Lecture Notes in Statistics

Edited by J. Berger, S. Fienberg, J. Gani,
K. Krickeberg, and B. Singer

55

L. McDonald B. Manly
J. Lockwood J. Logan (Eds.)

Estimation and Analysis
of Insect Populations

Proceedings of a Conference held in Laramie,
Wyoming, January 25–29, 1988

Springer-Verlag

Berlin Heidelberg New York London Paris Tokyo Hong Kong

Editors

Lyman L. McDonald
Departments of Statistics and Zo
University of Wyoming
Laramie, WY 82071, USA

Bryan F.J. Manly
Department of Mathematics and
University of Otago
Dunedin, New Zealand

Jeffrey A. Lockwood
Department of Plant, Soil and Ins
University of Wyoming
Laramie, WY 82071, USA

Jesse A. Logan
Departments of Entomology and
Virginia Polytechnic Institute
Blacksburg, VA 24061, USA

Mathematical Subject Classification Codes: 62P10

ISBN 0-387-96998-5 Springer-Verlag New York Heidelberg Berlin
ISBN 3-540-96998-5 Springer-Verlag Berlin Heidelberg New York

© Springer-Verlag Berlin Heidelberg 1989
Printed in Germany

Printing and binding: Druckhaus Beltz, Hemsbach/Bergstr.
2847/3140-543210

Preface

The papers in this volume were presented at a symposium/workshop on "The Estimation and Analysis of Insect Populations" that was held at the University of Wyoming, Laramie, in January, 1988. The meeting was organized with financial support from the United States — New Zealand Cooperative Science Program and the University of Wyoming. The purpose was to bring together approximately equal numbers of quantitative biologists and biometricians in order to (1) provide a synthesis and evaluation of currently available methods for modeling and estimating parameters of insect population, and to (2) stimulate research into new methods where this is appropriate.

The symposium/workshop attracted 46 participants. There were 35 papers presented in four subject areas: analysis of stage—frequency data, modeling of population dynamics, analysis of spatial data, and general sampling and estimation methods. New results were presented in all these areas. All except one of the papers is included in the present volume.

There were ten papers concerned with the analysis of stage—frequency data. Manly began the workshop with his review of past work in this area. His paper also discusses a new model for the estimation of the parameters of a single generation of a population based on the concept of a development variable. New single generation models were also described by Braner and Hairston, Faddy, Kemp et al., Munholland et al., and Pontius et al. Bellows et al. address the difficult problem of estimating losses from parasitism using stage—frequency and other data in a host—parasite system. Caswell and Twombly consider the estimation of stage transition matrices for cases where generations cannot be separated. Here estimates are naturally ill—conditioned and there is interest in techniques such as ridge regression as possible ways to improve their properties. Morton and Vogt consider the problem of estimating relative recapture rates for blowflies of different ages and development stages. Again ill-conditioning problems are apparent. Schneider proposes a computer—intensive approach for estimating parameters for a nematode population.

There were 12 papers on various aspects of modeling of population dynamics. Manly reviews development in key factor analysis and proposed a test for delayed density—dependent mortality with data consisting of stage—based life tables from a series of successive generations. Allen shows that natural enemy populations can easily display chaotic behavior. Carey presents a demographic framework for the analysis of cohort life histories, and discusses the modeling of harvesting and two sexes. Dennis reviews the use of stochastic differential equations for population modeling. Dianmo and Chang review a number of approaches for modeling single species populations. Fargo and Woodson note the potential use of the computer language SLAM for simulating insect populations. Lih and Stephen describe SPBMODEL, a computer simulation model that predicts infestation rates of the southern pine beetle. Lockwood and Lockwood show that some aspects of grasshopper population dynamics can be described by a cusp catastrophe. Logan discusses the composite modeling approach to population analysis which involves first developing a complex realistic model which is then simplified until it is amenable to a mathematical analysis. Nordheim et al., consider stochastic Leslie matrix models for population development, noting that their behavior may be somewhat different from that of the equivalent deterministic models. Schaalje and van der Vaart, and Smerage address the relationship between alternative models for the development of populations with a stage structure.

There were five papers concerned with various aspects of the analysis of spatial data. Hutcheson and Lyons discuss the analysis of quadrat counts in two papers. Johnson describes two modeling methods for testing hypotheses concerning the relationship between grasshopper densities and soil type in southern Alberta. Liebhold and Elkinton show how classical key factor analysis can be extended for studying life tables determined for different locations in space as well as different generations. Willson and Young present a model for arthropod movement that accounts for the negative binomial distributions often observed for counts.

The final section consists of eight papers on various types of statistical analysis and sampling problems. Blough discusses the analysis of the time series of trapping records of the pink bollworm moth from a cotton field in southern Arizona, in terms of the relationship of this series with amounts of irrigation water and the times when insecticide was applied. Buonaccorsi and Liebhold review statistical methodologies appropriate for estimating gypsy moth numbers from the ratio of mean frass drop from a forest canopy to the mean frass production from caged larvae. Burnham reviews recent developments in the analysis of population parameters from capture–recapture data. Cooper, Ellington et al., and Morrison et al., address various aspects of sampling designs. Gates describes a computer program DISCRETE that can be used to fit a variety of discrete distributions to data. McDonald and Manly discuss ways in which biased sampling procedures can be calibrated, as in cases where the sampling fraction differs for the various development stages in a population.

Lyman L. McDonald
Bryan F. J. Manly
Jeffrey A. Lockwood
Jesse A. Logan

ACKNOWLEDGEMENTS

In addition to the authors of the papers in this volume, a number of other people have made valuable contributions, and we would like to express our gratitude for their assistance. Our thanks are due to referees, to the Department of Statistics, University of Wyoming for both financial and moral support, to Charles Cooper and Ester Fouts, Conferences and Institutes, University of Wyoming, for hosting the conference, to Brenda Shriner and Julie Williams for clerical assistance, and especially to Lisa Eckles, Secretary for the symposium and publication of the proceedings.

CONTENTS

SECTION I

ANALYSIS OF STAGE–FREQUENCY DATA 1

SECTION IV
GENERAL SAMPLING AND ESTIMATION METHODS . . . 387

SECTION I

ANALYSIS OF STAGE—FREQUENCY DATA

A Review of Methods for the Analysis of Stage–Frequency Data

Bryan F.J. Manly[1]

ABSTRACT Stage–frequency data consists of counts or estimates of the numbers of individuals in different development stages in a population at a series of points in time as one generation is developing. Ecologists collect such data because in principle it can be used to estimate population parameters such as survival rates and durations of stages. This paper is a review of the various methods that have been proposed for this estimation. Several examples are presented to highlight the fact that there can be important differences between sets of stage–frequency data that appear superficially to be similar. These differences are then used as a basis for a classification scheme involving three types of data: multi–cohort, where entries into stage 1 take place over an appreciable part of the sampling period; single cohort, where all entries to stage 1 occur at approximately the same time; and reproducing, where the individuals entering stage 1 are produced by the individuals already in the population. Most of the paper consist of a review of methods for analyzing these three types of data using computer generated examples with known parameters.

The Classification of Stage–Frequency Data

Definition of Stage–Frequency Data

Stage–frequency data consists of counts or estimates of the numbers of individuals in q development stages in a population, or in a fraction of a population, at n points in time, $t_1 < t_2 ... < t_n$. Here we are only concerned with certain special types of situation that occur in biology. Interest is restricted to cases where individuals enter a population in stage 1, and then gradually progress through to stage q, unless they die before that happens. Thus all losses from stage q are by death whereas in the earlier stages there are losses to the next highest stage. Individuals cannot miss stage i by passing directly from stage i−1 to stage i+1. A definite maturation process is undergone in stages 1 to q−1 so that the probability of moving to stage i+1 depends upon how long an individual has been in stage i. In this paper the description "stage–frequency data" will generally imply that all these conditions hold.

It is often the case that the only practical way to study the dynamics of a population is in terms of life stages, no other information being readily available on the ages of individuals. Consequently, it is important for the biologist to have a range of straightforward and reliable methods for estimating population parameters from stage–frequency data. Some or all of the following are likely to be of interest:

---

[1]Department of Mathematics and Statistics, University of Otago, P.O. Box 56, Dunedin, New Zealand. On leave in the Deparment of Statistics, University of Wyoming, Laramie, Wyoming, 82071, for 1988.

(a)   the total number of individuals entering each stage;

(b)   the average time spent in the stages 1 to q−1;

(c)   the probabilities of surviving each of the stages 1 to q−1 (the stage–specific survival rates);

(d)   the mean times of entry to stages; and

(e)   unit time survival rates.

The examples that will now be described are intended to illustrate the characteristics that are important for modelling data. They show that although two sets of stage–frequency data appear to be superficially the same, there may be differences that will have to be taken into account when deciding on an appropriate method of analysis.

Examples

To begin with, consider the data in Table 1 which shows the stage–frequency counts obtained by Qasrawi (1966) when he sampled a population of the grasshopper *Chorthippus parallelus* on East Budleigh Common, Devonshire, in 1964. *C. parallelus* hatches from eggs and then passes through four instars before reaching the adult stage. Since Qasrawi sampled a fraction 0.002 of his defined site (except as noted in the table), population frequencies are estimated as the sample frequencies multiplied by 500.

These results are typical of what is often obtained by sampling insect populations. Simply inspecting the frequencies is quite instructive. Thus we see that when sampling began on 20 May all of the grasshoppers found were in instar 1. Subsequently, the numbers in this stage increased to a peak on 3 June and then declined to zero from 13 July onward. Grasshoppers were first seen in instar 2 on 3 June. Numbers in this stage peaked on 25 June, and declined down to zero from then on. The other two instar stages also show this pattern of increasing numbers followed by a decrease to zero. When sampling ceased on 23 September the few grasshoppers still alive were in the adult stage.

The sample frequencies are quite small and very likely contain substantial sampling errors. Nevertheless, there is a clear picture of the individuals in the population beginning to hatch just before the 20 May sample and continuing to hatch until about the middle of June. After hatching the grasshoppers proceeded to pass gradually through the instar stages to the adult stage. The individuals that hatched first reached the adult stage early in July while the late hatching individuals reached the adult stage by about the middle of August. Ashford et al. (1970) argued that migration across the boundaries of the study area can be neglected because of the limited mobility of the grasshoppers and the similarity of the surrounding area. On that assumption, losses from the adult must be due to death, and losses from the other stages must be due to death or advancement to the next stage.

For a second example consider Table 2, which shows Rigler & Cooley's (1974) estimates of the total numbers of individuals in six naupliar stages (N1 to N6), five copepodite stages (CI to

Table 1. Sample counts of grasshoppers *Chorthippus parallelus* in four instar stages and the adult stage at East Budleigh Common in 1964. The sampling fraction was 0.002 for 29 May to 23 September. The sampling fraction was 0.00143 on 20 and 25 May so the sample counts on these two days have been multiplied by the ratio 0.002/0.00143 = 1.4 to give adjusted counts that are comparable with the other ones.

Date		Day	Instar 1	Instar 2	Instar 3	Instar 4	Adult	Total
May	20	4	7.0*					7.0
	25	9	8.4*					8.4
	29	13	14					14
June	3	18	10	1				11
	10	25	7	5	1			13
	15	30	1	10				11
	18	33	1	8	1	1		11
	22	37	3	8	4	2		17
	25	40	7	12	6			25
	29	44	7	6	6			19
July	2	47	1	1	6	4	1	13
	9	54	1	3	2	1	7	14
	13	58		4	4	4	5	17
	16	61			1	3	2	6
	20	65		1	1	5	6	13
	24	69		1	1	2	5	9
	27	72					6	6
	31	75					6	6
Aug.	5	81				1	6	7
	11	87				1	1	2
	14	90					3	3
	19	95					3	3
	24	100					5	5
	28	104					3	3
Sept.	2	109					4	4
	8	115					2	2
	11	118					2	2
	17	124					2	2
	23	130					1	1

* adjusted counts

Table 2. Estimated population frequencies (millions in the entire lake) of *Skistodiaptomus oregonensis* in Teapot Lake, Ontario, in 1966.

Day	N1	N2	N3	N4	N5	N6	CI	CII	CIII	CIV	CV	A	Total
	96	26	4										30
101	81	38											119
106	161	237	45	3									446
109	27	176	182	58	15	1							459
112	33	68	192	330	258	46	2						929
115	25	55	68	119	243	311	94	14					929
118	37	51	53	68	112	205	203	30	2				761
122	16	54	54	51	84	165	177	102	47	17	3		770
126	18	49	41	51	45	94	97	90	105	104	48	1	743
130	9	16	37	23	49	63	98	64	96	113	122	9	699
133	17	23	49	41	40	82	110	121	160	173	248	138	1202
137	37	16	13	25	44	54	62	57	96	115	101	301	921
140	78	145	41	12	16	19	33	44	67	78	53	172	758
147	181	437	538	168	69	30	15	11	52	101	54	196	1897
154	111	186	294	238	309	455	289	56	42	83	50	201	2314
161	14	15	84	144	125	142	222	322	126	157	88	255	1694
167	17	46	23	39	29	64	156	300	439	271	61	143	1588
175	57	133	100	54	34	20	60	132	506	237	50	96	1479
181	59	88	165	136	50	49	64	77	308	465	160	121	1742
189	33	66	98	42	67	54	107	75	412	669	353	147	2123
199	47	82	101	57	20	3	11	6	81	508	261	78	1255
203	100	179	183	111	50	21	18	10	64	963	317	50	2066
210	124	264	370	149	50	9	10	10	21	365	143	84	1599
217	203	293	250	236	85	55	27	9	11	109	146	87	1511
224	117	242	219	130	131	77	45	11	10	125	430	238	1775
236	68	173	200	108	57	21	32	38	70	87	268	325	1447
249	42	128	130	71	40	7	9	1	5	17	51	67	568
256	133	203	160	124	99	36	9	11	20	138	145	125	1203
268	72	220	130	119	60	33	26	22	22	36	137	137	1014
281	40	73	52	35	38	57	45	23	9	27	37	379	725
293		2	11	18	21	16	10	13	15	16	19	193	334
308						1	3	5	5	6	11	219	250
318							3	3	10	6	162		184

(Population killed by anearobiosis shortly after day 318)

CV), and the adult stage for the zooplankter *Skistodiaptomus oregonensis* in Teapot Lake, Ontario, for the spring to autumn period in 1966. This is similar to the previous example, except that there are more stages, and recruitment to stage 1 occurs for almost the entire sampling period. Also, the sampling variation seems to be rather large. For example, the population size is estimated to have increased by 72% between days 130 and 133. However, this does not seem to have been a genuine change since the increase is primarily in the more advanced stages. Obviously any new entries to the population should show up first in the early stages.

An important factor in the dynamics of the *S. oregonensis* population comes from the fact that the females lay two different types of egg. Some are diapausing ones that lay dormant over the winter and then hatch in the spring to regenerate the population. Others are normal eggs that develop and hatch immediately. When sampling began in the spring of 1966 there were individuals in stages N1 and N2. These had developed from diapausing eggs laid in the previous autumn. All of the individuals in the population up to day 133 came from the same source. However, by day 133 the adults in the population had begun to lay eggs that could hatch to produce entries to N1. Hence from day 133 onward the population was reproducing itself to some extent.

Stage–frequency data can be generated in the laboratory, and in some cases laboratory populations have been allowed to reproduce. Lefkovitch (1964a, experiment II) gives the results for a population of the cigarette beetle *Lasioderma serricorne* for which this is the case. Counts of eggs, larvae, pupae and adults are given in the Table 3 at three weekly intervals up to 48 weeks. An important aspect of this example is that the adults produce the eggs. Hence the relationship between the four stages is circular: eggs –> larvae –> pupae –> adults –> eggs –> larvae, etc. When he designed his experiment, Lefkovitch wanted to generate a laboratory population that was equivalent as far as possible to a natural population with unlimited food but limited space.

The stages in a population are usually defined in terms of qualitatively different parts of the life cycle. However, this is not a necessity and van Straalen (1982) has suggested that there are advantages in defining stages in terms of a continuous variable such as size. To illustrate this approach he used his own data on the Collembola species *Orchesella cincta* shown here in Table 4. This has a similar form to the data in earlier tables but the 'stages' are size classes. For example, stage 1 consists of individuals with a length of between 0.73 and 0.90 mm. As individuals age they move progressively through the size classes until they die. There is no difficulty in analyzing data like this using any of the methods that have been proposed for stage–frequency data in general. In addition, models can be constructed that incorporate equations for body growth with age. Note, however, that this type of data is only stage–frequency data as defined above providing that all individuals will grow into the final size class if they survive long enough.

The data shown in Table 5 were obtained by Bellows & Birley (1981) for a laboratory population of *Callosobruchus chinensis*, a bruchid pest of stored pulses. By sampling the population they were able to estimate the number of individuals in different stages each day for 26 days after eggs were laid. The numbers shown correspond to 100 hatched eggs. In the population the total number of surviving individuals must have either remained constant or declined as time increased. The increases shown on some days in Table 5 must therefore be an artifact of the sampling programme. Clearly, the daily survival rate was quite high. What makes this example different from the previous ones is that it refers to a single cohort: the stage–frequencies on any day all relate to individuals that entered stage 1 at approximately the same time. When cohort stage–frequencies like this are available it becomes possible to estimate more population parameters than would otherwise be the case. Thus Bellows & Birley (1981) were able to fit distributions to the times spent in different stages rather than just estimate the mean time in each stage.

Table 3. Stage–frequencies for a population of the cigarette beetle *Lasioderma serricorne* set up as Experiment II by Lefkovitch (1964), and allowed to develop for 105 weeks. These are totals from four replicates started with 20 adults each.

Week	Eggs	Larvae	Pupae	Adults	Total
0	0	0	0	80	80
3	0	1355	0	0	1355
6	4671	53	23	1008	5755
9	3	7545	4	12	7562
12	117	1586	661	783	3147
15	1196	4749	13	855	6813
18	55	2838	551	135	3579
21	2023	2488	54	704	5269
24	450	4117	82	328	4977
27	1001	2901	288	659	4849
30	847	3916	54	446	5263
33	534	1834	254	516	3138
36	1177	3883	112	526	5698
39	156	2937	278	297	3668
42	1206	3168	122	594	5090
45	389	5258	129	567	6233
48	358	3514	218	275	4365

Table 4. Data for a generation of *Orchesella cincta* in forest near Dronten, the Netherlands, published by van Straalen (1982). The 'stages' are seven size classes with the lower limits 0.73, 0.90, 1.35, 1.80, 2.25, 2.70 and 3.15 mm.

				Stage				
Week	1	2	3	4	5	6	7	Total
1	34							34
3	588	124	47	1				760
5	378	199	55	33	1	2		668
7	175	302	193	73	11	2		756
9	15	89	140	86	42	7		379
11	25	53	85	137	117	42	1	460
13		12	24	51	89	65	3	244
15		2	14	44	25	0		85
17				27	31	34	6	98
19				18	34	52	13	117
21				7	2	5	1	15
25				4	2	3	1	10
29						1	1	2

Table 5. Cohort stage–frequency data for the number of *Callosobruchus chinensis* per 100 hatched eggs. Stage I to IV are larvae, A1 is adults still in beans, A2 is adults after they have emerged from beans.

		Stage							
Day	Eggs	I	II	III	IV	Pupae	A1	A2	Total
1	100.0								100.0
2	100.0								100.0
3	100.0								100.0
4	59.0	41.0							100.0
5		96.8							96.8
6		91.8							91.8
7		23.1	71.8						94.9
8		4.8	74.2	17.7					96.8
9		1.1	57.9	40.0					98.9
10		1.9	3.7	98.8	0.9				96.3
11			4.8	41.7	48.8				95.2
12			0.0	3.5	90.6				94.1
13			1.2	0.0	96.3				97.5
14				0.0	100.0				100.0
15				1.2	82.6	15.1			98.8
16					37.7	58.0			95.7
17					10.6	74.2			84.9
18					10.5	81.4		1.2	93.0
19					2.1	70.8	18.8	0.0	91.7
20						22.6	72.6	1.2	96.4
21						8.2	75.3	16.4	100.0
22						4.8	38.6	53.0	96.4
23							12.7	85.9	98.6
24							8.5	91.5	100.0
25							2.0	88.2	90.2
26								88.4	88.4

Complications and Special Cases

There are many complications that can occur with stage–frequency data, particularly when it is collected in the field from natural population. For instance, it may not be possible to sample all stages equally well. The adults, in particular, may be more mobile than the individuals in the other stages so that a completely different sampling method has to be employed. In that case it may well turn out that the estimated adult numbers cannot be reconciled with the estimated numbers in the other stages. The adults may even disperse so that they cannot be counted at all.

Many models for stage–frequency data assume that the daily survival rate is constant either for all individuals, or for all individuals in the same development stage. Although these assumptions are often fair enough, situations do arise where they are clearly untenable. For example, if an insect population is sprayed with an insecticide while a sampling programme is being carried out it is almost certain that one of the effects will be temporary low survival. Another reason why daily survival rates and other population parameters may not be constant is that they are linked to environmental conditions, particularly ambient temperatures. A common situation is that physiological time passes slowly at low temperatures and quickly at high temperatures. It may then be appropriate to replace 'ordinary' time with 'physiological' time before starting to analyze data.

Stage–frequency data is most easy to analyse if recruitment to the first stage takes place over a relatively short period of time and sampling is continued until almost all individuals have died. The data in Tables 1 and 4 come closest to this ideal situation although in both cases the sampling errors in the data are larger than is desirable.

A major complication occurs if recruitment to the first development stage continues over the entire sampling period so that there is no time when stage–frequencies are changing only through deaths and individuals developing through stages. Under these conditions an analysis can only be made if it is either (a) possible to relate the recruitment to stage 1 to the numbers already in the population, or (b) the entry distribution is simple enough to be described using a relatively small number of parameters.

A Classification Scheme

Consideration of the examples given above indicates that there are various properties that can be used to characterize stage–frequency data, and a large number of special cases that are possible. In practice however the majority of examples seem to fall within one of the following categories:

(a) Multi–cohort stage–frequency data: Entries to the population occur for the first part of the observation period, these entries coming from a source that is independent of the stage–frequencies in the population; only single stage transitions are possible in the time between observations; the data available consists of either population counts or sample estimates of these.

(b) Single cohort stage–frequency data: This is similar to multi–cohort data except that all entries to the population occur at the same time.

(c) Reproducing stage–frequency data: Observations are either total counts or sample estimates of total counts for a reproducing population so that the stage–frequencies at an observation time are determined (apart from stochastic variation) by the stage–frequencies at earlier times.

The data in Tables 1, 2 and 4 fall within the first category, the data in Table 5 falls within the second category, and the data in Table 3 falls within the third category.

Analysis of Multi–Cohort Data

General Considerations and Notation

The various methods have been proposed for analyzing multi–cohort stage–frequency data fall into three groups: methods that require no information in addition to the stage–frequencies; methods that assume that the survival rate per unit time varies from stage to stage, and which require an independent estimate of at least one population parameter; and methods that assume that the survival rate changes with time, and which require independent estimates of some population parameters.

For discussing the various methods it is useful to have some data for which true parameters are known. To this end, two sets of data have been simulated on a computer. Data set 1 has four stages, 20 samples taken at 'daily' intervals, a constant daily survival probability of 0.7, and other parameters as shown in Table 6. Data set 2 has five stages, 20 'daily' samples, daily survival rates varying from stage to stage, and other parameters as shown in Table 7. The first of these two sets of data is a 'good' set in the sense that it can produce estimates of population parameters that are close to the true values. The second set of data is not easy to analyse because of the variation in the survival rates in different stages.

It is convenient at this point to define some of the notation that will be used in describing methods for the analysis of data. It is assumed that there are q development stages in a population, and that n samples are taken from the population at times $t_1 < t_2 < ... < t_n$. For stage j, let:

M_j = the total number of individuals entering;

a_j = the mean duration;

s_j = the standard deviation of the duration;

w_j = M_{j+1}/M_j = the stage–specific survival rate;

ϕ_j = $\exp(-\Theta_j)$ = the survival rate per unit time;

μ_j = the mean time of entry;

σ_j = the standard deviation of the time of entry;

$g_j(x)$ = the probability density function for the time of entry of individuals into the stage (which has the mean of μ_j and standard deviation of σ_j).

Since losses from stage q are by deaths only, a_q and s_q are not defined and w_q is zero. For some methods of estimation it is assumed that the survival rate per unit time is the same in all stages. In that case this rate is taken as $\phi = \exp(-\Theta)$.

Table 6. First computer simulated test data, with four stages and 20 samples. The generated stage–frequencies were determined so as to be equivalent to values that would be obtained by sampling a small fraction of a large population. Entries to the population took place up to the time of the fifth sample. For the fraction of the population sampled, the entries between times 0 and 1, 1 and 2, ..., 4 and 5 were 25, 200, 150, 75 and 50, respectively, with a total of 500. The durations of the first three stages in the population were set at 2, 3 and 2 days, respectively. A daily probability of survival of 0.7 was imposed, which resulted in stage–specific survival rates of 0.49, 0.34 and 0.49, respectively, for the first three stages.

Time	Stage			
	1	2	3	4
1	10	0	0	0
2	164	0	0	0
3	252	4	0	0
4	181	90	0	0
5	107	121	0	0
6	26	96	3	0
7	1	81	27	0
8	0	41	41	1
9	0	8	20	18
10	0	1	24	24
11	0	0	3	20
12	0	0	0	13
13	0	0	0	14
14	0	0	0	8
15	0	0	0	6
16	0	0	0	5
17	0	0	0	4
18	0	0	0	1
19	0	0	0	2
20	0	0	0	0

It is necessary to distinguish between the stage–frequencies in the population (the 'true' frequencies) and the data stage–frequencies. To this end let:

$f_j(t)$ = the number of individuals in the population in stage j at time t;

$F_j(t)$ = the number of individuals in the population in stage j or a later stage at time t;

f_{ij} = the data value for the number of individuals in stage j in the sample taken at time t_i; and

F_{ij} = the data value for the number of individuals in stage j or a later stage in the sample taken at time t_i.

Often the data frequencies are supposed to relate to a certain fraction of a total population. In that case the population frequencies $f_j(t)$ and $F_j(t)$ are to be understood to relate to the same fraction. In general it will be assumed that f_{ij} is an estimate of $f_j(t_i)$ and F_{ij} is an estimate of $F_j(t_i)$.

Table 7. Second computer simulated test data with five stages, and 20 samples. As for the first set of data, the generated stage–frequencies were determined so as to be equivalent to values that would be obtained by sampling a small fraction of a large population. Entries to the population took place up to the time of the tenth sample. For the fraction of the population sampled, the entries between times 0 and 1, 1 and 2, ..., 9 and 10 were all equal to 20, with a total of 200. The durations of the first four stages in the population were set at 1.5, 2, 2 and 2 days, respectively. The daily survival probabilities were set at 0.7, 0.75, 0.8, 0.7 and 0.6, for stages 1 to 5, respectively. This resulted in stage–specific survival rates of 0.59, 0.56, 0.64 and 0.49 for stages 1 to 4, respectively.

	Stage				
Time	1	2	3	4	5
1	7	0	0	0	0
2	17	5	0	0	0
3	18	7	0	0	0
4	22	25	2	0	0
5	20	21	6	0	0
6	31	21	8	3	0
7	22	12	11	5	0
8	34	20	9	2	0
9	21	17	14	6	2
10	24	24	11	6	2
11	13	13	8	2	1
12	0	13	9	5	4
13	0	3	11	7	2
14	0	0	10	11	3
15	0	0	3	2	5
16	0	0	0	4	4
17	0	0	0	0	1
18	0	0	0	0	2
19	0	0	0	0	0
20	0	0	0	0	1

Methods of Analysis Based only on Stage–Frequency Data

The methods for estimating population parameters that will be considered first have the common characteristic of not requiring any information in addition to the data stage–frequencies. They will be reviewed roughly in the order of their mathematical complexity.

The Richards & Waloff (1954) Method

Richards & Waloff (1954) used a regression method for estimating the numbers entering stages and the survival rate per unit time. They assumed that the survival rate per unit time is the same for all stages so that an appropriate equation for the decline in the number in stage j or higher inclusive is $F_j(t) = F_j(0)\phi^t$. Hence there is the equation

$$\log\{F_j(t)\} = \log\{F_j(0)\} + t\log(\phi)$$

which can be estimated by fitting a linear regression of the data values $\log(F_{ij})$ against time t_i, using only the samples after F_{ij} reached a peak. If time is measured with t=0 when peak numbers occur, then $F_j(0)$ corresponds approximately to the number entering stage j (Fig. 1). Dempster (1956) suggested that if survival rates vary between stages then estimation will be improved if the regression for a stage is carried out only using data for sample times during which the stage is present. This is sensible providing that the regression for each stage can still be based on a reasonable number of data points.

The simplicity of Richards & Waloff's method makes it appealing. However, with many sets of data its use is limited because of a prolonged period during which entries to a stage take place. New entries will occur after the time at which the numbers reach a peak. Hence if the peak time is taken as the starting time for the regression then some deaths will be offset by new entries. Consequently, the estimated survival rate will tend to be higher than the real rate. Also, the regression estimate of the population size at the time when entries to stage j begin will be an over–estimate of the number that enter stage j because some of the individuals counted will die before they leave stage j–1.

For the first set of simulated data shown in Table 6 there are the following estimates from Richards & Waloff's method:

| Stage | Number entering | | Daily survival | |
	Estimate	True	Estimate	True
1	865	500	0.70	0.70
2	343	245	0.71	0.70
3	157	84	0.70	0.70
4	53	41	0.73	0.70

The survival estimates are excellent but (as expected) the numbers entering stages have been overestimated. If Dempster's suggestion of restricting the regression data for a stage to samples when the stage is present is followed then the estimates become much worse. For example 763 are estimated to enter stage 1 but 911 to enter stage 3.

If the same calculations are carried out on the second set of simulated data (Table 7) then the results obtained are as shown in the following table:

Stage	Number entering		Daily survival	
	Estimate	True	Estimate	True
1	4658	200	0.64	0.70
2	1658	117	0.66	0.75
3	4066	66	0.58	0.80
4	463	42	0.63	0.70
5	28	21	0.73	0.60

No comment is required here! The estimates are even worse if Dempster's modification is used.

The Rigler & Cooley (1974) Method

The method for analyzing stage–frequency data that was proposed by Rigler & Cooley (1974) has been popular with limnologists, but unfortunately there are a number of problems with it that have been exposed recently (Hairston & Twombly, 1985; Aksnes & Hoisaeter, 1987; Hairston et al., 1987; Saunders & Lewis, 1987). The principle difficulty is that certain key equations are only valid when there is no mortality.

The Manly (1974a) Method

This method can be thought of as a generalization of the Richards & Waloff (1954) approach since it involves also regressing stage–frequencies against time. However all the accumulated stage–frequencies are used rather than just the ones after the peak (Fig. 1). It is assumed that the times that individuals enter a development stage follow a distribution with frequency function $g(x)$, and that the survival rate per unit time is $\exp(-\Theta)$ in all stages. It then follows that the expected number of individuals in the stage, or a later stage, at time t is given by the equation

$$F(t) = M \int_{-\infty}^{t} \exp\{-\Theta(t-x)\}\, g(x)\, dx,$$

where M is the total number entering the stage. If the entry time distribution is assumed to be normal with mean μ and standard deviation σ then this becomes

$$F(t) = M^*\exp(-\Theta t)/\sqrt{(2\pi)} \int_{-\infty}^{(t-\mu^*)/\sigma} \exp(-x^2/2)\, dx,$$

where $M^* = \exp\{\Theta(\mu+\Theta\sigma_2/2)\}M$ and $\mu^* = \mu + \Theta\sigma_2$. This is the equation for a non–linear regression of stage–frequencies $F(t)$ on time t, with unknown parameters M, μ, σ and Θ that can be estimated using a standard computer program. Once M, μ and Θ are estimated for a stage they can be used to estimate other parameters of interest. For stage j the duration is $a_j = \mu_{j+1} - \mu_j$, the stage–specific survival rate is $w_j = M_{j+1}/M_j$, and the daily survival rate is $\phi_j = \exp(-\Theta_j)$.

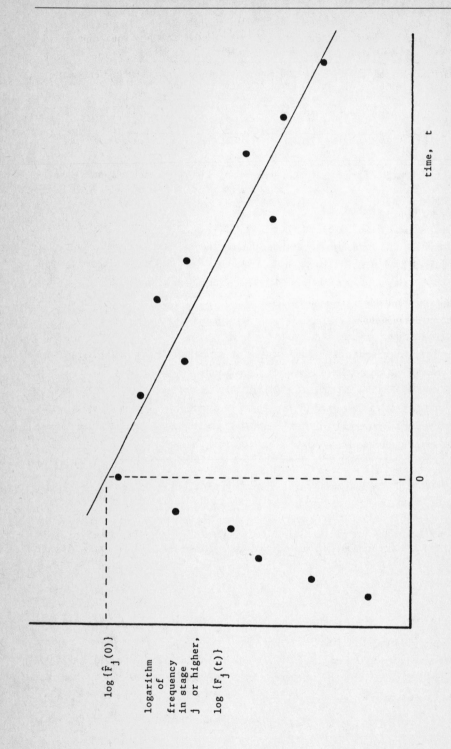

Fig. 1. The principle behind Richards and Waloff's regression method of estimation. The slope of the regression line estimates $\log(\phi)$ while $\hat{F}_j(0)$ corresponds to the number entering stage j.

Substituting the regression estimates into the right—hand sides of these equations provides estimates of these extra parameters.

The regression estimates obtained for the computer simulated test data in Table 6 are as follows:

Stage	Number entering		Duration		Daily survival	
	Est	True	Est	True	Est	True
1	495	500	1.7	2.0	0.71	0.70
2	189	245	3.6	3.0	0.77	0.70
3	95	84	1.3	2.0	0.67	0.70
4	34	41	—	—	0.76	0.70

These estimates are fairly reasonable. Each stage provides a separate estimate of the daily survival rate $\exp(-\Theta)$, although the assumption is being made that this is constant. The estimate from stage 1 is the most reliable if this assumption is true.

For the second set of simulated data in Table 7 the estimates come out to be:

Stage	Number entering		Duration		Daily survival	
	Est	True	Est	True	Est	True
1	401	200	1.5	1.5	0.52	0.70
2	326	117	2.0	2.0	0.43	0.75
3	176	66	1.8	2.0	0.40	0.80
4	80	42	1.6	2.0	0.38	0.70
5	33	21	—	—	0.31	0.60

These are clearly unsatisfactory.

The Kiritani—Nakasuji—Manly Method

When Kiritani & Nakasuji (1967) first described their method for estimating stage—specific survival rates it had the restriction of requiring samples to be taken at fixed intervals of time for the entire period that a population is present. Subsequently, this requirement was relaxed, and their approach was extended to provide estimates of other population parameters (Manly, 1976, 1977, 1985b). The method assumes that the duration of stage j is a_j for all individuals and that the survival rate per unit time is the same for all stages at all times.

Let A_j^* be the area under the stage—frequency curve $F_j(t)$ for stages j, j+1, ..., q combined, and D_j^* is the area under corresponding time—stage—frequency curve $tF_j(t)$. Then there are the relationships

$$A_j^* = M_j/\Theta,$$

and

$$D_j^* = (\mu_j + 1/\Theta)A_j^* .$$

It follows that $w_j = A_{j+1}^*/A_j^* = M_{j+1}/M_j$, is the stage–specific survival rate for stage j; $\Theta = -\log_e(A_q/A_1)/(B_q^* - B_1^*)$, is the survival parameter; $a_j = -\log_e(w_j)/\Theta$, is the duration of stage j; and $M_j = A_j^*\Theta$, is the number entering stage j, where $B_j^* = D_j^*/A_j^*$. Hence population parameters can be determined from areas under accumulated stage–frequency and time–stage–frequency curves.

The values A_j^* and D_j^* can be estimated from data using the trapezoidal rule. This is done most easily when observation times are equally spaced at t_1, $t_2 = t_1 + h$, $t_3 = t_1 + 2h$, ..., $t_n = t_1 + (n-1)h$, and all stage–frequencies are zero at times t_1 and t_n so that $F_{1j} = F_{nj} = 0$. In that case the trapezoidal rule gives the areas under the $F_j(t)$ and $tF_j(t)$ curves to be $A_j^* = h \Sigma F_{ij}$ and $D_j^* = h \Sigma t_i F_{ij}$, respectively (Fig. 2). If data are not equally spaced then equally spaced data can be obtained by interpolating between the frequencies at observation times.

The trapezoidal rule does not provide valid estimates of the areas under population curves unless the stage–frequencies are zero for times t_1 and t_n. Thus it is necessary for a population to be observed from the time that individuals first enter stage 1 until all individuals are dead in order to use the above equations. In some cases this condition can be met by adding 'dummy' samples before and after the actual sampling period. However, if the first sample from a population shows many individuals in stage 2 or a higher stage, or the last observation includes many individuals still alive, then an iterative method for adjusting the data is available (Manly, 1985b). Experience has indicated that this usually works with data that are incomplete due to sampling terminating too soon. However, convergence problems are apt to occur when sampling begins too late. The method can also be modified for situations where no frequencies are available for some stages, providing that information is available for three consecutive stages (Manly, 1977).

When the above equations are applied to the first set of test data in Table 6, the results shown in the following table are obtained. A dummy sample with zero frequencies was added at time 0 to complete the data and avoid iteration.

Stage	Number entering		Duration		Daily survival	
	Est	True	Est	True	Est	True
1	536	500	2.0	2.0	0.69	0.70
2	256	245	2.8	3.0	0.69	0.70
3	89	84	1.9	2.0	0.69	0.70
4	44	41	—	—	0.69	0.70

These estimates are very reasonable. Estimates for the second set of test data in Table 7 are shown in the next table. In this case dummy samples with zero frequencies were added at times 0 and 21 to complete the data.

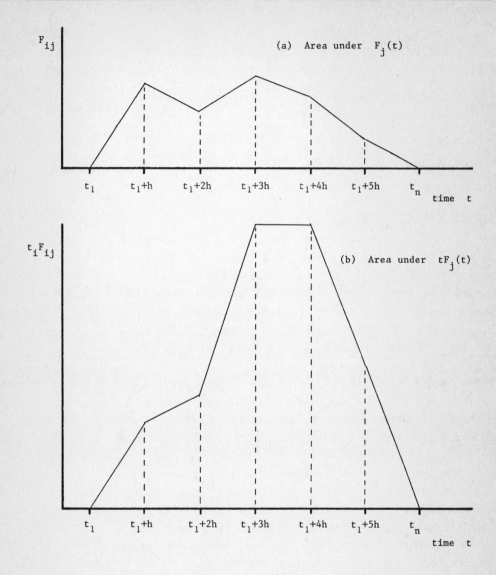

Fig. 2. Areas under the $F_j(t)$ and $tF_j(t)$ curves estimated by the
trapeizoidal rule.

Stage	Number entering Est	Number entering True	Duration Est	Duration True	Daily survival Est	Daily survival True
1	310	200	0.9	1.5	0.59	0.70
2	190	117	1.3	2.0	0.59	0.75
3	95	66	1.6	2.0	0.59	0.80
4	42	42	2.1	2.0	0.59	0.70
5	14	21	–	–	0.59	0.60

The (incorrect) assumption of the same survival rate in all stages has resulted in some large discrepancies between the estimates and the true values.

The Read & Ashford (1968) Method

The first stochastic model for stage–frequency data was provided by Read & Ashford (1968), and discussed further by Ashford et al. (1970). They assumed Erlangian distributions for the times spent in different development stages, and determined maximum likelihood estimates of population parameters for two sets of data. This model will not be considered here, although it is important as a precursor to the next two models to be considered. The particular assumptions made by Read and his colleagues result in equations that are not particularly suitable for computational purposes.

The Kempton (1979) Method

Kempton (1979) noted that models must include three components. The first concerns the survival rates in different stages. He suggests that a constant survival rate per unit time may be too simplistic, although this may be an acceptable assumption within one life stage. Assuming a gamma, Weibull or lognormal distribution for lengths of life may be more satisfactory. The second component of a model concerns times of entry to stage 1. For this, he suggested that it is realistic to conceive of development starting at a time 0 with all individuals in a stage 0. The time of entry to stage 1 for an individual then corresponds to the duration of stage 0 for that individual. A normal or a gamma distribution may be realistic for the distribution of this duration. The third component of a model concerns the durations of stages 1, 2, ... q. In this case Kempton pointed out the advantages of assuming that durations are (a) normally distributed, (b) gamma distributed, (c) inverse normally distributed, or (d) constant.

As far as the gamma distribution is concerned, the advantages of its use come about because of an 'additive' property. Let the gamma distribution with scale parameter b and shape parameter k be denoted by $G(k,b)$. Note that the mean of this distribution is $\mu = k/b$. Suppose that the time of entry to stage 1 follows a $G(k_0,b)$ distribution, and the duration of stage j follows a $G(k_j,b)$ distribution, for $j = 1, 2, ..., q-1$, with all these gamma distributions being independent. Then the time of entry to stage j will follow a $G(c_j,b)$ distribution, where $c_j = k_0 + k_1 + ... + k_{j-1}$. The mean time of entry to stage j will therefore be $\mu_j = c_j/b$. Also, the mean duration of stage j is $a_j = k_j/b$, so that $\mu_j = \mu_1 + a_1 + a_2 + ... + a_{j-1}$.

Assuming a gamma distribution for the time of entry to stage 1 will be realistic for situations where there is a single peak in the curve of the numbers entering into a population. Also, a gamma distribution for the duration of a life stage is a reasonable assumption in the absence of any knowledge about the true form of the distribution. Read & Ashford (1968) and Ashford et al. (1970) used a gamma model for which k values were constant between stages and b values varied since the Erlangian distribution that they used is a special case of the gamma distribution with an integer value of k.) Whatever assumptions are made about the form of distributions, the probability that an individual is in stage j at time t, $p_j(t_i)$, can be written as in terms of the various parameters involved providing that it is assumed that the probability of an individual surviving depends only on time, and that the times that an individual spends in different stages are independently distributed.

For his examples Kempton assumed that the sample count of the number of individuals in stage j in the sample at time t_i was a random value from the Poisson distribution with mean $M_0 p_j(t_i)$, where M_0 is the total number of individuals at time 0. The Poisson assumption is reasonable if M_0 is large and $p_j(t_i)$ is small, which will usually be the case. He maximized the likelihood function numerically using a general purpose maximum likelihood program. An alternative fitting process is possible using the computer program MAXLIK (Reed and Schull, 1968; Manly, 1985b, p. 433).

The MAXLIK program was used to fit the model described above to the test data in Tables 6 and 7, assuming the same survival rates in each stage. For the first of these two sets of data the following table shows rather satisfactory results:

Stage	Number entering Est	Number entering True	Duration Est	Duration True	Daily survival Est	Daily survival True
1	469	500	2.4	2.0	0.72	0.70
2	220	245	3.2	3.0	0.72	0.70
3	79	84	2.1	2.0	0.72	0.70
4	41	41	–	–	0.72	0.70

For the second set of data the estimates are not so good, with the following results:

Stage	Number entering Est	Number entering True	Duration Est	Duration True	Daily survival Est	Daily survival True
1	172	200	2.1	1.5	0.75	0.70
2	105	117	3.0	2.0	0.75	0.75
3	53	66	3.5	2.0	0.75	0.80
4	24	42	4.2	2.0	0.75	0.70
5	11	21	–	–	0.75	0.60

The Van Straalen (1982, 1985) Method

Van Straalen's (1982) contribution to the modelling of stage–frequency data was the introduction of the notion of a continuous stage–variable. This is a measurable variable y such as body size that increases with age and can be used as an indicator of the development stage of individuals. The concept can be used with qualitative development stages if individuals in stage j are thought of as having a stage–variable between j and j+1. As individuals develop their underlying stage–variable is increasing steadily. The outward manifestation consists of integer jumps as higher stages are reached. In his original paper van Straalen (1982) allowed for the possibility of linear growth for the stage variable y but later (van Straalen, 1985) considered also linear, power, logistic, Gompertz, and von Bertalanffy growth.

Van Straalen developed his model as a solution to a partial differential equation for the number of individuals with a size y at time t. Assuming a normal distribution for the times at which individuals enter stage 1 he was able to determine the expected frequency in stage i at time t as a function of the parameters M_j, Θ, μ_j and σ_j, together with the parameters for whichever of the growth relationships is assumed. If the stage–frequency data are assumed to consist of counts with Poisson distributions then the model can be fitted to data using the MAXLIK algorithm mentioned above.

Although this model has a certain appeal, it has not yet been shown to be appropriate for any data. Van Straalen (1982) used the example shown in Table 4 for a population of the litter inhabiting spring–tail *Orchesella cincta*, with body length as the size variable, assuming the linear growth function. The fit of the model to the data was very poor. Later van Straalen (1985) tried all of the growth equations on the same data, and another similar set but the fit was still not good. He suggested that the lack of fit arises mainly because of the assumption of a constant daily survival rate. However the problem may be due to the imposition of a definite growth function on the population which, in effect, determines stage durations.

The Manly (1987) Method

The final method to be discussed in the present section is based on a regression model which allows the survival rate per unit time to vary from stage to stage. It is assumed that observations are taken at unit intervals of time i = 1, 2, ..., n. The key equation is

$$F_{i1} \approx B_{1i-1} + f_{i-11}\phi_1 + f_{i-12}\phi_2 + ... + f_{i-1q}\phi_q,$$

for i = 2, 3, ..., n, where B_{1i-1} is the number of entries to stage 1 between times i–1 and i that are still alive at time i. That is to say, the count in all stages at time i, F_{i1}, consists approximately of survivors from the number entering stage 1 between times i–1 and i, plus the survivors from different stages that were in the population at time i–1. Here ϕ_j is the probability of surviving from one sample time to the next for an individual that is in stage j at the time of the first sample, which is not quite the same as the survival rate per unit time for individuals in stage j.

Suppose it is known that $B_{i1} = 0$ for i>r because, for example, there are no individuals in stage 1 in samples after the rth one. Then the above equation can be written as

$$F_{i1} \approx B_{11}X_{i1} + ... + B_{r1}X_{ir} + \phi_1f_{1\text{-}11} + ... + \phi_qf_{i\text{-}1q},$$

where $X_{ij} = 0$ unless $j = i{-}1$, in which case $X_{ii\text{-}1} = 1$. Hence it is possible to estimate the constants $B_{11}...B_{r1}$, ϕ_1, ..., ϕ_q using standard multiple regression calculations.

This method breaks down if entries to stage 1 occur for almost all of the samples because the number of parameters to be estimated (r+q) exceeds the number of regression points (n−1). In fact generally it may lead to an excessive number of entry parameters having to be estimated. This suggests that the distribution of the numbers entering stage 1 should be defined by a smaller number of parameters. Thus entry numbers E_1, E_2, ..., E_p can be defined at p equally spaced points in time between 1 and r, inclusive, so that $B_{11} = E_1$ and $B_{r1} = E_p$. The samples 2 to p−1 all then fall at times that are between the times of two entry parameters E_j and E_{j+1}, for some value of j. Hence B_{i1} for $1 < i < r$ can be found by interpolating between the two E values that enclose it (Fig. 3). Using this type of p−point entry distribution means that $B_{i\text{-}11}$ can be expressed as a weighted average of two of the E values, or, more generally, as $B_{i\text{-}11} = E_1Z_{i1} + ... + E_pZ_{ip}$, with values of Z that are determined from the interpolation. Hence, the multiple regression takes the form

$$F_{i1} \approx E_1Z_{i1} + ... + E_pZ_{ip} + \phi_1f_{i\text{-}11} + ... + \phi_qf_{i\text{-}1q}.$$

One advantage of using this equation is that it is relatively straightforward to set a survival rate equal to a fixed value (such as 1, if the regression estimate exceeds 1). Also, it is a simple matter to allow several successive stages to have the same survival parameter. It is just necessary to define a new stage comprised of individuals from any one of these stages. Estimates are then based on the frequencies for the composite stage.

Once estimated survival rates $\hat{\phi}_1,..., \hat{\phi}_q$ have been found by regression, estimates of other parameters can be determined. An estimate of the number of entries to stage j between time i and time i+1 that are still alive at time i+1 is

$$B_{ij} = F_{i+1j} - \hat{\phi}_jf_{ij} - \hat{\phi}_{j+1}f_{ij+1} - ... - \hat{\phi}_qf_{iq}.$$

Hence estimates of the total number entering stage j that survive until a sample time and the mean time of entry to stage j, are

$$M_j = \sum_{j=1}^{r_j} B_{ij} + F_{1j}$$

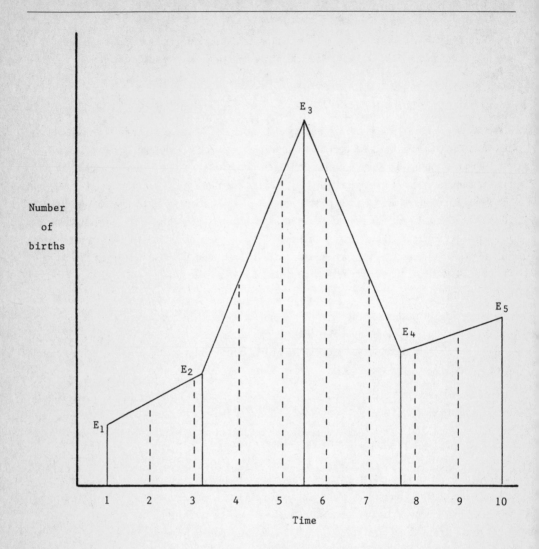

Fig.3. An example of approximating an entry distibution with p
parameters. In this example entry numbers for ten samples
(indicated by broken vertical lines) are determined by interpolating between numbers E_1 to E_5 at equally-spaced times
between and including times 1 and 10 .

and

$$m_j = \sum_{i=1}^{r_j} {}_iB_{ij}/M_j,$$

respectively, where r_j is the last sample to contain new entries to stage j. Here F_{1j} is included in M_j to account for those in stage j, or a higher stage, in the first sample. These are considered to enter at time 0, the B_{1j} to enter at time 1, etc.

If sampling covers the period from when individuals begin to enter stage 1 until all individuals have entered the final stage the estimated entry distributions will be complete for all stages. Then the mean duration of stage j can then be estimated from $a_j = \mu_{j+1} - \mu_j$, and the stage–specific survival rate for stage j can then be estimated from $w_j = M_{j+1}/M_j$.

If the entries to a stage occur at a uniform rate between sample times then it can be shown that the probability of an individual that enters the stage surviving until a sample time is $(\phi-1)/\log_e(\phi)$, where ϕ is the probability of surviving the full time between samples. This implies the relationship $M = M^*(\phi-1)/\log_e(\phi)$ between M^*, the total number entering the stage, and M the number that enter and survive until a sample time. Hence when there is appreciable mortality between samples it may be appropriate to correct the estimate of M_j given by the regression to the value $M^*_j = M_j\log_e(\phi)/(\phi-1)$, using a suitable value for ϕ. Unfortunately, using the regression estimates of ϕ values pertaining to particular stages tends to produce unsatisfactory corrections with this equation because the correction factors can differ greatly from stage to stage. It therefore seems best to use the same ϕ value to correct all entry numbers, and make this equal to the value that reduces M_1 to M_q in the time from μ_1 to μ_q, which is $\phi = (M_q/M_1)^{1/(\mu_q-\mu_1)}$.

For the data in Table 6 the above equations produce very poor estimates when it is assumed that each stage has a different survival rate. More realistic estimates are obtained when the survival rate is (correctly) assumed to be the same in all stages. The resulting estimates are then as follows:

stage	Number entering		Duration		Daily survival	
	Est	True	Est	True	Est	True
1	581	500	2.2	2.0	0.64	0.70
2	284	245	2.8	3.0	0.64	0.70
3	96	84	1.6	2.0	0.64	0.70
4	43	41	–	–	0.64	0.70

For the second set of simulated data in Table 7 the estimates are also completely unsatisfactory when a different survival rate is assumed for each stage. Assuming the same survival rate in each stage and a five parameter entry distribution produces the following estimates that are obviously still not very satisfactory.

Stage	Number entering		Duration		Daily survival	
	Est	True	Est	True	Est	True
1	284	200	1.1	1.5	0.60	0.70
2	174	117	2.5	2.0	0.60	0.75
3	100	66	1.5	2.0	0.60	0.80
4	45	42	0.9	2.0	0.60	0.70
5	13	21	–	–	0.60	0.60

A Development Variable Model

The final model to be discussed in this section is somewhat similar to van Straalen's one discussed above. However, it differs because the development variables for individuals (the continuous stage variables for van Straalen's model) are not assumed to be directly observable. Closely related models have been used by various authors for the analysis of stage–frequency data (Osawa et al., 1983; Stedinger & Shoemaker, 1985; Dennis et al., 1986; Kemp et al., 1986; Shoemaker et al., 1986).

Different times of entry to stage 1 are allowed for by assuming that all individuals enter an unobservable stage 0 (i.e., begin developing) at time zero. Samples from the population are taken at later times $t_1 < t_2 \ldots < t_n$. It is assumed that each individual has a value for a development variable $D(t)$ associated with it at time t such that this determines the stage that the individual is in at that time. Specifically, if $y_{i-1} < D(t) < y_i$ then the individual is in stage i, with $y_{-1} = -\infty$ and $y_q = +\infty$. The 'cut–points' on the development scale $y_0, y_1, \ldots, y_{q-1}$ are the same for all individuals at all times. It is also assumed that at time t the development variable values $D(t)$ are normally distributed with mean $\mu(t)$ and standard deviation $\sigma(t)$. The development scale is fixed by assuming that there is a standard normal distribution at the time of the first sample so that $\mu(t_1) = 0$ and $\sigma(t_1) = 1$. With these assumptions, the probability of an individual being in stage i at time t is

$$P_i(t) = \int_{y_{i-1}}^{y_i} \{\sqrt{(2\pi)}\sigma(t)\}^{-1} \exp[-_1/_2\{x-\mu(t)\}^2/\sigma(t)^2]dx.$$

The model is illustrated graphically on Fig. 4. It is essentially Thurstone's (1925) mental age model (Togerson, 1958, p. 393), and also a model for the 'method of successive intervals' used by psychometricians. Schonemann & Tucker (1967) have provided equations for the maximum likelihood estimation of the unknown parameters for the case where all stages are observable and the sample sizes at different times are arbitrary (rather than having expected sizes determined by the survival of the individuals in the population). Their model is applicable to some types of single cohort data (see section 3.5). An application of it to analyzing data on the intellectual development of New Guinean and Australian school children is given by Manly & Shannon

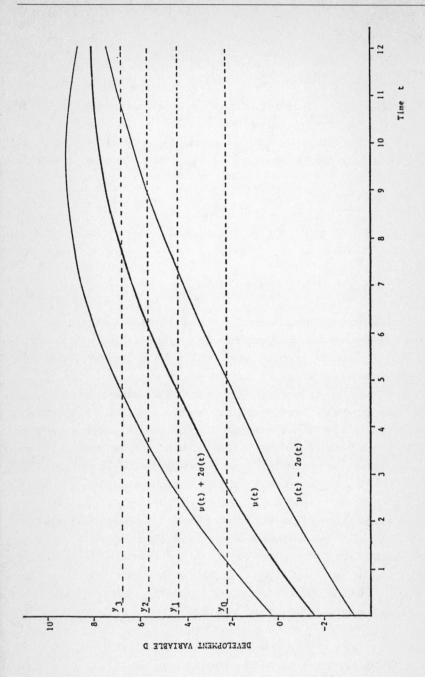

Fig. 4. The development variable model. The broken lines indicate development levels for changing stages, y_1 to y_4; $\mu(t)$ indicates the development progress for average individuals; $\mu(t) \pm 2\sigma(t)$ indicates the development progress for individuals two standard deviations from the mean.

(1974). Dennis et al. (1986) and Kemp et al. (1986) use the model as defined here except that the normal distribution is replaced by the logistic distribution. This change is probably only of minor importance.

With stage–frequency data it is reasonable to put more structure into the model in order to reduce the number of parameters to be estimated, and take into account the fact that the mean μ_t can be expected to increase with time and the standard deviation σ_t to change slowly with time. Any assumptions made to achieve this end are rather arbitrary. Here quadratic time changes in the mean and linear time changes in the standard deviation of the development variable are assumed so that

$$\mu(t) = \alpha_1(t-t_1) + \alpha_2(t-t_1)^2$$

and

$$\sigma(t) = 1 + \beta(t-t_1).$$

The distribution is still standard normal at the time t_1 of the first sample. Other authors have assumed that the mean and variance are both proportional to physiological time (Osawa et al., 1983; Stedinger et al., 1985; Dennis et al., 1986; Kemp et al., 1986) on the basis of models of constant development rates on this scale.

The model is not complete as yet since nothing has been said about mortality. If $F_0(0)$ is the size of the population at time zero, and the constant unit time survival rate is ϕ, then the expected number to be seen in stage i at time t is $F_0(0)\phi^t P_i(t)$, with $P_i(t)$ given by the integral above. Hence the MAXLIK algorithm can be used to estimate the unknown parameters ϕ, α_1, α_2, β, y_1, ..., y_{q-1}, assuming that stage frequencies are sample counts with Poisson distributions.

If all individuals develop at the same rate then $\sigma(t)$ will remain constant and the durations of stages will be the same for all individuals. On the other hand, if $\sigma(t)$ is changing then this implies that the durations of stages vary from individual to individual. If $\sigma(t)$ decreases with time then this suggests that individuals with extreme development values (high or low) are becoming more like average individuals (fast developers are slowing down and slow developers are catching up). If $\sigma(t)$ increases with time then the gap between fast and slow developers is increasing. The implicit assumption in these interpretations of changes in $\sigma(t)$ is that the order of the individuals on the development scale remains the same, so that an individual that is x standard deviations above the mean in one sample is the same number of standard deviations above the mean in all other samples. In that case, an individual with development value $D(t_1)$ at time t_1 (when the population mean and standard deviation are zero and one, respectively) has the value

$$D(t) = \mu(t) + D(t_1)\sigma(t)$$

at time t.

Once estimates of parameters are obtained it is possible to estimate the development value at any time for an individual with the value $D(t_1)$ at time t_1. It is therefore possible to estimate when this individual enters each stage, and hence to estimate the durations of stages. Since the $D(t_1)$ are assumed to follow a standard normal distribution this makes it possible to calculate distributions for stage durations as well as mean durations.

Obviously the assumption that individuals remain in order with regard to their development is questionable. However, if some changes of order do occur then this implies that the durations of stages are more variable than the present model suggests. Hence estimated distributions of durations from the model are in a sense 'minimum variation' estimates. That is to say, the true distributions of durations will tend to have higher variances than the estimated ones if this assumption does not hold.

The first set of test data shown in Table 6 was not generated using the model just described. The procedure adopted was somewhat simpler. (See the description of 'method 3' simulation given above.) The development variable model would have been approximately correct if the distribution of entry times to stage 1 had been made normal, which it was not.

The estimates obtained using the MAXLIK algorithm are shown below, together with the standard errors that are produced as part of the estimation process.

	Estimate	Std. err.
ϕ	.70	0.01
α_1	1.13	0.16
α_2	0.13	0.03
β	0.34	0.04
y_0	2.19	0.10
y_1	6.11	0.14
y_2	13.77	0.53
y_3	19.83	1.12

The true value of ϕ was 0.70 so the estimate of the daily survival rate is correct to two decimal places.

The estimates of α and β values are not themselves particularly informative. What is interesting are the means and standard deviations of development values that they provide at different times, and the population sizes (including those in the unobservable stage 0. These are as follows: —

t	Mean	SD	Size	t	Mean	SD	Size
1	0.00	1.00	885	2	1.26	1.34	621
3	2.77	1.68	436	4	4.54	2.01	306
5	6.56	2.35	215	6	8.84	2.69	151
7	11.38	3.03	106	8	14.17	3.37	74
9	17.22	3.71	52	10	20.53	4.04	37
11	24.09	4.38	26	12	27.91	4.72	18
13	31.98	5.06	13	14	36.31	5.40	9
15	40.89	5.73	6	16	45.74	6.07	4
17	50.83	6.41	3	18	56.19	6.75	2
19	61.80	7.09	2	20	67.66	7.42	1

The population mean is estimated to be increasing with time, as is obviously expected. The estimated standard deviation also increases, which can be interpreted as a spreading out of the population development variables as time goes on.

A simple way to determine the distributions and means of stage durations that are implied by the model involves calculating times of changing stages for individuals with different development values at the time of the first sample, approximating the standard normal distribution at that time with a discrete distribution. This idea may become clearer by considering Table 8 where the results of doing this are shown. Differences between the estimated times of starting stages estimate stage durations. These are shown in Table 9.

A comparison between the mean durations of stages used for generating the data (2, 3 and 2 days for stages 1 to 3, respectively) and the estimated means shown in Table 8 (2.1, 3.1, and 1.9, respectively) shows a good agreement. The stage durations were constant for the generated data but the model has estimated a small amount of variation with non–zero standard deviations for stage durations. This is inevitable, given the model.

The number entering stages 1 to 4 can be estimated using the discrete distribution approximation of Tables 8 and 9. For each of the 13 values of the development variable estimated times of entries to stages 1 to 4 are known, and hence survival rates from the first sample time can be determined. Numbers surviving can then be estimated from the estimated number with each development value at the time of the first sample (with a total of 885), and these can be added to give estimates of the total numbers entering stages. The estimates obtained in this way are 511.4, 246.4, 83.5 and 42.7, respectively. The corresponding true values are 500, 245, 84 and 41. Overall, the development variable model has performed very well with this example.

Table 8. Estimated times of entering stages for individuals with different development parameters in sample 1.

Development parameter	Probability	Stage			
		1	2	3	4
−3.00	0.002	6.9	9.0	12.0	13.9
−2.50	0.009	6.0	8.1	11.2	13.1
−2.00	0.027	5.2	7.3	10.4	12.4
−1.50	0.065	4.4	6.6	9.7	11.7
−1.00	0.121	3.7	5.9	9.0	11.0
−0.50	0.176	3.1	5.3	8.4	10.3
0.00	0.200	2.6	4.7	7.8	9.7
0.50	0.176	2.1	4.2	7.3	9.2
1.00	0.121	1.7	3.7	6.7	8.6
1.50	0.065	1.4	3.3	6.3	8.1
2.00	0.027	1.1	2.9	5.8	7.6
2.50	0.009	*	2.6	5.4	7.2
3.00	0.002	*	2.3	5.0	6.8
	1.000				

* Not entered during the sampling period (in stage 1 at time 1).

Table 9. Estimated durations of stages calculated by taking differences between times of entry to stages shown in Table 8, where this is possible. Means and standard deviations are for completed stages only, using conditional probabilities. Stage 0 refers to the time of entry to stage 1.

Development parameter	Probability	Stage			
		0	1	2	3
−3.00	0.002	6.9	2.1	3.0	1.9
−2.50	0.009	6.0	2.1	3.1	1.9
−2.00	0.027	5.2	2.1	3.1	2.0
−1.50	0.065	4.4	2.2	3.1	2.0
−1.00	0.121	3.7	2.2	3.1	2.0
−0.50	0.176	3.1	2.2	3.1	1.9
0.00	0.200	2.6	2.1	3.1	1.9
0.50	0.176	2.1	2.1	3.1	1.9
1.00	0.121	1.7	2.0	3.0	1.9
1.50	0.065	1.4	1.9	3.0	1.8
2.00	0.027	1.1	1.8	2.9	1.8
2.50	0.009	*	*	2.8	1.8
3.00	0.002	*	*	2.7	1.8
Mean		2.7	2.1	3.1	1.9
Sd		1.0	.1	.1	.1

* Cannot be determined because individuals were in stage 1 at the time of the first sample.

When the same analysis was carried out on the second set of test data shown in Table 7, the estimates shown in the following table were obtained: —

Stage	Number entering		Duration		Daily Survival	
	Est	True	Est	True	Est	True
1	242	200	1.2	1.5	0.67	0.80
2	150	117	1.8	2.0	0.67	0.75
3	73	66	2.1	2.0	0.67	0.80
4	33	42	2.5	2.0	0.67	0.70
5	12	21	–	–	0.67	0.60

These estimates are fairly reasonable considering the poor data and the incorrect assumption that the survival rate was constant.

Estimation with Additional Information: Parameters Constant in Time

Many of the methods that have been suggested for estimating population parameters from stage–frequency data require more information to be available than just these stage–frequencies. The types of additional information that are required varies from method to method, but in most cases is one or more of (a) the number of individuals entering stage 1, (b) the distribution of entry times to stage 1, or (c) the mean durations of stages. Generally speaking, the reason for incorporating extra information into the estimation process is in order to allow more population parameters to be estimated. For example, if mean stage durations are known accurately then it seems likely that this will make the estimation of a different daily survival rate for each stage easier than would otherwise be the case.

There are various reasons why extra information might be available on a population, over and above sample stage–frequencies. In laboratory experiments, it could be that one or more parameters have been fixed by the experimenter. On the other hand, it could be that certain parameters (particularly the mean durations of stages) can be estimated from previous experiments.

The present section reviews the methods that require extra information and are based on the assumption that survival parameters are constant over time. This includes methods that allow for a different daily survival rate in each stage, providing that these survival rates are assumed to be constant during the entire sampling period.

The Richards et al. (1960) Method

Richards et al. (1960) suggested the use of the equation

$$A = M(\phi^a - 1)/\log_e(\phi)$$

to estimate the survival rate per unit time, ϕ, in a stage where M is the known number entering the stage, a is the known duration of the stage, and A is the area under the sample stage–frequency curve. This is the same as one of the equations for the Kiritani–Nakasuji–Manly method of estimation but with the survival rate per unit time set equal to ϕ rather than $\exp(-\Theta)$. Richards et al. pointed out that if M, a and A are known then the equation can be solved to find ϕ for a stage. Different stages will have their own particular values of M, a and A. Consequently, they will have different estimated survival rates.

With samples taken at unit intervals of time during the entire period during which individuals are alive in stages 1 to q (using interpolated stage–frequencies if necessary), the trapezoidal rule makes the area under the sample stage j frequency curve equal to the sum $A_j = \Sigma f_{ij}$ of the stage frequencies. Once A_1 has been determined it can be substituted, together with M_1, the number entering stage 1, and a_1, the duration of stage 1, into the above equation, which can then be solved to find the estimated daily survival rate ϕ_1 for stage 1. Having estimated ϕ_1, the stage–specific survival rate is estimated as $w_1 = \phi_1^{a1}$ so that the number entering stage 2 is estimated as $M_2 = M_1 w_1$. The daily survival rate in stage 2, ϕ_2, can then be found by solving the equation using this value for M_2 together with appropriate values for a_2 and A_2. Having determined an estimate ϕ_2 for ϕ_2, and estimate of M_3 is $M_2 \phi_2^{a2}$, which can be used to determine ϕ_3, etc. Thus, starting with stage 1 it is possible to successively estimate a daily survival rate for all stages.

A simple way to solve Richards et al.'s equation involves rewriting it as

$$\phi = \exp\{-M(1 - \phi^a)/A\},$$

and using it iteratively. Thus $\phi = 0$ can be substituted into the right had side to get a new value for ϕ. This new value can then be used to get another new value. If the process converges then the value obtained for ϕ will be the required solution of the equation. If $\phi = 1$ then all individuals spend exactly a time a in stage i, which implies than $A = aM$. This is therefore the largest possible area under the stage–frequency curve that is allowed. If A is larger than aM then the equation does not have a solution with $\phi < 1$.

As an example, consider again the first set of computer generated stage–frequency data shown in Table 6. There were $M_1 = 500$ individuals that entered stage 1, the mean duration of the stage was $a_1 = 2.0$ days and the area under the sample stage–frequency curve is $A_1 = 741$. The estimation equation therefore becomes $\phi_1 = \exp\{-500(1 - \phi_1^2)/741\} = \exp\{-0.6747(1 - \phi_1^2)\}$. Putting $\phi_1 = 0$ in the right hand side produces the new value $\phi_1 = 0.509$. Substituting this into the right hand side gives $\phi_1 = 0.607$. Continuing in this way it is found that iterations stabilize on the value $\phi_1 = 0.729$. This is therefore the Richards et al. estimate of the daily survival rate in stage 1. The stage–specific survival rate is then estimated as $w_1 = 0.729^2 = 0.531$ and the number entering stage 2 is estimated as $M^2 = w_1 M_1 = 265.72$.

Treating the other stages in a similar way leads to the following comparison between estimates and true values for all the stages: –

Stage	Daily survival Estimate	True	Stage specific survival Estimate	True	Number entering stage Estimate	True
1	0.73	0.70	0.53	0.49	–	500.0
2	0.64	0.70	0.27	0.34	265.7	245.0
3	0.83	0.70	0.69	0.49	70.7	84.0
4	0.66	0.70	–	–	48.6	41.2

The estimates are fairly reasonable.

Applying the Richards et al. equations to the second set of test data (Table 7) produces the following results: –

Stage	Daily survival Estimate	True	Stage–specific survival Estimate	True	Number entering stage Estimate	True
1	0.69	0.70	0.57	0.59	–	200
2	0.79	0.75	0.62	0.56	114	117
3	0.71	0.80	0.50	0.64	71	66
4	0.74	0.70	0.55	0.49	35	42
5	0.49	0.60	–	–	19	21

Again, these are fairly reasonable estimates.

The Southwood and Jepson (1962) Method

In his review of methods for the analysis of stage–frequency data, Southwood (1978, p. 358) describes Southwood and Jepson's (1962) graphical method as the crudest and simplest one available, which provides an overestimate of the number alive mid–way through the stage. Sawyer & Haynes (1984) also noted the simplicity of the method but claimed that it is applicable to realistic situations in which the daily survival rate varies from stage to stage, or when data are not available for all stages. To use the method it is only necessary to know the mean durations of stages. This is an apparent advantage over Richard's et al.'s method which requires a knowledge of the number entering stage 1 as well.

If a is the duration of a stage and A is the area under the stage–frequency curve, then the number entering the stage is estimated by

$$M = A/a.$$

Since the area A under a stage–frequency curve is equal to the number entering the stage multiplied by the mean time spent in the stage, the estimate M will only be unbiased if all individuals spend a time a in the stage. Therefore, using the equation amounts to assuming that all the mortality in the stage takes place at the end of the stage. Non—survivors are assumed to remain in the stage for a time a and then die rather than passing on to the next stage.

Sawyer & Haynes (1984) considered the biases involved in using Southwood and Jepsons's equation and suggested a 'correction' process for improving estimates. Unfortunately, the 'corrected' estimates will not always be better than the original ones. On the whole, it does not seem to be a good idea to use a method of estimation that is based on an equation that is known to be incorrect, and not necessarily even a good approximation.

The Dempster (1961) Method

The method proposed by Dempter (1961) requires a knowledge of the distribution of times of entry to stage 1. It is based on the equation

$$F_{i+11} - F_{i1} = P\alpha_i - 1/2(f_{i1}+f_{i+11})\Theta_1 - 1/2(f_{i2}+f_{i+12})\Theta_2 - ... -$$
$$1/2(f_{iq}+f_{i+1q})\Theta_q$$

for the change in the total sample frequencies between sample i and sample i+1. Here the left hand side is the change in the sample cumulative frequency, which is attributed to the net result of $P\alpha_i$ new entries to stage 1, and deaths in the stages 1, 2, ..., q. The value P represents the total number entering stage 1; α_i is the proportion of these entries to occur between samples i and i+1, which is assumed known. The term involving Θ_j on the right hand side represents the deaths of individuals in stage j. The number of individuals at risk in this stage in the time interval being considered it taken as $1/2(f_{ij}+f_{ij+1})$, the average of the observed frequencies in the stage. The daily mortality rate in stage $_i$ is Θ_j.

It is possible to set up a multiple regression for this model. The left hand side of the equation gives the dependent variable, the independent variables are α_i, and the coefficients of Θ_1, Θ_2, ..., Θ_q. Fitting the multiple regression model is then equivalent to estimating P and the Θ values by Dempster's least squares method.

The Kobayashi (1968) Method

Kobayashi (1968) put forward a method for estimating the numbers in a population entering different stages which does not require a knowledge of stage durations and allows daily survival rates to be different in different stages. It is necessary to know the number entering stage 1.

The method involves a two part process. First, the deaths are estimated for each stage between each pair of successive samples by apportioning the total deaths between samples according to the numbers in stages mid—way between samples. The deaths in each stage are then added up to get first approximations to the total numbers dying in each stage. Second, the numbers of deaths in different stages between successive samples are 'corrected' by making them proportional to the product of stage—frequencies mid—way between samples and overall deaths in a stage divided by the total stage—frequency. It is by no means clear from Kobayashi's explanation why the second part in this process should lead to improved estimates of the total number of deaths in stages. Only one equation is provided to justify the calculations, and the connection between the equation and the calculations is somewhat obscure.

The Lakhani & Service (1974) Method

Lakhani & Service (1974) set out to use Dempster's (1961) method to estimate mortalities in the immature stages of the mosquito *Aedes cantans*, but were unable to determine the required information on the emergence time distribution. Instead, they ended up with estimates of the durations of stages from laboratory experiments, and an estimate of the proportion of 1st instar larvae that survive to become adults. From this information, and their sample stage–frequencies, they proposed three methods for estimating mortality rates in different development stages. Methods A and C are graphical, and partly subjective; they will not be considered here.

Method B is based on Richards et al.'s (1960) equation. Writing $\phi_j = \exp(-\Theta_j)$, this becomes

$$A_j = M_j\{1 - \exp(-\Theta_j a_j)\}/\Theta_j ,$$

where

$$M_j = \exp(-\Theta_1 a_1 - \Theta_2 a_2 - ... - \Theta_{j-1} a_{j-1}) ,$$

Hence if the A_j values are estimated for the stages 1 to q−1, then there are q equations in q unknowns

$$A_1 = M_1\{1 - \exp(-\Theta_1 a_1)\}/\Theta_1,$$
$$A_2 = M_1 \exp(-\Theta_1 a_1)\{1 - \exp(-\Theta_2 a_2)\}/\Theta_2,$$
$$\cdot$$
$$\cdot$$
$$\cdot$$
$$A_{q-1} = M_1 \exp(-\Theta_1 a_1 - \Theta_2 a_2 - ... - \Theta_{q-2} a_{q-2})\{1 - \exp(\Theta_{q-1} a_{q-1})\}/\Theta_{q-1},$$

and

$$P = \exp(-\Theta_1 a_1 - \Theta_2 a_2 - ... - \Theta_{q-1} a_{q-1}),$$

where P is the probability of an individual that enters stage 1 surviving to enter stage q. The unknowns are the Θ values and M_1, all other parameters being assumed to be known. The equations can be solved by a standard computer program for sets of non–linear equations.

Lakhani & Service describe an approximate method of solution that involves changing the equations to linear ones. Nowadays there is no need to adopt this type of expediency.

For the test data in Table 6 the equations are satisfied with the estimates $\Theta_1 = 0.375$, $\Theta_2 = 0.386$, $\Theta_3 = 0.296$ and $M_1 = 526.8$. The Θ values used for generating the data were 0.357 for all stages and the true value for M_1 was 500. The estimates are therefore good. For the test data of Table 7 the estimates obtained by solving the equations are as follwos, with true values in parethesis: $\Theta_1 = 0.464$ (0.357), $\Theta_2 = 0.159$ (0.288), $\Theta_3 = 0.443$ (0.223), $\Theta_4 = 0.185$ (0.357) and $M_1 = 211.9$ (200). Here the estimates of the survival parameters are not very good.

The Birley (1977) Method

Birley (1977) noted four problems that had not been properly addressed with previously proposed methods for analyzing stage–frequency data. First, many insect species migrate shortly after becoming adults and the daily survival rate for the sampled population should fall abruptly to zero when this happens. Second, sources of mortality associated with specific dates, such as the application of insecticide need to be allowed for. Third, survival rates are likely to vary with stage and age, and this variation is of crucial importance in understanding population dynamics. Lastly, the distribution of recruitment time to the first stage may well be a complicated function of temperature and other environmental variables.

Birley proposed a method for overcoming these problems, assuming that the distributions of times of entry to stages are known. This is based on the equation.

$$f_j(t) = M_j \sum_{i=0}^{D_j} P_j(t-i)S_{ij},$$

for the population frequencies in stage j at the discrete points in time 1, 2, 3, ... n. Here M_j is the total number entering the stage, D_j is the maximum duration of the stage, $P_j(t)$ is the fraction of the population entering the stage just before time t, and S_{ij} is the number of individuals that are still in the stage at a time i after entering, with $S_{oj} = 1$. The equation says that the $f_j(t)$ consists of the sum of contributions from individuals entering the stage at various times before t. The first contribution, $M_j P_j(t)S_{oj} = M_j P_j(t)$, is from individuals that enter just before time t. The second contribution, $M_j P_j(t-1)S_{1j}$, comes from the survival in the stage of the $M_j P_j(t-1)$ individuals that entered just before time t−1. The last contribution, $M_j P_j(t-i)S_{ij}$, where i = D_j, comes from the survival of the $M_j P_j(t-D_j)$ individuals that entered just before time t−D_j. The maximum duration of the stage ensures that individuals that entered before time t−D_j will have left by time t.

Birley suggested several approaches for estimating the parameters of the model, assuming that the proportions $P_j(1)$, $P_j(2)$, $P_j(3)$, ... are known. The only sure restriction on the survival function is that $S_{oj} \geq S_{1j} \geq S_{2j} \geq S_{3j} ... \geq 0$. One possibility therefore involves estimating each of the survival parameters S_{ij} independently, subject to these constraints, using the principle of least squares. Another possibility for estimation involves assuming an appropriate parametric form for the survival function and estimating by non–linear regression. If the stage duration D_j is unknown then this can be estimated by finding the value that produces the minimum sum of squares. It is easy to incorporate a 'catastrophe' into the survival function if the time of this is known. The main problem with using this model is the need to know the distribution of entry times to stages.

The Bellows & Birley (1981) Method

Bellows & Birley (1981) extended Birley's (1977) method by developing equations for estimation that allow for a different survival rate for each stage but only need a knowledge of the distribution of times of entry to stage 1 in addition to sample stage–frequencies. Although Bellows & Birley only provided equations for the development of the individuals in a single cohort entering stage i at the same time, it is fairly straightforward to extend these equation to cover the development of several cohorts entering a stage at different times.

The first important equation is

$$f_j(t) = M_1 \sum_{i=0}^{t} P_j(i) \phi_j^{t-i} \{1 - H_j(t-i)\},$$

which says that the frequency in stage j at time t consists of contributions from those entering at times 0, 1, 2, ..., t. The number entering at time i is the total number of individuals entering stage 1, M_1, multiplied by the proportion of these entering stage j at this time, $P_j(i)$. These individuals will still be in the stage at time t if they survive, with probability ϕ_j^{t-i}, and do not pass into stage j+1, with probability $1 - H_j(t-i)$. Thus the contribution to $f_j(t)$ from entries at time i is given by the ith term in the summation on the right hand side of the equation.

An important aspect of this model is the assumption of a distribution for the duration of a stage. In the above equation, $H_j(t)$ denotes the probability that an individual spends t or less units of time in stage j. Hence if $h_j(t)$ denotes the probability of a stage duration of t time units, then

$$H_j(t) = h_j(0) + h_j(1) + ... + h_j(t).$$

A second important equation is

$$P_{j+1}(t) = \sum_{i=0}^{t} P_j(i) \phi_j^{t-i} h_j(t-i),$$

which shows the proportion of the population entering stage j=1 at time t as a function of the proportions entering stage j at different times. It says that the entries to stage j+1 at this time is made up of individuals entering stage j at times 0, 1, 2. ..., t that survive to leave stage j at time t. Thus the ith term in the summation is the proportion entering stage j at time i, $P_j(i)$, multiplied by the probability of surviving the time interval i to t, $\phi_j{}^{t-i}$, multiplied by the probability of a stage duration of t–i, $h_j(t-i)$.

To complete the model it is necessary to make appropriate assumptions about the distributions of durations of stages. Bellows & Birley suggest that it is realistic to assume that the development rate in a stage is normally distributed between individuals so that the duration of a stage is the reciprocal of a normally distributed variable. This reflects the distributions of durations that have been observed experimentally. Another possibility is to assume that the logarithm of the stage duration is normally distributed.

Bellows & Birley suggested using non–linear regression to estimate the parameters of their model. However, if the data stage–frequencies are regarded as Poisson variates then the MAXLIK algorithm can be used instead. An attempt was made to estimate parameters for the test data using this algorithm but unfortunately convergence difficulties were experienced, apparently because of high correlations between estimates of means and the standard deviations. This type of problem is often an indication of an over–parameterized model. It can probably be overcome by reducing the number of parameters to be estimted, for example by setting one or more standard deviations to zero and estimating the remaining parameters. However, this possibility has not been pursued.

The Bellows & Birley model is considered further below for the analysis of single cohort data. It is for this application that it seems most useful.

The Mills (1981a, 1981b) Methods

In his first paper, Mills (1981a) considered how to estimate the mean durations of stages. With arguments based upon particular numerical examples, he demonstrated that in the absence of mortality the relationship

$$2(B_{j+1} - B_j) = a_j + a_{j+1}$$

holds, where B_j is the 'mean' time of the stage frequency curve for stage j,

$$b_j = \sum_{i=1}^{n} jf_{ij} \Big/ \sum_{i=1}^{n} f_{ij},$$

assuming that samples are taken daily over the full development time of the population. Mills then went on to suggest a method for 'correcting' this equation to allow for mortality. The corrected equation can then be used to estimate values for mean stage durations providing at least one duration, or the ratio of two durations is known.

In his second paper, Mills (1981b) tackled the problem of estimating the numbers entering different stages assuming that stage durations are known. He noted that the area under the stage j frequency curve can be expressed as the sum of a contribution from those that survive to enter stage J+1, and a contribution from those that die in stage j. That is,

$$A_j = M_j w_j a_j + M_j(1-w_j)x_j a_j,$$

$$= m_j\{(1-x_j)a_j w_j + x_j a_j\},$$

where M_j is the number entering stage j, w_j is the proportion surviving stage j, a_j is the duration of stage j, and x_j is the fraction of the stage duration a_j that non–survivors of the stage are present for. The problem with using this equation is that the value x_j is unknown. Mills suggested taking $x_j = 0$ (all deaths occurring at the start of the stage), so that $A_j/a_j = M_{j+1}$, and $x_j = 1$ (all deaths occurring at the end of the stage), so that $A_j/a_j = M_j$. This shows that A_j/a_j is an upper limit for the number entering stage j+1 and a lower limit for the number entering stage j. An 'average' estimate of M_j is then

$$M_j = (A_{j-1}/a_{j-1} - A_j/a_j)/2.$$

This equation cannot be used with stage 1 since A_0 and a_0 have no meaning. Mills suggested using $M_1 = M_1/w_2$ in this case, with w_1 being determined in some other way. This seems most unsatisfactory.

Estimation with Additional Information: Parameters Varying with Time

The previous section reviewed eight different methods that have been proposed for estimating population parameters from stage–frequency data augmented with some extra information. For all of these methods a prime consideration was to avoid the assumption of the same daily survival rate in each stage. However, all model parameters were assumed to remain constant with time. It is in this respect that the methods to be considered in the present section are different. For the methods that are to be reviewed now, a primary consideration of the authors was the need to allow survival rates to change with time.

The Ruesink (1975) Method
Ruesink (1975) suggested that it is often the case that the mortality in a population occurs mainly around the time when individuals are transfering from one stage to the next. On this basis he proposed a model from which it is possible to estimate stage–specific survival rates that are changing with time. The mean durations of all stages have to be known.

Ruesink noted that the total number of individuals that have entered stage j by time t is given by the equation

$$C_j(t) = f_j(t) + D_j(t)$$

where $f_j(t)$ is the number of individuals still in stage j and $D_j(t)$ is the number that have passed through the stage at this time. If there are no deaths within stage j then the relationship

$$D_j(t) = C_j(t-a_j)$$

will also apply, where a_j is the duration of the stage j for all individuals: all individuals that entered the stage at or before time $t-a_j$ will have passed on to the next stage by time t.

These two equations can be used to calculate $C_j(t)$ and $D_j(t)$ values for a range of values of t. Linear interpolation can be used to evaluate $C_j(t-a_j)$ if $t-a_j$ does not correspond to one of these t values. The survival rate in passing from stage j to stage j+1 can then be estimated at time t by

$$w_j(t) = \{C_{j+1}(t+\delta) - C_{j+1}(t-\delta)\}/\{D_j(t+\delta) - D_j(t-\delta)\}.$$

This is the estimated number entering stage j+1 in the time interval from $t-\delta$ to $t+\delta$, divided by the estimated number leaving stage j in the same interval. The value of δ is at choice. Increasing δ has the effect of increasing the 'smoothing' of survival estimates.

Table 10 shows the $C_j(t)$ and $D_j(t)$ values that are obtained from the test data in Table 6 when the equations are applied. Ruesink suggested that sampling errors may make it necessary to adjust the $C_j(t)$ and $D_j(t)$ values before using them in the last equation since it is important to ensure that these are both non–decreasing functions of time. For the present example all that was done was to make $C_j(t)$ equal to $C_j(t-1)$ in cases where otherwise $C_j(t) < C_j(t-1)$. The stage durations used in the calculations were $a_1 = 2.0$, $a_2 = 3.0$, $a_3 = 2.0$ and $a_4 = 2.8$. The latter figure is the mean survival time in the adult stage based on a daily survival rate of 0.70. The need to use this figure highlights a problem with Ruesink's method that is not discussed in his paper. If the final stage studied is the adult stage then it is necessary to assume that all adults live exactly the same time before leaving the stage (i.e., before dying). This is obviously a rather questionable approximation.

Table 10 also shows the stage–specific survival estimates obtained using $\delta = 2$. For example, it is estimated that at time $t_6 = 6$, the probabilities of an individuals surviving stages 1 to 3 were $w_1(6) = 0.42$, $w_2(6) = 0.39$ and $w_3(6) = 0.67$. In fact, the stage–specific survival rates were constant for the population at $w_1 = 0.49$, $w_2 = 0.34$ and $w_3 = 0.49$. The estimates from Ruesink's equations are very variable and include an impossible value exceeding 1. It seems clear that an attempt is being made to estimate far too many parameters from the amount of data available. The value of $\delta = 2$ is arbitrary. However, using values of δ equal to 1 and 3 gave somewhat similar (unsatisfactory) results.

The C values $C_1(19) = 371.0$, $C_2(19) = 172.0$, $C_3(19) = 68.0$ and $C_4(19) = 45.2$ for the last sample are estimates of the total numbers entering the four stages. We can expect that these estimates will be more stable than the large number of very variable $w_j(t)$ estimates. In fact, the numbers entering the four stages were 500.0, 245.0, 84.0 and 41.2, respectively, so the estimates are not particularly good.

The estimates from Ruesink's method for the second set of test data (Table 7) are shown in Table 11. Here again the time varying stage–specific survival estimates are extremely variable and not clearly related to the true values. The C values for the final sample are calculated as

$C_1(19) = 148.9$, $C_2(25) = 108.0$, $C_3(25) = 53.0$, $C_4(25) = 31.0$ and $C_5(25) = 16.2$. Again, these should be estimates of the total numbers entering stages, for which the true values were 200.0, 117.1, 65.9, 42.2 and 20.7, respectively.

In assessing these examples it must be kept in mind that the computer model used to generate the data is not Ruesink's model. Mortality was inflicted on a daily basis rather than at the conclusion of each stage.

Table 10. Calculation of stage specific survival rates using Ruesink's (1975) method on the first set of test data. The C_i values are estimates of the total numbers that have entered a stage by the sample time shown; the D_i values are estimates of the total numbers that have left the stage at the sample time. These are both rounded to integers for display. The δ value for time has been set at 2.

Time	Total entries to time				Total losses by time				Stage-specific survival rates		
	C_1	C_2	C_3	C_4	D_1	D_2	D_3	D_4	w_1	w_2	w_3
2	164	0	0	0	0	0	0	0	–	–	–
3	252	4	0	0	0	0	0	0	0.48	–	–
4	345	90	0	0	164	0	0	0	0.35	0.75	–
5	359	121	0	0	252	0	0	0	0.47	0.30	–
6	371	121	3	0	345	4	0	0	0.39	0.36	0.33
7	371	171	27	0	359	90	0	0	0.42	0.39	0.67
8	371	171	44	1	371	121	3	0	1.96	0.39	0.55
9	371	171	47	18	371	121	27	0	0.08	0.51	0.52
10	371	172	68	24	371	171	44	0	–	0.48	0.48
11	371	172	68	24	371	171	47	4	–	0.41	0.49
12	371	172	68	32	371	171	68	19	–	–	0.59
13	371	172	68	38	371	172	68	24	–	–	0.72
14	371	172	68	38	371	172	68	26	–	–	–
.											
.											
20	371	172	68	45	371	172	68	43	–	–	–
True value	500	245	84	41	500	245	84	41	0.49	0.34	0.49

The Bellows et al. (1982) Method

An extension to the use of Bellows & Birley's (1981) model was suggested by Bellows et al. (1982) for situations where there are marked changes in the survival rate at particular points in time. They suggested a two stage estimation procedure. At the first stage only total population counts are considered. The time trend of these is used to estimate two or more survival rates that apply at different intervals of time. Once these survival rates have been determined, the second stage involves using them with Bellows & Birley's model to determine the distribution of stage durations. Thus suppose that the total population size at time 0 is N_0, and decreases from then on because of deaths. If the survival rate from time t−1 to time t is $\exp(-\Theta_t)$ in all stages, then at time t the population size will be

$$N_t = N_0 \exp(-\Theta_1 - \Theta_2 - ... - \Theta_t).$$

Hence the logarithm of the population size at time t will be

$$Y_t = \log_e(N_t) = \log_e(N_0) - \Theta_1 - \Theta_2 - ... \Theta_t.$$

Table 11. Estimates of stage–specific survival rates for the second set of test data, as calculated using Ruesink's (1975) method with $\delta = 2$.

Time	w_1	w_2	w_3	w_4
2	–	–	–	–
3	0.84	0.87	–	–
4	0.93	0.33	1.50	–
5	0.74	0.57	0.83	–
6	0.65	0.37	0.50	–
7	0.59	0.57	0.65	0.40
8	0.60	0.51	0.47	0.42
9	0.65	0.54	0.32	0.27
10	0.59	0.48	0.52	0.77
11	0.83	0.43	0.41	0.52
12	0.49	0.51	0.76	0.64
13	1.00	0.38	0.74	0.92
14	–	1.00	.079	0.44
15	–	0.23	0.79	0.51
16	–	–	0.31	0.41
17	–	–	1.33	0.36
18	–	–	–	0.75
19	–	–	–	–
20	–	–	–	–
True values	0.59	0.56	0.64	0.49

If the daily survival rate is constant from time 0 to time τ, changes, and remains constant from time $\tau+1$ on, then this implies that $\Theta_1 = \Theta_2 = ... \Theta\tau$, and $\Theta\tau_{+1} = \Theta\tau_{+2} = \Theta\tau_{+3} = ...$ Given sample estimates of the Y_t values, this model can be estimated by a multiple regression, for any given τ. If τ is not known then the regression can be carried out for all possible values of τ, and the one that minimizes the residual sum of squares can be chosen.

This regression approach for estimating time varying survival parameters can be generalized easily enough for situations where there are three or more Θ's. Bellows et al. describe an alternative way of doing essentially the same calculations. Given the ready availability of multiple regression programs, the approach suggested here seems easier to use.

The Derr & Ord (1979) Method

Derr & Ord (1979) method bears a superficial resemblance to Ruesink's (1975) method. Both assume that the durations of stages are known, and both methods result in estimates of stage–specific survival rates that change with time. They also share the assumption that most mortality occurs at the point of transition from one stage to the next. However, the calculations for the two methods are based on different principles.

Derr & Ord developed the equation

$$w_j(t) = \frac{\sum\limits_{v=0}^{d_j-1} f_{j+1}(t-v)\exp(-\Theta v) + f_{j+1}(t-d_j)\exp(-\Theta d_j)\delta_j}{\sum\limits_{u=0}^{d_{j+1}-1} f_j(t-u-a_j)\exp(-\Theta u) + f_j(t-a_j-d_{j+1})\exp(-\Theta d_{j+1})\delta_{j+1}}$$

for time–dependent stage–specific survival rates, where d_j is the integer part of a_j and $\delta_j = a_j - d_j$. Assuming that Θ is known, sample stage frequencies can be used in place of the population values on the right hand side of this equation, and hence an estimate of $w_j(t)$ can be determined. It will usually be necessary to interpolate to determine sample values for $f_j(t)$ when t is not an integer. In practice, Θ will not usually be known. However, Derr & Ord suggest that the estimate of $w_j(t)$ will not be strongly effected by small changes in Θ so an approximate value will be all that is needed. In their own example they take $\Theta = 0$.

For the test data of Table 6, the durations of stages are $a_1 = 2.0$, $a_2 = 3.0$ and $a_3 = 2.0$ days. using these values together with the sample stage–frequencies (interpolated as required), produces the estimates of stage–specific survival rates that are shown in Table 12. The calculations have been done for $\Theta = 0$, 0.2 and 0.4. (The true Θ value was 0.36, corresponding to a 70% survival rate per day.) The true stage–specific survival rates, which were constant, are also shown in the table. Clearly, the time varying estimates of stage–specific survival rates have large components of random sampling variation. An attempt is being made to estimate too many parameters from the available data. Also, the estimates are not insensitive to fairly large changes in the assumed Θ value. Indeed, the sequences of estimates of w_3 have changed dramatically with Θ. What is more, these sequences are showing trends although the true value of this parameter was constant.

For the second set of test data of Table 7 estimates are shown in Table 13. In this case the daily survival rate in the population did vary, but this was from stage to stage rather than with time. Estimates are shown for $\Theta = 0$ and 0.4. True Θ values varied from 0.51 in stage 5 to 0.22 in stage 3. Again, the estimates are extremely variable and the Θ value used is important.

Table 12. Estimates of stage–specific survival rates for the first set of test data, as calculated using Derr & Ord's (1979) method with $\Theta = 0$, 0.2 and 0.4.

Time	$\Theta = 0$			$\Theta = 0.2$			$\Theta = 0.4$		
	w_1	w_2	w_3	w_1	w_2	w_3	w_1	w_2	w_3
3	0.02	–	–	0.02	–	–	0.02	–	–
4	0.22	–	–	0.24	–	–	0.25	–	–
5	0.35	–	–	0.39	–	–	0.43	–	–
6	0.40	0.03	–	0.46	0.03	–	0.52	0.03	–
7	0.56	0.14	–	0.68	0.15	–	0.81	0.16	–
8	0.91	0.33	0.03	1.14	0.33	0.03	1.43	0.34	0.03
9	1.81	0.50	0.27	2.28	0.45	0.29	2.87	0.41	0.31
10	9.00	0.70	0.46	11.26	0.63	0.53	14.16	0.59	0.06
11	–	0.96	0.38	–	0.87	0.47	–	0.79	0.56
12	–	3.00	0.28	–	2.46	0.41	–	2.01	0.57
13	–	3.00	0.23	–	2.46	0.42	–	2.01	0.73
14	–	–	0.19	–	–	0.40	–	–	0.84
15	–	–	0.12	–	–	0.32	–	–	0.82
16	–	–	0.09	–	–	0.31	–	–	0.97
17	–	–	0.08	–	–	0.31	–	–	1.17
18	–	–	0.04	–	–	0.20	–	–	0.88
19	–	–	0.03	–	–	0.16	–	–	0.95
20	–	–	0.02	–	–	0.11	–	–	0.71
True values	0.49	0.34	0.49	0.49	0.34	0.49	0.49	0.34	0.49

The Relative Merits of Different Methods of Analysis

The previous three section have reviewed 21 different methods that have been proposed for the analysis of multi–cohort stage–frequency data. It is now possible to state some general conclusions, although it is clear that far more work is required to properly assess and compare the most promising of these methods.

To begin with, certain of the methods have been shown to be based upon models that are either unlikely to be correct or of doubtful validity. Obviously these methods should not normally be used. Within this category there are the methods of Rigler & Cooley (1974), Southwood and Jepson (1962), Kobayashi (1968), and Mills (1981a,b).

In assessing the other methods two factors are crucial. First, it makes a difference whether the survival rate per unit time is (i) constant, (ii) varies from stage to stage but is constant with time, or (iii) varies with time but is the same in all stages. Second, it makes a difference whether

Table 13. Estimates of stage–specific survival rates for the second set of test data, as calculated using Derr & Ord's (1979) method with $\Theta = 0$ and 0.4.

Time	$\Theta = 0$				$\Theta = 0.4$			
	w_1	w_2	w_3	w_4	w_1	w_2	w_3	w_4
2	0.71	–	–	–	0.86	–	–	–
3	0.61	–	–	–	0.60	–	–	–
4	0.97	0.17	–	–	1.07	0.19	–	–
5	0.89	0.25	–	–	0.93	0.25	–	–
6	0.77	0.30	0.38	–	0.81	0.32	–	0.41
7	0.48	0.45	0.57	–	0.48	0.47	0.58	–
8	0.50	0.61	0.37	–	0.55	0.63	0.33	–
9	0.50	0.72	0.40	0.20	0.52	0.71	0.45	0.30
10	0.59	0.68	0.52	0.25	0.64	0.67	0.50	0.32
11	0.50	0.46	0.32	0.14	0.51	0.43	0.30	0.18
12	0.48	0.46	0.37	0.21	0.52	0.49	0.41	0.44
13	0.38	0.77	0.71	0.21	0.39	0.78	0.72	0.38
14	0.23	1.31	0.90	0.14	0.23	1.48	0.92	0.29
15	–	4.33	0.62	0.17	–	4.83	0.54	0.33
16	–	–	0.46	0.18	–	–	0.55	0.45
17	–	–	1.33	0.09	–	–	1.33	0.25
18	–	–	–	0.06	–	–	–	0.27
19	–	–	–	0.04	–	–	–	0.20
20	–	–	–	0.02	–	–	–	0.22
True values	0.59	0.56	0.64	0.49	0.59	0.56	0.64	0.49

data are complete or not. Here 'complete' means that observations start either before or at the time when individuals enter stage 1, and continue until all individuals are dead.

With a survival rate of type (i) and complete data the Kiritani–Nakasuji–Manly (KNM) method of estimation has a great deal to recommend it in comparison with the possible alternatives. A simulation study (Manly, 1974b) has indicated that it produces better estimates than the methods of Richards & Waloff (1954), Richards et al. (1960), Dempster (1961) and Manly (1974a). It has the advantage over Kempton's (1979) method, van Straalen's (1982, 1985) method, and the development variable method of not requiring as many assumptions and involving calculations that are far more straightforward.

The situation with a constant survival rate and incomplete data is far less clear. Experience has shown the the iterative method of Manly (1985a) for applying the KNM method to incomplete data usually works if observations start when individuals first enter stage 1 but end before all individuals are dead. However, if observations start with individuals in stages higher than 1 then convergence problems often arise. It is probably fair to say that the KNM method is a good choice for the analysis of incomplete data providing that it is only incomplete because observations stopped shortly before all individuals were dead. In other cases Kempton's method or the development variable method should be considered.

If the survival rate is of type (ii) then it is desirable to estimate a separate survival parameter for each stage. However, in many cases this will be impractical because the data are not sufficient to allow good estimates of so many parameters. Given good data, the regression method of Manly (1987) provides a convenient analysis. It does not require independent estimates of any population parameters. The calculations are straightforward and it is easy to allow several successive stages to share the same survival parameter.

If the survival rate is of type (iii) then several possible methods of analysis are available. It appears the Ruesink's (1975) method and Derr & Ord's (1979) method will be unsatisfactory except with very good data sets since they attempt to estimate so many parameters. However, Kempton's (1979) method can incorporate age—dependent survival and may enable an appropriate analysis to be carried out. An interesting point here is that Kempton's method can be modified in a straightforward way to allow for survival to depend on calendar time but but stage durations to depend on physiological time. The same is true for the development variable method. Alternatively, if the survival rate changes abruptly at certain points in time then Bellow's et al (1982) method will be better.

Another general point to be kept in mind when comparing methods is the best balance between assumptions and mathematical complexity. The methods that have been described vary considerable in this respect. At one extreme there are the very simple approaches of Richards & Waloff (1954) and Southwood and Jepson (1962). At the other extreme there are the sophisticated stochastic models of Read & Ashford (1968), Kempton (1979), van Straalen (1982, 1985), and the development variable model. If estimated by maximum likelihood, these latter models can be expected to produce the best possible estimates from data providing that the assumptions are correct. Unfortunately, the assumptions will often (or usually) not be correct. The optimality of these models may then be seriously upset. It is for this reason that less complicated methods of estimation must often be preferable.

Assessing Estimates Using Simulation.

Stage—frequency data is derived under many different circumstances. As a consequence, it is difficult to discuss the determination of the accuracy of estimates and the testing of assumptions without referring to particular examples. However, it is appropriate to say a few words here about simulation, which often provides the simplest and most versatile method for addressing these questions.

If the mechanism generating data is known then many artificial sets of data can be produced on a computer with parameter values equal to those estimated from a particular set of data. These sets can be analyzed in order to generate approximate sampling distributions for estimates of population parameters. Biases and standard errors can then be determined. Goodness of fit statistics can also be calculated both for the observed data and for the computer generated data.

If the observed statistic falls well within the distribution for the generated data then the model can be considered to give a reasonable fit to the data. If the observed statistic is very large when compared to the generated distribution then the model is questionable. If the observed statistic is very small when compared to the generated distribution then this suggests that the model includes too many sources of variation.

There are several different approaches to simulation that can be used. Obviously, if the data stage–frequencies have been determined by some rather special method such as capture–recapture sampling then the simulation will have to attempt to model this process. However, in many cases the data values are simply obtained by sampling a small fraction of a large population. Assuming random sampling, the sample frequencies will then be independent Poisson variates with mean values equal to the population stage–frequencies multiplied by the sampling fraction. In that case, four possible methods of simulation can be suggested.

Method 1 involves simulating the population individual by individual, and then giving each live individual a probability of being 'captured' at each sample time. This has the advantages of allowing any desired distributions for the time of entry to stage 1 and the durations of stages, and making it possible to study the variation in estimates due to stochastic effects in the population being sampled as well variation due to sampling errors. The main disadvantage is that generating samples from large populations takes a long time. This was the approach used by Manly (1974a).

Method 2 is attractive for simulating samples from a large population. It involves using a specific model such as that of Bellows & Birley (1981) to calculate expected population stage–frequencies, and then determining sample stage–frequencies by generating independent Poisson variates with these expected values. Since the expected frequencies are determined as functions of the entry distribution to stage 1 and the distributions of stage durations, this method assumes that the population being sampled is large enough to make stochastic variation in population values negligible.

Method 3 is very simple. It involves assuming that the duration of stages is the same for all individuals. The individuals that enter between two sample times can then be allowed to enter at (say) ten equally–spaced times between the samples and it is quite easy to calculate their expected contributions to stage–frequencies from then on, taking into account the survival rates in different stages and the durations of stages. Once expected frequencies are determined, sample frequencies can be taken as independent Poisson variates with these means.

Method 4 is even simpler still. This just involves generating a simulated set of data by replacing each observed stage–frequency with a value from the Poisson distribution with a mean equal to this observed value. This is then equivalent to random sampling from a population with expected stage–frequencies equal to the observed ones. The main advantage of this method of simulation is that no particular model has to be assumed for the population. Nevertheless, the simulated data should display the amount of variation that is to be expected from sampling.

Analysis of Single Cohort Stage–Frequency Data

Single cohort data often looks rather similar to multi–cohort data. The only difference is that with single cohort data it is known that all individuals entered stage 1 at the same time. Indeed, some of the methods for analyzing multi–cohort data are applicable to single cohort data without any changes. It might therefore seem that there is no point in looking in detail at the special case of single cohort data. However, there are two reasons why this is worthwhile.

First, single cohort data makes it possible to estimate the distributions of durations of stages, rather than just mean durations. This is because all of the overlap between stages can be attributed to variation from individual to individual in the durations of stages. Estimating distributions of durations is difficult with multi–cohort data since variation in durations is confounded with variation in times of entry to stage 1, although the methods of Kempton (1979) and the development variable model do attempt this.

Second, it is sometimes the case with single cohort data that there are virtually no deaths. This means that differences between the total sample stage–frequencies at different times can be attributed to sampling variation and other factors that are not of any particular importance. This is different from the situation for most sets of multi–cohort data for which an increase in total stage–frequencies with time indicates new entries to stage 1, and a decrease in numbers indicates deaths.

Two sets of computer simulated data are provided in Tables 14 and 15 for use as examples. The first set of data is for five stages and 20 'daily' samples, with 'daily' survival rates ranging from 0.7 in stage 1 to 0.9 in stages 3 and 4. The second set of data is for five stages and daily samples taken from day 2 to day 12 after the cohort entered the population. In this case there were no deaths.

The analysis of data when there are deaths

As mentioned above, many of the methods of analysis that have been proposed for multi–cohort data can be used without change with single cohort data. The Richards & Waloff (1954), Kiritani–Nakasuji–Manly, Manly (1987), Richards et al. (1960), Lakhani & Service (1974), Birley (1977) and Bellows & Birley (1981) methods all fall into this category. Other methods require minor changes to take into account the single entry time to stage 1. Manly's (1974) method reduces to Richards & Waloff's method for stage 1, since there is a single time of entry to the stage, but for later stages it works the same for single or multi–cohort data. Kempton's (1979) model and the development variable model already involves having a single cohort in an (unobservable) stage 0 at time 0. This has to be changed to an observable stage. Van Straalen's (1982, 1985) model assumes a normal distribution for the time of entry to stage 1, with mean μ and standard deviation σ. For a single cohort these parameters become $\sigma = 0$ and μ equal to the time that the cohort entered stage 1.

Having said that many different methods can be used to analyze single–cohort data, it must be noted that in practice most of them must be ruled out since they only provide estimates of mean stage durations. This is not good enough if one of the reasons for collecting single cohort data is to estimate distributions for the durations of stages. It is true that Kempton's (1979) method estimates gamma distributions for stage durations, but it assumes that the scale parameter is the same for all these distributions. This assumption is of no real concern with multi–cohort data where the main interest is in getting good estimates of mean durations. It may be too much of a restriction with single cohort when good estimates are wanted for the complete distributions. Another limitation with this method is to cases with constant or age–dependent survival rates.

If the possibility of a different survival rate in each stage is to be allowed, then the only immediately suitable method for the analysis of single cohort data is that of Bellows & Birley (1981). However, the development variable model for single cohort data is also discussed here since this model is being presented for the first time in the present paper.

Bellows & Birley's method when there are deaths

The main drawback with Bellows & Birley's (1981) method when applied to multi–cohort data (the need to know the distribution of entry times to stage 1) does not apply with single cohort data. Bellows & Birley suggested fitting the model to data using non–linear regression but the MAXLIK algorithm is more appropriate if the data consist of stage–frequency counts from a small sample from a population.

When Bellow's & Birley's model is fitted to the data in Table 14 using the MAXLIK algorithm, assuming lognormal distributions for stage durations, the estimated population parameters compare with the true values as shown in the following table: –

Stage	Mean duration		Std. dev. of duration		Daily survival rate		Number entering stage	
	Est	True	Est	True	Est	True	Est	True
1	2.1	2.0	0.7	0.7	0.71	0.70	1032	1000
2	1.4	2.0	0.8	0.7	0.82	0.80	439	488
3	2.8	3.2	1.0	1.1	0.91	0.90	299	309
4	2.9	3.2	1.5	1.1	0.86	0.90	218	220
5	–	–	–	–	0.82	0.80	131	156

The estimates are fairly reasonable. The full estimated distributions are compared with the true distributions in Table 16. Here again the estimates are quite good.

Table 14. Computer simulated single cohort data with five stages and 20 samples. The daily survival rates for stages 1 to 5 were 0.7, 0.8, 0.9 and 0.8, respectively. In stages 1 and 2 the probabilities of durations within the ranges 0–1, 1–2, 2–3, and 3–4 'days' were 0, 0.6, 0.3 and 0.1, respectively. In stages 3 and 4 the probabilities for durations in the ranges 0–1, 1–2, 2–3, 3–4, 4–5 and 5–6 were 0, 0.1, 0.4, 0.3, 0.1 and 0.1, respectively. The data were generated to correspond to samples from a large population. For the sampling fraction the numbers entering stages 1 to 5 were 1000.0, 487.9, 309.3, 220.0 and 156.4, respectively.

Day	Stage 1	2	3	4	5
1	1008	0	0	0	0
2	716	0	0	0	0
3	221	299	0	0	0
4	44	265	59	0	0
5	0	139	190	0	0
6	0	44	243	15	0
7	0	3	204	48	1
8	0	1	125	95	5
9	0	0	47	138	12
10	0	0	19	123	38
11	0	0	6	95	52
12	0	0	1	44	71
13	0	0	0	23	88
14	0	0	0	18	78
15	0	0	0	5	60
16	0	0	0	0	67
17	0	0	0	1	51
18	0	0	0	0	39
19	0	0	0	0	35
20	0	0	0	0	20

Table 15. Computer generated single cohort data with five stages and 12 samples, and no deaths. The distributions of stage durations were fixed as shown in Table 17. The generated data represent samples of a fixed fraction of a large population such that the expected frequency at time 1 was 500.

Day	Stage 1	2	3	4	5
1	506	0	0	0	0
2	412	109	6	0	0
3	218	254	37	2	0
4	44	317	135	34	2
5	0	207	199	76	13
6	0	93	158	141	61
7	0	18	124	207	145
8	0	2	61	161	267
9	0	0	13	110	360
10	0	0	1	33	441
11	0	0	0	8	502
12	0	0	0	0	536

The development variable model when there are deaths

It is assumed with Bellows & Birley's model that the times spent in successive stages by an individual are independent variables. This means, for example, that if an individual spends an unusually long time in stage 1 then it is just as likely as any other individual to spend a long time in stage 2. Put in this way, it seems unlikely that the assumption of independence is completely true, but without following the history of individuals it does not seem to be possible to determine what the true situation is. What can be done is to analyze data using a model that does not make the assumption and see what difference this makes to estimates. One such model is the development variable model discussed above. Some minor modifications are needed to apply the model with single cohort data since there is no unobservable stage 0.

Thus suppose that the individuals in a population enter stage 1 (i.e., begin developing) at time zero. Samples from the population are taken at times t_1, t_2, ..., t_n. Each individual has a value for a development variable $D(t)$ associated with it at time t such that this determines the stage that the individual is in at that time. Specifically, if $y_{i-1} < D(t) < y_i$ then the individual is in stage i. The 'cut–points' on the development scale $y_0 = -\infty$. y_1, ..., $y_q = +\infty$ are the same for all individuals at all times. Suppose, further that at time t the development variable values $D(t)$ are normally distributed with mean $\mu(t)$ and standard deviation $\sigma(t)$, with the development scale being fixed by assuming that this is a standard normal distribution at the time of the first sample so that $\mu(t_1) = 0$ and $\sigma(t_1) = 1$. The probability of an individual being in stage i at time t is then the integral from y_{i-1} to y_i for a normal distribution. The only difference between this and the multi–cohort model of Fig. 4 is the labelling of the cut points from y_1 to y_{q-1} instead of from y_0 to y_{q-1}.

Following the multi–cohort model, quadratic time changes in the mean and linear changes in the standard deviation can be assumed so that

$$\mu(t) = \alpha_1(t-t_1) + \alpha_2(t-t_1)^2,$$

and

$$\sigma(t) = 1 + \beta(t-t_1).$$

Other authors have assumed that the mean and variance are proportional to physiological time.

With a constant survival rate of ϕ per unit time and expected sample sizes being proportional to survival rates, the expected sample number in stage i at time t is equal to the probability of being in this stage at this time multiplied by the probability of surviving to that time. It is therefore straightforward to use the MAXLIK algorithm to estimate the unknown parameters ϕ, α_1, α_2, β, y_1, ..., y_{q-1}, assuming that stage frequencies are sample counts with Poisson distributions.

Still following the multi–cohort model, the distributions of stage durations can be estimated by assuming that the order of the individuals on the development scale remains the same, so that an individual that is x standard deviations above the mean in one sample is the same number of standard deviations above the mean in all other samples, and approximating the standard normal distribution for development variable at time t_1 with a 13 point discrete distribution. The numbers entering stages can also be determined using the same discrete approximation.

Since the test data shown in Table 14 were generated using a Bellows & Birley type of model, with a different survival rate in each stage, it is unrealistic to expect the development variable model to provide good estimates of the parameters. Nevertheless, it is of some interest to see what results are obtained from this model. A summary is shown below: —

Stage	Mean duration		Std. dev. of duration		Daily survival rate		Number entering stage	
	Est	True	Est	True	Est	True	Est	True
1	2.1	2.0	0.6	0.7	0.84	0.70	1055	1000
2	1.9	2.0	0.4	0.7	0.84	0.80	722	488
3	3.3	3.2	0.5	1.1	0.84	0.90	520	309
4	3.5	3.2	0.5	1.1	0.84	0.90	292	220
5	–	–	–	–	0.84	0.80	160	156

Table 16. Comparison between estimated and true distributions of stage durations for the data in Table 14. Estimates were obtained using the original Bellows & Birley model with lognormal distributions for stage durations. Mean values are calculated by assigning durations from i to i+1 the value i+^{1}/$_2$.

| | Stage 1 | | | | | Mean |
| | Probability of duration between | | | | | |
	0–1	1–2	2–3	3–4	4–5	Duration
Estimated	0.01	0.50	0.42	0.07	0.00	2.1
True	0.00	0.60	0.30	0.10	0.00	2.0

| | Stage 2 | | | | | Mean |
| | Probability of duration between | | | | | |
	0–1	1–2	2–3	3–4	4–5	Duration
Estimated	0.28	0.55	0.14	0.03	0.00	1.4
True	0.00	0.60	0.30	0.10	0.00	2.0

| | Stage 3 | | | | | | Mean |
| | Probability of duration between | | | | | | |
	0–1	1–2	2–3	3–4	4–5	5–6	Duration
Estimated	0.00	0.18	0.45	0.26	0.08	0.02	2.8
True	0.00	0.10	0.40	0.30	0.10	0.10	3.2

| | Stage 4 | | | | | | | Mean |
| | Probability of duration between | | | | | | | |
	0–1	1–2	2–3	3–4	4–5	5–6	6–7	Duration
Estimated	0.02	0.27	0.33	0.20	0.10	0.04	0.02	2.9
True	0.00	0.10	0.40	0.30	0.10	0.10	0.00	3.2

There are very reasonable estimates for the mean durations of stages. However, the estimated standard deviations are too small. As was explained with the multi–cohort model, this under–estimation of variation is exactly what is expected. Generally the development variable model will indicate less variation for the durations of stages 2 to q–1 than Bellows & Birley's model. Better agreement is expected for stage 1.

The analysis of data when there are no deaths

It is no accident that most of the models that have been proposed for stage–frequency data arising in biological contexts have allowed for mortality. It is almost inevitable that wild populations will lose a substantial proportion of their members before they reach the mature stage. Nevertheless, there are situations where stage–frequency data is generated without mortality taking place. In these cases all that can be estimated is the distribution of stage durations.

The first attempt at modelling single cohort stage–frequency data without mortality seems to have been made by Aitchison & Silvey (1957). They assumed that the time spent in stage i is normally distributed with mean μ_i and standard deviation σ_i, with the durations of successive stages being independently distributed for an individual. They proposed estimation by the method of maximum likelihood.

Normally distributed stage durations imply that negative durations are possible but this can be avoided if another distribution is used in place of the normal. If the gamma distribution is used with the same scale parameter for all stages then this gives Kempton's (1979) model with the survival parameter Θ set at zero. Some other possiblities are discussed by McCullagh (1985).

Fitting Bellows & Birley's (1981) model to data involves no special problems when there are no deaths. It is simply necessary to make the observed and expected sample sizes agree for each sample time when using the MAXLIK algorithm. Doing this with the data in Table 15, assuming lognormal distributions for stages, results in the estimated distributions in Table 17. These compare very well with the true distributions.

There are also no particular difficulties with fitting the development variable model. Again, the observed and expected sample sizes should be made to agree at each sample time rather than to be determined by the survival rate. Using the MAXLIK algorithm results in the estimate distributions of stage durations shown in Table 18 using the 13 point discrete approximation for the distribution of the development variable at the time of the first sample.

A comparison between the mean durations of stages used for generating the data (1.8, 1.8, 1.5 and 1.5 for stages 1. to 4, respectively) and the estimated means (1.8, 2.0, 1.5 and 1.6, respectively) shows reasonable agreement. However, a comparison between the true and estimated standard deviations (0.9, 0.9, 0.4 and 0.4 compared to 1.0, 0.2, 0.2 and 0.3) does not show good agreement, except with stage 1. For stages 2 to 4 the estimated distributions of stage durations show much less variation than actually occurred as expected since the data were generated using Bellows & Birley's model.

Reproducing Stage–Frequency Data

The important characteristic of the situations that will be considered now is that the population being studied is reproducing: all entries to stage 1 are generated by the individuals that are already in the population. An example of such a situation is provided by Lefkovitch's laboratory population of the cigarette beetle *Lasioderma serricorne* (Table 3). In this case the individuals in the population developed through the stages eggs → larvae → pupae → adults, with losses through deaths in all stages. The adults produce the eggs, and the cycle continues.

Most of the mathematical models that have been developed to account for the dynamics of reproducing popualations assume that the ages of individuals are known. When this is true, there are two alternative approaches that have been used. The first is the continuous time integral equation method pioneered by Sharpe & Lotka (1911). The second uses grouped age intervals and a matrix formulation as was first proposed independently by Bernardelli (1941), Lewis (1942) and Leslie (1945). Both of these approaches are reviewed by Pollard (1973). In the context of the present paper it is the matrix approach that is most relevant since several people have developed the ideas of Bernardelli, Lewis and Leslie to allow for stage rather than age grouping of individuals.

Table 17. Comparison between observed and estimated distributions of stage durations for the data of Table 15. Estimates are from Bellows & Birley's method, assuming lognormally distributed stage durations.

| | Stage 1 Probability of duration between | | | | | | |
	0–1	1–2	2–3	3–4	4–5	Mean	SD
Estimated	0.17	0.52	0.22	0.06	0.02	1.8	1.0
True	0.20	0.40	0.30	0.10	0.00	1.8	0.9

| | Stage 2 Probability of duration between | | | | | | |
	0–1	1–2	2–3	3–4	4–5	Mean	SD
Estimated	0.29	0.50	0.15	0.04	0.01	1.5	0.9
True	0.20	0.40	0.30	0.10	0.00	1.8	0.9

| | Stage 3 Probability of duration between | | | | | |
	0–1	1–2	2–3	3–4	Mean	SD
Estimated	0.57	0.38	0.05	0.00	1.0	0.6
True	0.10	0.80	0.10	0.00	1.5	0.4

	Stage 4 Probability of duration between					
	0–1	1–2	2–3	3–4	Mean	SD
Estimated	0.50	0.42	0.06	0.00	1.1	0.7
True	0.10	0.80	0.10	0.00	1.5	0.4

Table 18. Estimated durations of stages determined from fitting the development variable model to the data in Table 15.

Development parameter	Probability	Duration of stage			
		1	2	3	4
−3.00	0.002	5.0	2.5	2.0	0.0
−2.50	0.009	4.5	2.3	1.9	2.6
−2.00	0.027	4.0	2.2	1.8	2.2
−1.50	0.065	3.4	2.2	1.7	1.9
−1.00	0.121	2.9	2.1	1.6	1.8
−0.50	0.176	2.4	2.0	1.5	1.7
0.00	0.200	1.8	2.0	1.5	1.5
0.50	0.176	1.3	2.0	1.3	1.4
1.00	0.121	0.8	1.9	1.3	1.3
1.50	0.065	0.3	1.8	1.3	1.2
2.00	0.027	0.1	1.4	1.2	1.3
2.50	0.009	0.1	0.8	1.2	1.2
3.00	0.002	0.1	0.3	1.1	1.1
Mean		1.86	1.98	1.45	1.56
Std. dev.		1.01	0.20	0.16	0.25

Table 19 shows a computer generated set of data that will be used for example calculations. This was generated using a model described below.

The Bernardelli–Leslie–Lewis model

With the Bernadelli–Leslie–Lewis model a population is considered at discrete points in time 0, 1, 2, etc, with individuals in age groups 0, 1, 2, etc. Age group x comprises all individuals with an age from x until just less than x+1. For convenience only females are considered. The following notation will be used: −

$n(x,t)$ = the number of females in the age group x at time t;

$p(x)$ = the probability that a female in the age group x at time t will survive to be in the age group x+1 at time t+1; and

$B(x)$ = the average number of female offspring born to females aged from x to x+1 in a unit period of time that survive to the end of that period of time.

The number of females in age group x at time t+1 will be the sum of the offspring from females of different ages:

$$n(0,t+1) = B(0)n(0,t) + B(1)n(1,t) + ... + B(k)n(k,t)$$

where k+1 is the maximum possible age. Also,

$$n(x+1, t+1) = p(x)n(x,t)$$

for x = 0, 1, ..., k−1. These equations can be written together as the matrix equation

$$
\begin{bmatrix}
n(0,t+1) \\
n(1,t+1) \\
\cdot \\
\cdot \\
\cdot \\
n(k,t+1)
\end{bmatrix}
\begin{bmatrix}
B(0) & B(1) & ... & B(k-1) & B(k) \\
p(0) & 0 & ... & & 0 \\
\cdot & & & & \\
\cdot & & & & \\
\cdot & & & & \\
0 & 0 & ... & p(k-1) & 0
\end{bmatrix}
\begin{bmatrix}
n(0,t) \\
n(1,t) \\
\cdot \\
\cdot \\
\cdot \\
n(k,t)
\end{bmatrix}
$$

or

$$N_{t+1} = MN_t .$$

It then follows that

$$N_t = M^t N_0 .$$

The matrix M, whose elements are the fecundity rates B(x) and the survival probabilities p(x), is usually called a Leslie matrix. The last equation shows that the numbers in different age groups at an arbitrary time t are determined by the numbers in the age groups at time zero (N_0) and the Leslie matrix raised to the power t. Subject to certain mild assumptions it is possible to show that a population will eventually reach a stable distribution for the relative numbers of individuals with different ages, and be growing or declining at a constant rate. In fact, the long term behavior of the population is determined by the dominant eigenvalue of the Leslie matrix. See Pollard (1973, Chapter 4) for a discussion of the various theoretical results that are known.

Lefkovitch's model for populations grouped by life stages

Lefkovitch (1965) modified the Bernardelli–Leslie–Lewis model to allow a population to be grouped by life stages rather than by age. He did this by the simple expedient of allowing the number in stage i at time t+1 to depend on the numbers in all previous stages at time t. Thus if f(i,t) is the number of individuals in stage i at time t then for q stages his model is , in matrix notation,

Table 19. Simulated data from a population with reproduction. The frequencies in stages 1 to 4 in the population at time t+1 were determined from the frequencies at time t by $F_{t+1} = MF_t$ using the transition matrix

$$M = \begin{bmatrix} 0.72 & 0 & 0.22 & 0.14 \\ 0.67 & 0.06 & 0 & 0 \\ 0 & 0.78 & 0.65 & 0 \\ 0 & 0 & 0.20 & 0.09 \end{bmatrix} .$$

The elements in the vector F_{t+1} were then individually multiplied by independent normally distributed random variables with mean 1 and standard deviation 0.05 to simulate population disturbances. This produced population expected frequencies. Sample values were generated as independent Poisson distributed random variables with means equal to population expected frequencies. Sample sizes were set so that the first sample was expected to yield 100 individuals in stage 1 only.

Sample	1	2	3	4	Total
1	91	0	0	0	91
2	68	59	0	0	127
3	55	47	58	0	160
4	35	37	91	9	172
5	47	38	61	11	157
6	48	38	80	20	186
7	44	29	71	16	160
8	64	44	77	11	196
9	65	35	87	15	202
10	65	27	80	23	195
11	85	52	91	13	241
12	47	44	103	15	209
13	68	56	107	16	247
14	82	57	92	27	258
15	80	45	123	14	262
16	100	51	118	31	300
17	94	63	112	34	303
18	94	59	105	22	280
19	107	61	125	24	317
20	81	62	139	29	311

$$\begin{bmatrix} f(1,t+1) \\ f(2,t+1) \\ \cdot \\ \cdot \\ \cdot \\ f(q,t+1) \end{bmatrix} = \begin{bmatrix} m_{11} & m_{12} & ... & m_{1q} \\ m_{21} & m_{22} & ... & m_{2q} \\ \cdot & & & \\ \cdot & & & \\ \cdot & & & \\ m_{q1} & m_{q2} & ... & m_{qq} \end{bmatrix} \begin{bmatrix} f(1,t) \\ f(2,t) \\ \cdot \\ \cdot \\ \cdot \\ f(q,t) \end{bmatrix}$$

or,

$$F_{t+1} = MF_t \ ,$$

so that

$$F_t = M^t F_0 \ .$$

Here M_{ij} is a constant which reflects how the number in stage i at time t+1 depends on the number in stage j at time t.

The matrix M of the last equation does not have the simple structure of a Leslie matrix. consequently, Lefkovitch's model is not as straightforward as the Bernardelli–Leslie–Lewis model to study from a theoretical point of view. Nevertheless, as noted by Lefkovitch (1965), the long term behavior of the population will be determined by the eigenvalue of M with maximum modulus, with its corresponding eigenvector. To be precise, let τ_1 be the eigenvalue of the matrix M with the largest absolute value, with corresponding eigenvector v_1. Then the proportions in different stages in the population should tend towards the ratios in the vector v_1, and the numbers in each stage should increase by the factor τ_1 per generation.

If the numbers in the different stages are known for a series of at least q^2 equally spaced sample times then the constants in the matrix M can be determined by regression. For stage j

$$f(j,t+1) = m_{i1}f(1,t) + m_{i2}f(2,t) + ... + m_{iq}f(q,t).$$

Estimates of $m_{i1}, m_{i2}, ..., m_{iq}$ can therefore be found by a multiple regression of $f(j,t+1)$ values on the numbers in the different stages at time t. The regression should not include a constant term. Combining the regression estimates of m values for different stages leads to the estimate of M suggested by Lefkovitch (1964b), although this is not immediately obvious.

There are practical difficulties with using this multiple regression method for estimation. The number of parameters to be estimated may be rather large for available data so that estimates are subject to large sampling errors. This may then mean that estimates are not biologically meaningful. In particular, negative values will be difficult to interpret. Lefkovitch

(1965) suggested overcoming this problem by: (a) setting m_{ij} values to zero rather than estimating them whenever biological knowledge allows this; (b) setting negative estimates of m_{ij} to zero, and reestimating the remaining parameters; and (c) replacing estimates that are not significantly different from zero with zero, and reestimating the remaining parameters. There is a problem with (c) since the usual assumptions of multiple regression do not apply which means that determining the significance levels for estimates is not a simple matter.

When Lefkovitch's regression method is applied to the data of Table 19 without worrying about negative estimates of transition coefficients, the estimate of the matrix M is found to be

$$M = \begin{bmatrix} 0.68 & -0.09 & 0.23 & 0.38 \\ 0.53 & 0.07 & 0.11 & -0.20 \\ 0.11 & 0.90 & 0.51 & 0.04 \\ 0.04 & -0.06 & 0.21 & 0.02 \end{bmatrix},$$

which has a resemblance to the true matrix

$$M = \begin{bmatrix} 0.72 & 0 & 0.22 & 0.14 \\ 0.67 & 0.06 & 0 & 0 \\ 0 & 0.78 & 0.65 & 0 \\ 0 & 0 & 0.20 & 0.09 \end{bmatrix},$$

but differs substantially in some ways. Setting the negative estimates to zero and reestimating the other ones produces

$$M = \begin{bmatrix} 0.65 & 0 & 0.20 & 0.40 \\ 0.52 & 0.09 & 0 & 0 \\ 0 & 0.90 & 0.54 & 0 \\ 0.02 & 0 & 0.20 & 0.05 \end{bmatrix}.$$

If it is assumed that positions of the zero coefficients in M are known then the estimates of the other coefficients are little changed, these being

$$M = \begin{bmatrix} 0.65 & 0 & 0 & 0.14 \\ 0.54 & 0.18 & 0 & 0 \\ 0 & 1.02 & 0.54 & 0 \\ 0 & 0 & 0.20 & 0.05 \end{bmatrix}.$$

The most interesting question now is how this matrix compares with the true one as far as the long term behavior of the population is concerned. Answering that is not as simple as comparing coefficients and will not be gone into here.

Usher's model

The main problem with using Lefkovitch's model is the large number of coefficients to be estimated in the transition matrix. However, this can be overcome to some extent if the time between samples is small so that the possibility of an individual developing through more than one stage in this time can be discounted. A model based upon this assumption has been developed by Usher (1966, 1969) in the context of the management of a forest. The population then consists of trees in development stages that are determined by their size. However, the model can be applied with other definitions for stages.

Usher assumed that the population devleopment equation can be simplified to

$$
\begin{bmatrix}
f(1,t{+}1) \\
f(2,t{+}1) \\
f(3,t{+}1) \\
\cdot \\
\cdot \\
f(q{-}1,t{+}1) \\
f(q,t{+}1)
\end{bmatrix}
=
\begin{bmatrix}
B_1 & B_2 & B_3 & \cdots & B_{q-1} & B_q \\
b_1 & a_2 & 0 & \cdots & 0 & 0 \\
0 & b_2 & a_3 & \cdots & 0 & 0 \\
\cdot & & & & & \\
\cdot & & & & & \\
0 & 0 & 0 & \cdots & a_{q-1} & 0 \\
0 & 0 & 0 & \cdots & b_{q-1} & a_q
\end{bmatrix}
\begin{bmatrix}
f(1,t) \\
f(2,t) \\
f(3,t) \\
\cdot \\
\cdot \\
f(q{-}1,t) \\
f(q,t)
\end{bmatrix}
$$

Here B_i is the contribution to the number in stage 1 at time $t{+}1$ that comes from those in stage i at time t. For $i = 1$ this contribution comes from those that remain alive but do not develop to stage 2, and also possibly from the reproduction of stage 1 individuals. For $i > 1$ the contribution is from reproduction only. Also, a_i is the probability that an individual in stage i at time t is still in stage i at time $t + 1$, while b_i is the probability that an individual in stage i at time t moves to stage $i + 1$ by time $t + 1$. The sum $a_i + b_i$ gives the survivial rate between two sample times for an individual in stage i at the first of these times.

In principle the parameters B_i, ..., B_q, a_2, ..., a_q, b_1, ..., b_{q-1} can be estimated by regressing the frequencies at time $t + 1$ on the frequencies at time t in a similar manner to the method of estimation that has been described for Lefkovitch's model. In fact, the data of Table 19 satisfy Usher's model so that an example of fitting the model has in effect already been given in the last section. A simpler method was used by Usher (1966) for his forest example, but this requires more data than just the stage–frequencies at equal intervals of time.

Brown (1975) used the Usher model in a study of the population dynamics of the freshwater pulmonate snail *Physa ampullacea*. He noted the need for non–negative coefficients and used linear programming to estimate the coefficients in row 1 of the transition matrix. This involved an arbitrary choice of coefficients in an objective function to be macimized, which seems to make it a generally unsatisfactory method. He used a form of least squares for estimating the other coefficients.

Sampling variation and other sources of errors

At best the models described above will only be approximations to reality. Also, data stage–frequencies will usually only be estimates of population values so that even if a model is quite correct the data will not be fitted exactly. Hence there is a need for model descriptions to incorporate allowances for random variation in population stage–frequencies and sampling errors. One plausible model says that for the population the relationship between the stage–frequencies at time t and at time t + 1 is

$$
\begin{bmatrix} f(1,t+1) \\ f(2,t+1) \\ . \\ . \\ f(q,t+1) \end{bmatrix} = \begin{bmatrix} m_{11} & m_{12} & ... & m_{1q} \\ m_{21} & m_{22} & ... & m_{2q} \\ . & & & \\ . & & & \\ m_{q1} & m_{q2} & ... & m_{qq} \end{bmatrix} \begin{bmatrix} f(1,t) \\ f(2,t) \\ . \\ . \\ f(q,t) \end{bmatrix} \begin{bmatrix} e_{1t+1} \\ e_{2t+1} \\ . \\ . \\ e_{qt+1} \end{bmatrix}
$$

or

$$
F_{t+1} = MF_t + E_{t+1} \ .
$$

This is Lefkovitch's model with an 'error' term E_{t+1} which takes into account random disturbances in the system. It is Usher's model if the appropriate transition coefficients are set at zero. Generally it can be expected that the magnitude of the disturbances will increase with the population size. This can be accounted for by making the standard deviation of e_{ij} equal to $\alpha f(i,j)$, where α is a constant.

The last equation is for population stage–frequencies $f(i,j)$. If the observed stage–frequencies are sample counts f_{ij} of a small fraction of the population then they can be expected to be Poisson distributed about the population values. The estimate of e_{ij} from the data will therefore be

$$
\begin{aligned}
e_{ij} &= f_{ij} - m_{i1}f_{1j-1} - m_{i2}f_{2j-1} - ... - m_{iq}f_{qj-1} \\
&= e_{ij} + \{f_{ij} - f(i, j)\} + m_{i1}\{f_{1j-1} - f(1,j-1)\} + \\
&\quad m_{i2}\{f_{2j-1} - f(2,j-1)\} + ... + m_{iq}\{f_{qj-1} - f(q,j-1)\} \ .
\end{aligned}
$$

The terms within curly brackets on the right–hand side of this equation are differences between observed stage–frequencies and their expected population values. Assuming that these are all independently distributed as Poisson random variables it follows that

$$\begin{aligned}
\text{Var}(e_{ij}) \quad &= \quad \text{Var}(e_{ij}) + \text{Var}(f_{ij}) + m_{i1}{}^2\text{Var}(f_{ij\text{-}1}) + m_{i2}{}^2\text{Var}(f_{2j\text{-}1}) \\
&\quad + \ldots + m_{iq}{}^2\text{Var}(f_{qj\text{-}1}) \\
&= \quad \alpha^2 f(i,j)^2 + f(i,j) + m_{i1}{}^2 f(1,j{-}1) + m_{i2}{}^2 f(2,j{-}1) \\
&\quad + \ldots + m_{iq}{}^2 f(q,j{-}1)
\end{aligned}$$

It can also be established easily enough that covariance between e_{ij} values are all zero except that

$$\text{Cov}(e_{ij}, e_{ij\text{-}1}) = m_{ii} f(i,j) \ .$$

The last two equations can be made on the basis of a generalized least–squares method for estimating the coefficients m_{ij} from data (Manly, 1985, p. 414), an idea which will not be pursued here. Instead it will merely be noted that $\text{Var}(e_{ij})$ gives the expected contribution to the residual sum of squares from the regression of the stage–frequency f_{ij} on the stage–frequencies at time $j{-}1$. Adding up all these contributions (replacing the population frequencies $f(i,j)$ with their sample estimates) and equating the total to the observed sum of squares provides an estimate of α.

For example, the total residual sum of squares for the last estimates of transition coefficients discussed above for the data in Table 19 (with no negative values) is 69274.2. Equating this to the sum of the variances for all stage–frequencies gives the equation.

$$8531.5 \approx 326479\alpha^2 + 6783.0 \ .$$

Hence $\alpha \approx 0.073$. The data were actually generated with $\alpha = 0.05$.

If the observed stage–freuqencies are population counts without sampling errors then

$$\text{Var}(e_{ij}) = \alpha^2 f(i,j)^2 \ .$$

Hence an estimate of α is given by equating the regression residual sum of squares to the sum of the right hand side of this equation over all stage–frequencies.

Simulation

Obtaining an estimate of the value of α is valuable for the understanding of the extent to which stage–frequencies are effected by random factors. It is also necessary to have an α value if a population is to be simulated in a way that is at all realistic. Simulation is very straightforward using the model of the previous section, at least for cases where the data stage–frequencies are either population counts or are obtained by the random sampling of a small fraction of a large population. Starting with a vector F_0 of population frequencies in stages 1 to q at time 0, the population frequencies at time 1 are found by first applying the transition matrix to get the vector of expected frequencies MF_0, and then disturbing this by multiplying each element in MF_0 by an independent, normally distributed random variable with mean 1 and standard deviation α.

Sample stage–frequencies can be found if necessary by generating independent Poisson variates with mean values equal to the population frequencies. Simulation of data at times 2, 3 and so on is carried out by simply repeating the same process. At all times the population frequencies at time t + 1 are obtained from the population frequencies at time t rather than the sample frequencies. The data in Table 19 were generated in this way with $\alpha = 0.05$.

Obviously, simulation in this simple manner is not appropriate when the data stage–frequencies are obtained by some complicated special sampling method. In that case simulations will have to be modified as approppriate.

Discussion

When Lefkovitch introduced his model for populations developing through stages he envisaged a situation where samples are taken at time intervals that are fairly large in comparison with the mean durations of stages. This results in a very flexible model. However, the meanings of the elements in the matrix become confused. Usher's model is much more satisfactory in this respect since this allows the elements to be interpreted in terms of birth rates and survival probabilities.

There is one problem with modeling populations using transition matrices that has not been mentioned so far. This concerns the assumption that the same transition matrix applies, irrespective of the age distribution of individuals within stages. This unfortunately contradicts one of the important conditions laid down for stage–frequency data in the introduction to this paper: a definite maturation process takes place in each stage, and the probability of an individual leaving a stage depends upon how long the individual has been in the stage. If a population is in a fairly stable state as far as the age distribution within stages is concerned then a transition matrix model may give a reasonable fit to data even though it ignores this important condition. However, the possibility of a bad fit to data because of a changing age distribution should always be kept in mind.

In concluding this section, mention must be made of some more elaborate methods that are available for modelling reproducing populations. First, there is Schneider & Ferris' (1986) computer intensive approach that can be applied over several generations providing that a fecundity rate is known. Second, there is the approach of Brillinger et al. (1980) which allows for death rates depending on population size. Certain generalizations along these lines were also discussed by Lefkovitch (1965). There are also numerous special purpose models (using simulation, differential equations, or difference equations) that have been developed to describe particular populations.

Acknowledgements

The work reported in this paper was supported by research grants from the United States — New Zealand Cooperative Science Project and the University of Otago Research Committee.

References Cited

Aitchison, J. & S. D. Silvey. 1957. The generalization of probit analysis to the case of multiple responses. *Biometrika* 44: 131–140.

Aksnes, D. L. & T. Hoisaeter. 1987. Obtaining life table data from stage–frequency distributional statistics. *Limnol. Oceanogr.* 32: 514–517.

Ashford, J. R., K. L. Q. Read & G. G. Vickers. 1970. A system of stochastic models applicable to animal population dynamics. *J. Anim. Ecol.* 39: 29–50

Bellows, T. S. & M. H. Birley. 1981. Estimating developmental and mortality rates and stage recruitment from insect stage–frequency data. *Res. Pop. Ecol.* 23: 232–244.

Bellows, T. S., M. Ortiz, J. C. Owens, & E. W. Huddleston. 1982. A model for analyzing insect stage–frequency data when mortality varies with time. *Res. Pop. Ecol.* 24: 142–156.

Bernardelli, H. 1941. Population waves. *J. Burma Res. Soc.* 31: 1–18.

Birley, M. 1977. The estimation of insect density and instar survivorship functions from census data. *J. Anim. Ecol.* 46: 497–510.

Birley, M. 1979. Estimating the developmental period of insect larvae with applications to the mosquito *Aedes aegypti L..* *Res. Pop. Ecol.* 21: 68–80.

Brillinger, D. R., J. Guckenheimer, P. Guttorp & G. Oster. 1980. Empirical modelling of population time series data: the case of age and density dependent vital rates. *Lect. Math. Life Sci.* 13: 65–90.

Brown, K. M. 1975. Estimation of demographic parameters from sampling data. *Amer. Midl. Natur.* 93: 454–459.

Dempster, J. P. 1956. The estimation of the number of individuals entering each stage during the development of one generation of an insect population. *J. Anim. Ecol.* 25: 1–5.

Dempster, J. P. 1961. The analysis of data obtained by regular sampling of an insect population. *J. Anim. Ecol.* 30: 429–432.

Dennis, B., W. P. Kemp & R. C. Beckwith. 1986. Stochastic model of insect phenology: estimation and testing. *Environ. Entom.* 15: 540–546.

Derr, J. A. & K. Ord. 1979. Field estimates of insect colonization. *J. Anim. Ecol.* 48: 521–534.

Hairston, N. G., M. Braner & S. Twombly. 1987. Perspective on prospective methods for obtaining life table data. *Limnol. Oceanog.* 32: 517–520.

Hairston, N. G. & S. Twombly. 1985. Obtaining life table data from cohort analyses: a critique of current methods. *Limnol. Oceanog.* 30: 886–893.

Kemp, W. P., B. Dennis & R. C. Beckwith. 1986. Stochastic phenology model for the western spruce budworm (*Lepidoptera: Tortricidae*). *Environ. Entom.* 15: 547–554.

Kempton, R. A. 1979. Statistical analysis of frequency data obtained from sampling an insect population grouped by stages. In *Statistical Distributions in Scientific Work* [eds.] J. K. Ord, G. P. Patil & C. Taillie, pp. 401–418. International Cooperative Publishing House, Maryland.

Kiritani, K. & F. Nakasuji. 1967. Estimation of the stage–specific survival rate in the insect population with overlapping stages. *Res. Pop. Ecol.* 9: 143–152.

Kobayashi, S. 1968. Estimation of the individual number entering each development stage in an insect population. *Res. Pop. Ecol.* 10: 40–44.

Lakhani, K. H. & M. W. Service. 1974. Estimating mortalities of the immature stages of *Aedes cantans* Mg. (*Diptera, Culicidae*) in a natural habitat. *Bull. Entom. Res.* 64: 265–276.

Lefkovitch, L. P. 1963. Census studies on unrestricted populations of *Lasioderma serricorne* F. (*Coleoptera: Anobiidae*). *J. Anim. Ecol.* 32: 221–231.

Lefkovitch, L. P. 1964a. The growth of restricted populations of *Lasioderma serricorne* F.) (*Coleoptera: Anobiidae*). *Bull. Entom. Res.* 55: 87–96.

Lefkovitch, L. P. 1964b. Estimating the Malthusian parameter from census data. *Nature* 204: 810.

Lefkovitch, L. P. 1965. The study of population growth in organisms grouped by stages. *Biometrics* 21: 1–18.

Leslie, P. H. 1945. On the use of matrices in certain population mathematics. *Biometrika* 33: 182–212.

Lewis, E. G. 1942. On the generation and growth of a population. *Sankhya* 6: 93–96.

McCullagh, P. 1985. Statistical and scientific aspects of models for qualitative data. In *Measuring the Unmeasurable* [eds.] P. Nijkamp, H. Leitner & N. Wrigley, pp. 39–49. Martin Nijhoff Publishers, Dordrecht, The Netherlands.

Manly, B. F. J. 1974a. Estimation of stage–specific survival rates and other parameters for insect populations passing through stages. *Oecologia* 15: 277–285.

Manly, B. F. J. 1974b. A comparison of methods for the analysis of insect stage–frequency data. *Oecologia* 17: 335–348.

Manly, B. F. J. 1975. A note on the Richards, Waloff and Spradbery method for estimating stage–specific mortality rates for insect populations. *Biom. J.* 17: 77–83.

Manly, B. F. J. 1976. Extensions to Kiritani & Nakasuji's method for the analysis of stage frequency data. *Res. Pop. Ecol.* 17: 191–199.

Manly, B. F. J. 1977. A further note on Kiritani & Nakasuji's model for stage–frequency data including comments on Tukey's jackknife technique for estimating variances. *Res. Pop. Ecol.* 18: 177–186.

Manly, B. F. J. 1985a. Further improvements to a method for analyzing stage–frequency data. *Res. Pop. Ecol.* 27: 325–332.

Manly, B. F. J. 1985b. *The Statistics of Natural Selection.* Chapman and Hall, London.

Manly, B. F. J. 1987. A regression method for analyzing stage–frequency data. *Res. Pop. Ecol.* 29: 119–127.

Manly, B. F. J. & A. G. Shannon. 1974. A analogical replication of a Piagetian experiment on New Guineans. *Malay. J. Educ.* 11: 77–82.

Mills, N. J. 1981a. The estimation of mean duration from stage frequency data. *Oecologia* 51: 206–211.

Mills, N. J. 1981b. The estimation of recruitment from stage frequency data. *Oecologia* 51: 212–216.

Osawa, A., C. A. Shoemaker & J. R. Stedinger. 1983. A stochastic model of balsam fir bud phenology utilizing maximum likelihood estimation. *Forest. Sci.* 21: 478–490.

Pollard, J. H. 1973. *Mathematical Models for the Growth of Human Populations.* Cambridge University Press, Cambridge.

Pontius, J. S., J. E. Boyer & M. L. Deaton. 1987. Estimation of stage transition time: application to ecological studies. *Ann. Ent. Soc. Amer.* (in press).

Qasrawi, H. 1966. *A Study of the Energy Flow in a Natural Population of the Grasshopper Chorthippus parallelus Zett. (Orthoptera Acridae).* Unpublished Ph.D. thesis, University of Exeter, U.K.

Read, K. L. Q. & J. R. Ashford. 1968. A system of models for the life cycle of a biological organism. *Biometrika* 55: 211–221.

Richards, O. W. & N. Waloff. 1954. Studies on the biology and population dynamics of British grasshoppers. *Anti–Locust Bull.* 17: 1–182.

Richards, O. W., N. Waloff & J. P. Spradbery. 1960. The measurement of mortality in an insect population in which recruitment and mortality widely overlap. *Oikos* 11: 306.

Rigler, F. H. & J. M. Cooley. 1974. The use of field data to derive populatin statistics of multivoltine copepods. *Limnol. Oceanogr.* 19: 636–655.

Ruesink, W. G. 1975. Estimating time–varying survival of anthropod life stages from populatin density. *Ecol.* 56: 244–247.

Saunders, J. F. & W. M. Lewis. 1987. A perspective on the use of cohort analysis to obtain demographic data for copepods. *Limnol. Oceonogr.* 32: 511–513.

Sawyer, A. J. & D. L. Haynes. 1984. On the nature of errors involved in estimating stage–specific survival rates by Southwood's method for a population with overlapping stages. *Res. Pop. Ecol.* 26: 331–351.

Schneider, S. M. & H. Ferris. 1986. Estimation of stage–specific development times and survivorship from stage–frequency data. *Res. Pop. Ecol.* 28: 267–280.

Schonemann, P. H. & L. R. Tucker. 1967. A maximum likelihood solution for the method of successive intervals allowing for unequal stimulus dispersion. *Psychometrika* 32: 403–417.

Sharpe, F. R. & A. J. Lotka. 1911. A problem in age distribution. *Philosoph. Mag.* 21: 435–438.

Shoemaker, C. A., G. E. Smith & R. G. Helgensen. 1986. Estimation of recruitment rates and survival fromcensus data with applications to poikilotherm populations. *Agricul. Sys.* 22: 1–21.

Southwood, T. R. E. 1978. *Ecological Methods with Particular Reference to the Study of Insect Populations.* Chapman & Hall, London.

Stedinger, J. R. & C. A. Shoemaker. 1985. A stochastic model of insect phenology for a population with spatially variable development rates. *Biometrics* 41: 691–701.

Straalen, N. M. van 1982. Demographic analysis of arthropod populations using a continuous stage–variable. *J. Anim. Ecol.* 51: 769–783.

Straalen, N. M. van 1985. Comparative demography of forest floor *Collembola* populations. *Oikos* 45: 253–265.

Thurstone, L. L. 1925. A method of scaling psychological and educational tests. *J. Educ. Psych.* 16: 433–451.

Togerson, W. S. 1958. *The Theory and Methods of Scaling.* Wiley, New York.

Usher, M. B. 1966. A matrix approach to the management of renewable resources, with special reference to selection forests. *J. Appl. Ecol.* 3: 355–367.

Usher, M. B. 1969. A matrix model for forest management. *Biometrics* 25: 309–315.

Life Tables and Parasitism: Estimating Parameters in Joint Host—Parasitoid Systems

T. S. Bellows, Jr.,[1] R. G. Van Driesche[2] and J. Elkinton[2]

ABSTRACT Estimating total losses from parasitism is a basic step in constructing life tables for host—parasitoid systems. Such estimates are not available in most cases directly from stage—frequency data or samples of percentage parasitism, particularly when recruitment or parasitism overlaps with advancement to the next developmental stage.

We present two approaches for calculating losses from parasitism. The first incorporates direct estimates of recruitment to the susceptible host stage and to the immature parasitoid stage. These data permit the direct calculation of total losses to parasitism and, when combined with stage density data for the parasitized stage and similar data for subsequent stages, permit thorough construction of life tables for both host and parasitoid which separately quantify host losses to parasitism, other host mortality, mortality in the parasitoid population, and time—specific mortality rates for both host and parasitoid populations during the course of the generation.

The second approach uses the graphical technique of Southwood & Jepson (1962) to estimate recruitment to both host and parasitoid populations. When total losses to parasitism are calculated from these estimates of recruitment, the resulting mortality estimate is subject to substantial biases. These biases arise from the impact of mortality on the accuracy of graphical estimates of recruitment. Consequently, this second approach is applicable only in cases where conditions indicate that the biases are minimal, specifically, when mortality in the popluation to which the graphical analysis is applied is less than 20%.

A central theme in the analysis of insect stage—frequency data is the estimation of numbers of individuals entering successive stages. This information is used to construct budgets or life tables and quantify stage—specific losses. Where these losses are associated with a specific cause such as insect parasitism, estimates of mortality can be assigned to particular factors in the insect's life table.

Almost ubiquitous among insect life tables are records of mortality due to insect parasitoids, and these range from minor to major sources of mortality. In spite of this abundance, however, methods to quantify specifically the impact of parasitoids in insect life tables remain relatively undeveloped. Mortality due to parasitism is often reported as percentage parasitism of samples of the host insect population, but as Van Driesche (1983) has pointed out, this approach will measure the total losses to parasitism only in exceptional cases. More typically, measures of percentage parasitism from samples do not estimate total losses to parasitism, and combinations of samples or selection of certain samples do not reconstruct the desired estimate of mortality.

Calculation of mortality due to parasitism in an insect life table requires estimates of numbers entering the susceptible host stage and estimates of the numbers of these host individuals

---

[1]Division of Biological Control, University of California, Riverside, CA 92521
[2]Department of Entomology, University of Massachusetts, Amherst, MA 01003

which subsequently become parasitized. Just as in the single species case where stage–frequency data are used to estimate numbers entering successive stages, for joint host–parasitoid systems stage–frequency data for host and parasitoid populations may be used together to estimate numbers of individuals entering the host stage and the number of these that subsequently become parasitized, or enter the parasitoid population.

Analysis of such a coupled, two–population system is somewhat more complicated than the analysis of a single population because there are more concurrent processes involved (Fig. 1). In the single species case, recruitment to a stage, mortality during a stage, and subsequent molting by survivors to the next stage are three processes which affect the number of individuals in each stage (Fig. 1a). In a coupled, parasitoid–host system, several additional processes are involved – parasitism (recruitment to the immature parasitoid stage), mortality of immature parasitoids (which may be the same as mortality to the parasitized hosts and is distinct from other non–parasitism mortality occurring in the host population) and recruitment of immature parasitoids to their next stage (Fig. 1b).

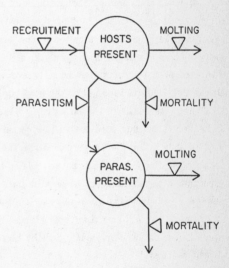

Fig. 1. (a) The pool of individuals present in a single–species population is the net result of three processes. (b) In the case of interacting host and parasitoid populations, the pools of individuals represent the net result of six processes, one which (parasitism) directly relates the two populations.

Estimates of numbers entering the several pools of individuals in such a coupled system may be obtained by analyzing stage–frequency data for both host and parasitoid populations together.

In this paper we consider two approaches to accomplish this. First, we consider the situation where information for both recruitment and stage density are available in the stage–frequency records for the host and parasitoid populations. This involves slightly more information than is customarily found in stage–frequency data but provides an immediate answer to the question of mortality due to parasitism and permits a detailed evaluation of the process of parasitism and its impact in the coupled system. Subsequently, we will consider extending an established method for estimating numbers entering a stage in the single–species case, the graphical technique of Southwood & Jepson (1962), to the case of two interacting populations and will discuss its potential applications and limitations. Both techniques are illustrated with references to populations of the cabbage butterfly, *Pieris rapae* (L.), parasitized by the braconid *Cotesia* (*Apanteles*) *glomerata* (L.).

The Methods

Use of recruitment data

In this approach to the problem of estimating mortality due to parasitism, we employ estimates of recruitment to both the host and parasitoid populations which are obtained by sampling the population directly. The usual stage–frequency data, consisting of numbers of individuals in separate developmental stages at successive points in time, is supplemented by estimates of the numbers entering these development stages during each sampling interval (Van Driesche & Bellows 1988). For the host population this is the number of individuals entering the host stage susceptible to parasitism. We must also measure the number of hosts which subsequently become parasitized, i.e., recruitment to the pool of parasitized hosts. For solitary parasitoids which do not superparasitize, the number of newly parasitized hosts is the same as the number of newly recruited parasitoids. However, in gregarious parasitoids or species that superparasitize, the number of new immature parasitoids (i.e., eggs) would be greater than the number of newly parasitized hosts. Superparasitism is not new host mortality and is not new recruitment to the parasitized host stage. We will use the term "parasitoid recruitment" as the equivalent of entry of previously healthy host individuals into the pool of immature parasitoids (or parasitized hosts).

Estimates of total recruitment can be obtained by adding together estimates of recruitment over all successive sampling intervals. Once estimates of total recruitment to both host and parasitoid populations are obtained, mortality due to parasitism (as a proportion or per capita rate) is estimated by

$$\text{mortality} = \frac{P_r}{H_r}, \tag{1}$$

where P_r is the total parasitoid recruitment and H_r is the total host recruitment. Equation (1) estimates parasitism as a proportion of the total number of hosts entering the susceptible host stage without recourse to the stage–frequency data. This provides a direct estimate of mortality due to parasitism and avoids the problems discussed by Van Driesche (1983).

The changes which occur in the host and parasitoid populations between sampling intervals may be described as follows:

$$H_i(t+1) = H_i(t) + H_{ri}(t) - H_{r(i+1)}(t) - H_{pi}(t) - \mu(t)H_i(t) \,, \qquad (2)$$

$$P_i(t+1) = P_i(t) + P_{ri}(t) - P_{r(i+1)}(t) - \mu'(t)P_i(t) \,. \qquad (3)$$

During the interval t to t+1, the number of hosts in stage i $[H_i(t)]$ is incremented by recruits to the stage $[H_{ri}(t)]$ and decremented by individuals leaving the stage either by molting (recruiting) to the next stage $[H_{r(i+1)}(t)]$, becoming parasitized $[H_{pi}(t)]$, or dying from causes other than parasitism $[\mu(t)H_i(t)]$. The immature parasitoid population at time t $[P_i(t)]$ similarly is incremented by recruits $[P_{ri}(t)]$ and decremented by recruits to the next stage $[P_{r(i+1)}(t)]$ and those dying $[\mu'(t)\,P_i]$. In cases where estimates of recruitment to the subsequent stages are not available, equations (4) and (5) can be used during that portion of the generation when no molting to the next stage is taking place, so that $H_{r(i+1)}$ and $P_{r(i+1)}$ are zero.

Where information on recruitment to the subsequent stage is available, equations similar to (2) and (3) can describe the dynamics of parasitism and other mortality over the entire span of the susceptible host stage:

$$H_{r(i+1)} = H_{ri} - H_{pi} - \mu H_{ri} \,, \qquad (4)$$

$$P_{r(i+1)} = P_{ri} - \mu' P_{ri} \,, \qquad (5)$$

where H_r is the total number of hosts recruited to the susceptible (i) and subsequent (i+1) stages and P_r is the total number of parasitoids recruited to the pool of immature (i) and the subsequent (i+1) parasitoid stage. H_p is the number of hosts dying due to parasitism; this number is considered equivalent to recruitment to the immature parasitoid stage so the $H_{pi} = P_{ri}$. The parameters μ and μ' are the per capita loss rates due to causes other than parasitism for the host and parasitoid respectively.

As discussed more fully by Van Driesche & Bellows (1988), estimates for $\mu(t)$ and $\mu'(t)$ are obtained from equations (2) and (3) by

$$\mu(t) = \frac{H_i(t) + H_{ri}(t) - H_{r(i+1)}(t) - H_{pi}(t) - H_i(t+1)}{H_i(t) + H_{ri}(t) - H_{r(i+1)}(t) - H_{pi}(t)}, \tag{6}$$

$$\mu'(t) = \frac{P_i(t) + P_{ri}(t) - P_{r(i+1)}(t) - P_i(t+1)}{P_i(t) + P_{ri}(t) - P_{r(i+1)}(t)}. \tag{7}$$

Similarly, estimates for μ and μ' over the span of the susceptible host stage are obtained from equations (4) and (5) by

$$\mu = \frac{H_{ri} - H_{pi} - H_{r(i+1)}}{H_{ri} - P_{ri}}, \tag{8}$$

$$\mu' = \frac{P_{ri} - P_{r(i+1)}}{P_{ri}}. \tag{9}$$

Estimates of per capita mortalities within sampling intervals (equations (6) and (7)) provide the ability to chronicle non—parasitism losses over the course of population development. They can be used to detect changes in mortality separately for both unparasitized hosts and immature parasitoids (parasitized hosts) over time. Equations (8) and (9) provide estimates of mortality other than parasitism to both healthy hosts and immature parasitoids across the span of the stage. Together these equations provide a powerful analytical approach to describe the dynamics of parasitism in an insect population.

Use of Graphical Estimates of recruitment

When sampling recruitment is not possible, an analytical technique must be applied to stage—frequency data to estimate the numbers entering various stages of the life table. A variety of such techniques exist for single—species systems (e.g. Southwood 1978). We will concentrate here on extending one such technique, the graphical technique of Southwood & Jepson (1962), to the two—species case.

In the single—species case, numbers entering a stage are estimated by plotting stage—frequency data against time, calculating the area enclosed by this plot and dividing this area by the developmental time of the stage. This yields an estimate of the number of individuals which entered the stage. Such an estimate is not without potential sources of error, but more importantly it is subject to consistent bias whenever mortality occurs during the stage (e.g. Southwood 1978; Sawyer & Haines 1984; Bellows et al. 1988), leading to underestimates of the number entering the stage. Graphical estimates ($\hat{N}_r$) are related to the actual number recruited (N_r) by, in the case of constant mortality during a stage,

$$\hat{N}_r = N_r \frac{S-1}{\ln S},$$ (10)

where S is overall stage survival. This bias, dependent on mortality acting on the host, increases as survival decreases and can be a source of significant bias in estimating the numbers entering the populations in a joint host–parasitoid system.

When graphical analyses are employed to provide estimates for life table construction for a host population subject to parasitism, the biases in the recruitment estimates of the host and parasitoid combine in complex ways (Fig. 2). In general, we may consider a host population to be surviving two sources of mortality, parasitism (denoted by the proportion surviving parasitism S_1) and non–parasitism sources (S_2), while parasitoid populations are subject to a third source of general mortality (the survival from which is S_3). Graphical analyses may be employed to estimate recruitment to either the host population, the parasitoid population, or both. When these graphical estimates are employed to calculate total mortality due to parasitism (equation (1)), the biases in the separate population estimates affect the accuracy of the estimate of proportionate mortality. Table 1 summarizes some of the findings of Bellows et al. (1988) regarding sources of bias when graphical estimates of recruitment are used to calculate mortality due to parasitism. This bias is quantified as the deviation from the actual percentage parasitism as a proportion of the actual value ($[\hat{P/H}-P/H]/[P/H]$), and generally increases as survival of either host or parasitoid decreases (Fig. 2). (Only healthy hosts were considered when plotting host stage–densities to estimate recruitment, although under some restrictions both healthy and parasitized individuals may may be included in the host plot (Bellows et al. 1988)). Deviation of the estimate of mortality due to parasitism ($\hat{P/H}$) from the actual value of P/H varies in a systematic way (Fig. 2). These biases are minimal when survival is greatest, and is less than 10% when survival of the host and parasitoid are both greater than 80% (Bellows et al. 1988).

Table 1. Application of graphical estimates of population recruitment for calculating mortality due to parasitism.

	Technique used to Estimate Recruitment to Stage		
Method	Host	Parasitoid	Source of bias in the ratio P/H
1	Direct measurement	Graphical	S_3
2	Graphical	Direct measurement	S_1, S_2
3	Graphical	Graphical	S_1, S_2, S_3

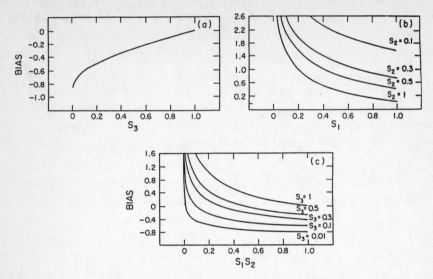

Fig. 2. Curves relating proportional bias in estimated percent parasitism $(\hat{P}/\hat{H})$ to survival of host population (S_1) = proporiton of hosts not parasitized, S_2 = proporiton of hosts surviving mortality factors toher than parasitism) and parasitoid population (S_3 = survival of immature parasitoids). (a), (b), and (c) are the methods (1), (2), and (3) of Table 2, respectively.

Application to *Pieris rapae* and *Cotesia glomerata*

As an example of the two methods described above, we will consider data for larvae of the cabbage butterfly, *P. rapae*, parasitized by the braconid *C. glomerata* (Fig. 3). Details regarding the collection of these data appear in Van Driesche & Bellows (1988) and Van Driesche (1988).

Recruitment to both host and parasitoid populations were recorded directly (Fig. 3a, b), and yielded a value for mortality due to parasitism (equation (1)) of 86.7%. Numbers entering the subsequent pupal stage of both host and parasitoid were estimated using the graphical technique of Southwood & Jepson (1962) and indicated losses in addition to parasitism of 98.2% to the host larvae (μ of equation (8)) and mortality of immature parasitoids (i.e., parasitized host larvae) of 98.4% (μ' of equation (9)). Data taken over intervals yielded time–specific mortality rates (equations (6) and (7)) for the host and parasitoid population, respectively (Fig. 3d). Mortality in the host population generally decreased as the population aged, while mortality rates in the parasitoid population increased slightly with age.

The plots of stage densities over time were used to provide graphical estimates of numbers entering the host and parasitoid stages. When these estimates were employed to calculate mortality due to parasitism (equation (1)), the values (Table 2) were significantly biased (with respect to the value calculated from recorded recruitment), consistent with the large degree of mortality acting on both host and parasitoid populations discussed above.

The underestimate of parasitism by method 1 reflects the underestimate of parasitoid recruitment in the numerator of equation (1) (see equation (10)), while the denominator is the actual recorded value of recruitment and hence is unbiased. Method 2 greatly overestimates parasitism because the graphical estimate of the number of hosts recruited underestimates actual recruitment. When graphical estimates are used for both populations (method 3), the resulting value for mortality lies between the previous two but is not particularly close to the actual value.

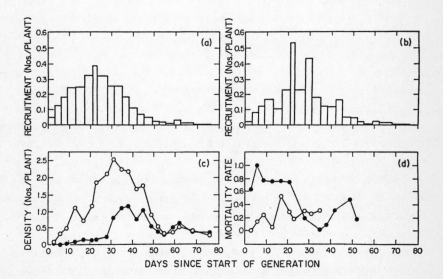

Fig. 3. (a) Average daily recruitment over 3 − 6 day intervals of first instar *P. rapae* larvae in the third generation of 1985 in a collard field in South Deerfield, Massachusetts. (b) Average daily recruitment over 3 − 6 day intervals of parasitized hosts. (c) Densities per plant of healthy *P. rapae* larvae (•) and larvae bearing the parasitoid = *C. glomerata* (o). (d) Interval−specific mortality rates of healthy (•) and parasitized (o) *P. rapae* larvae. The abscissal scales start at the beginning of the generation, where day 1 = 23 July 1988.

The deviations observed in this example would have been smaller if either mortality due to parasitism or other mortality acting in the system was substantially smaller. In addition, Bellows et al. (1988) have explored other approaches to employing graphical estimates of either host or

Table 2.   Mortality of *Pieris rapae* due to parasitism by *Cotesia glomerata* calculated using graphical estimates of numbers entering the host and parasitoid immature stages.

Method in Table 1	Estimated mortality due to parasitism (%)	Deviation from observed value (86.7%) (%)
1	35.3	−58.9
2	629.6	632.0
3	258.3	200.3

parasitoid recruitment to estimate total generational parasitism. Some of these variations reduce or limit the amount of bias in the resulting calculations, but their use implies strict assumptions about life histories of the species involved. Where these assumptions are appropriate, these alternative methods may be more applicable, and potentially less biased, than those presented here.

Discussion

Parasitism occurs widely in insect populations and is recorded frequently in insect life tables. Quantitatively estimating the impact of parasitism on a host population requires estimates of the numbers entering the susceptible host stage and those which subsequently become parasitized. In many insect populations these numbers cannot be determined by simple examination of single or serial samples of parasitized and healthy hosts. Such samples chronicle the proportion of parasitized individuals occurring in the population over time but do not relate simply to the total losses to parasitism in a generation (Van Driesche 1983).

The scoring of recruitment to the susceptible host stage and to the immature parasitoid stage (i.e., new parasitism of healthy hosts) provides a direct estimate of generational losses to parasitism. The scoring of mortality in this way records the parasitism as it occurs and avoids the significant risks of error associated with scoring parasitism by rearing individuals from late in the developmental stage when other mortality factors may have masked the actual impact of the parasitoid on the host population. In addition, it permits the host population to be represented in two categories: healthy hosts and those bearing parasitoids. Monitoring this second category of hosts, or more precisely the population of immature parasitoids (Varley et al. 1973) permits life table construction for the parasitoid population. Life tables for parasitoids are much less common than for their hosts but are much needed (Richards 1961) and can be used to great advantage in the quantitative study of host–parasitoid systems.

Measuring host and parasitoid recruitment at frequent intervals permits estimation of interval–specific mortality rates acting on the two separate populations. This is a particularly powerful tool through which age–related events can be recorded separately for parasitized and healthy hosts. Mortality rates can differ between healthy and parasitized individuals, and these differences can have significant bearing on the life tables constructed from such data and on inferences drawn from them.

Directly measuring recruitment when assessing losses from parasitism provides some analytical advantages over methods which estimate recruitment from stage–frequency data. Stage densities reflect the compound effects of recruitment, mortality, and advancement to the next stage and, when two populations are interacting, there are parallel processes in both. In contrast, recruitment estimates do not contain internal, counteracting processes and hence are conceptually simpler and analytically more tractable than density estimates alone. Recording recruitment permits the separate effects of these several processes to be easily distinguished and quantified.

There are cases, however, where recruitment to either or both the host or parasitoid population cannot be recorded, and in these cases we must turn to analysis of stage–density data to estimate losses to parasitism. A variety of techniques have been developed for single species systems, but few have been extended to the case of coupled host–parasitoid systems. The methods considered here, using graphical estimates of stage recruitment, are subject to significant biases when parasitism or other mortality is substantial. However, they may find application in systems where these biases are small. For example, method 1 (Table 1) may be applicable where direct measurement of host recruitment is possible and their is little mortality of the immature parasitoid; such cases may arise more commonly in systems with external parasitoids which paralyze their host or otherwise remove it from risk of other mortality factors. Any of the methods may be applicable in systems where parasitism and other mortality is less than 20%. Particular life histories also lend themselves to analysis by permitting special analytical assumptions which reduce the amount of bias in the calculations (Bellows et al. 1988).

Although the application of graphical estimates of recruitment for calculating mortality due to parasitism is limited, there are a number of other techniques for analysis of stage–frequency data which may have utility in addressing this problem. The regression techniques of Richards & Waloff (1954) and others may be modified to incorporate multiple interacting populations. Transfer–function techniques (e.g., Manly 1974, Bellows et al. 1982) may provide additional approaches. It remains to be seen, however, how the addition of another population to these techniques will affect their estimation abilities and general applicability. It seems likely that many of them will provide additional solutions to this important problem in the analysis of insect populations.

Acknowledgement

This study was supported in part (T.S.B.) by grants from the USDA (86–CRCR–1–2173) and NSF (BSR–8604546) and in part (R.G.V.D.) by grants from the USDA (84–CRCR–1–1385, 86–CRCR–1–2171).

References Cited

Bellows, T. S. Jr., M. Ortiz, J. C. Owens & E. W. Huddleston. 1982. A model for analyzing stage–frequency data when mortality varies with times. *Res. Pop. Ecol.* 24: 142–156.

Bellows, T. S. Jr., R. G. Van Driesche & J. S. Elkinton. 1988. Extensions to Southwood and Jepson's graphical method of estimating numbers entering a stage for calculating losses to parasitism. *Res. Pop. Ecol.*, in press.

Manly, B. F. J. 1974. Estimation of stage–specific survival rates and other parameters for insect populations developing through several stages. *Oecologia* 15: 277–285.

Richards, O. W. 1961. The theoretical and practical study of natural insect populations. *Ann. Rev. Entomol.* 6: 147–162.

Richards, O. W. & N. Waloff. 1954. Studies on the biology and population dynamics of British grasshoppers. *Anti–Locust Bull.* 17, 182 p.

Sawyer, A. J. & D. L. Haines. 1984. On the nature of errors involved in estimating stage–specific survival rates by Southwood's method for a population with overlapping stages. *Res. Pop. Ecol.* 26: 313–351.

Southwood, T. R. E. 1978. Ecological Methods. Chapman & Hall, London. 2nd ed.

Southwood, T. R. E. & W. F. Jepson. 1962. Studies on the populations of *Oscinella frit* L. (Diptera: Chloropidae) in the oat crop. *J. Anim. Ecol.* 31: 481–495.

Van Driesche, R. G. 1983. The meaning of "percent parasitism" in studies of insect parasitoids. *Environ. Entomol.* 12: 1611–1622.

Van Driesche, R. G. & T. S. Bellows, Jr. 1988. Host and parasitoid recruitment in quantifying losses from parasitism, with reference to *Pieris rapae* and *Cotesia glomerata*. *Ecol. Entomol.* 13: 215–222.

Van Driesche, R. G. 1988. Survivorship patterns of *Pieris rapae* (Lep.: Pieridae) larvae in Massachusetts collards with special reference to measurement of the impact of *Cotesia glomerata* (Hym.: Braconidae). *Bull. Entomol. Res.* 78: 199–208.

Varley, G. C., G. R. Gradwell & M. P. Hassell. 1973. Insect Population Ecology. Blackwell Scientific Publications, Oxford.

From Cohort Data to Life Table Parameters
Via Stochastic Modeling

Moshe Braner and Nelson G. Hairston, Jr.[1]

ABSTRACT We develop a new method through which parameters such as the duration of stages and the mortality rates within them can be deduced from data on abundance of the stages over time in one cohort of individuals. This method involves modeling of the development process, modeling of the sampling process with its inherent errors, a statistical approach based on the models, and a numerical algorithm designed to perform the statistical estimation on real (noisy) data. We discuss the meaning of *development time* in the face of mortality, and illustrate the use and the validity of the method with real data from cohorts where the development times were independently measured *in situ*.

Cohort analysis is the quantitative examination of stage–density data from one cohort for the purpose of uncovering its underlying demographic parameters. Several methods have been developed for cohort analysis, and their relative merits have been debated (e.g., Manly 1974b, 1977, 1985; Mills, 1981; Hairston & Twombly, 1985; Aksnes & Høisaeter, 1987; Hairston, Braner and Twombly, 1987). Most methods introduced in the past have made use of simple statistics derived from the data, such as the time of first appearance of an instar, mean or median time of sampling of individuals of each instar, and the area under the instar's time–density curve. A risk of using a few statistics derived from the data rather than the data themselves is the potential loss of information. Also, some model of cohort development is always present behind each method, even if not explicitly stated. The consequences of the model are transformed into equations relating the summary statistics to the estimated parameters. The hidden assumptions may be at odds with what is known about the developmental biology of the organisms, or have internal inconsistencies. For example, Rigler & Cooley (1974) used equations that implicitly assume that *no* mortality occurs in order to estimate mortality (see Hairston & Twombly, 1985).

Ideally, when estimating the demographic parameters from field population size estimates, one should take into account the random noise in the data. It is desirable to obtain not only a best–fit estimate but also some sort of a confidence interval around it. Doing that requires models for the development of the organism and for the sampling errors too. In a pioneering effort of this sort, Read & Ashford (1968; see also Ashford et al., 1970) clearly presented an explicit growth model together with its underlying assumptions. They treated the development time of each individual in each stage as a random variable, and proceeded to fit the model parameters to the data set as a whole, using a maximum–likelihood approach. Their method, however, did not

---

[1]Section of Ecology and Systematics, Cornell University, Ithaca, New York 14853.

immediately yield good results, especially when the data were noisy and when stage–specific mortality rates were allowed. Furthermore, the numerical effort was beyond the computing resources typically available at the time. Kempton's (1979) method is more practical due to simplifying assumptions. We follow his approach but attempt to remove some of the more restrictive assumptions. We also proceed further in modeling the sampling process.

In this paper we introduce a model of stage–structured cohort development, and a model of sampling errors. We derive from these models a new method of estimating the various growth parameters from field data. Before we can do that, however, we have to digress a bit into a discussion of the effect of mortality on the apparent development time and its philosophical and statistical consequences.

The Meaning of Development Time

We have to reevaluate the notion of "the time period an organism spends in development through a specific instar." If all individuals were to develop at the same rate then there would be no ambiguity. But in reality the development time is a random variable. The ambiguity arises when the process of mortality is superimposed on the development process. The chance of survival to the point of maturation into the next instar decreases with the time spent in development. Therefore, measurements of calculations made in the presence of mortality yield a biased sample of developmental times, with shorter development times overrepresented.

For example, suppose that N individuals start development at time 0, and that the development time, in the absence of mortality, is normally distributed, with mean m and variance σ^2. Under a constant mortality rate μ, the *rate* of maturation (the number of individuals leaving the instar alive per unit time) at time t, is:

$$X(t) = N \exp\{-\mu t\} \frac{1}{\sqrt{2\pi\sigma^2}} \exp\left\{ -\frac{(t-m)^2}{2\sigma^2} \right\}.$$

Combining the two exponents and completing the square, we get:

$$X(t) = N S \frac{1}{\sqrt{2\pi\sigma^2}} \exp\left\{ -\frac{(t-m')^2}{2\sigma^2} \right\},$$

where S is the probability of surviving through the instar:

$$S = \exp\{ -m\mu + \sigma^2\mu^2/2 \},$$

and m' is the mean time of maturation:

$$m' = m - \sigma^2\mu.$$

Thus, the rate of maturation is also normally distributed, with the same variance as the development time, but with a smaller mean. The difference between the means ($m - m' = \sigma^2\mu$) is proportional to both the variance in development time and the mortality rate (Fig. 1).

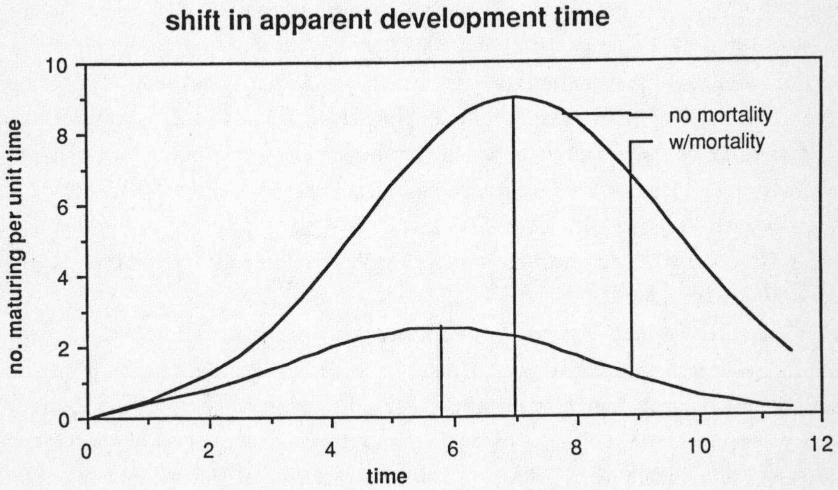

shift in apparent development time

Fig. 1. Rate of maturation as a function of time, without mortality (top curve) and after correcting for mortality (bottom curve), when development time is normally distributed. The vertical lines mark the mean development time in each case.

The significance of this observation is twofold. First, if one were to observe the maturation rates without measuring the mortality, there would be no way to disentangle the effects of m and $\sigma^2\mu$ from the observed m'. In other words, without further information there would be an inherent difficulty in estimating *both* development time *and* mortality from the data. Second, one might ask which is the *real* development time. The observed m' *is* the mean development time as it actually happens in the field. The biased sampling is not in our observation, but rather in the maturation process itself. One might say that the development time measured in the laboratory, under no—mortality conditions, is the biased sample! This redefinition of development time has the disadvantage that its value will vary with the severity of natural mortality, which itself is a random variable. This is no different, though, from the dependence of development time — under either definition — on other environmental variables such as temperature and food availability.

A Model of Cohort Development

We want to use a model of staged development that includes the essential ingredients necessary for the identification of the parameters of interest. We also would choose assumptions that allow an explicit mathematical expression of the growth model, an expression that is reasonably simple to transform into a computer algorithm. The model should not introduce more parameters than can be handled statistically.

Our model is conceptually based on inter–individual stochasticity in the developmental rates. However, we assume a large population, and thus the stochasticity at the individual level translates to deterministic distributions at the population level. The number of individuals sampled, though, is not large, and our model of sampling errors below is stochastic throughout.

We use a model of development in which the development time in an instar is normally distributed among individuals. This is of course biologically impossible, since it implies that some individuals finish their development before they start it, due to the indefinite extent of the left tail of the normal distribution. This problem is not of serious concern as long as that tail is small enough in area to be insignificant.

The time of entry into the first instar, we further assume, is also normally distributed. For the purpose of the mathematical derivation (but not in the practical data–fitting method) we assume that the development time in an instar is independent of the time of entry into that instar, and that the development time of an individual in an instar is independent of the development times of the same individual in other instars. Finally, we assume that the instantaneous mortality rate is constant throughout the development of an instar (but possibly different in different instars).

The details of the mathematical consequences of the above assumptions will be given elsewhere. Here we discuss some of the results. Let N denote the total number of individuals that enter an instar, and M the total number that pass into the next instar. Let a and u^2 denote the mean and variance of the time of entry into the instar, and let b and v^2 denote the mean and variance of the exit, or maturation, time. $P(t)$ will mean the number of individuals present in the instar at time t. Let m and σ^2 denote the mean and variance of the development time in an instar under the conditions of no mortality. Since we assume that development time is independent of entry time, and since both are assumed normal, the exit time — the sum of the two — is also normal, with the mean $a + m'$ and the variance $u^2 + \sigma^2$. That is (for each instar):

$$b = a + m - \sigma^2\mu$$

using the formula for m' developed above, where μ is the rate of mortality. The number present in the instar at time t is the difference between the number that have entered and the number that have left, corrected for mortality. Braner (1988) showed that to a good approximation

$$P(t) = N \exp\left\{ -\mu(t-a) + u^2\mu^2/2 \right\} \left[\Phi\left[\frac{t-a-u^2\mu}{u}\right] - \Phi\left[\frac{t-b-v^2\mu}{v}\right] \right],$$

where $\Phi\,()$ is the standard normal cumulative distribution function. This is an approximation since it takes the normality of the development times literally. If negative development times are omitted from integral ranges the formula becomes more complicated, but the relative numerical correction is small for typical parameter values. For more details see Braner (1988).

A Model of Sampling Error

Our goal is to develop a method for estimating the parameters in a model of cohort development from field data. But field data are noisy, and a perfect fit cannot be expected. In this section we develop an explicit model of the sources of error. This model will be used in the next section for the definition of the best fit.

The sampled population is frequently patchy over space and the patches move over time. Thus the success of the sampling effort is variable. Furthermore, the sampling effort is not precisely known.

For the purpose of modeling sampling error we assume unbiased sampling: the expected sample value equals the real mean of the population multiplied by the sampling effort. The expectation of the final number recorded as a sample point is the product of several error–prone numbers. On a logarithmic scale the process involves the summation of several independent random variables. But other considerations, such as patchiness of the sampled population, modify that description.

We chose to model the situation with the Gamma distribution because it is flexible enough to take on various shapes, including shapes similar to the lognormal and the normal, but involves only two parameters and is mathematically tractable in this context. It has the simple density form (Woodroofe, 1975, p. 124):

$$f(x) = \frac{\beta^\alpha x^{\alpha-1} e^{-\beta x}}{\Gamma(\alpha)},$$

where β is the scale parameter, α the shape parameter, and $\Gamma(\alpha)$ is the necessary normalization constant. The mean is α/β and the variance α/β^2. Thus the coefficient of variation (CV) is $1/\sqrt{\alpha}$, independent of β. For small values of α, the Gamma distribution resembles a lognormal distribution, while as $\alpha \to \infty$ the Gamma distribution converges to the normal distribution. We assume that the real sampling effort relative to the nominal effort is a factor that has a Gamma distribution with a mean of unity, i.e., $\beta = \alpha$. The now single parameter α can be varied to fit the CV of the process. If this CV is not known, it can be estimated from the data along with the growth parameters.

Given the actual sampling effort, there is yet another source of error. Individual organisms have individual chances of being caught in the net, and so the probability of counting k individuals is Poisson distributed (Woodroofe, 1975, p. 92):

$$p_k = \frac{e^{-\lambda}\lambda^k}{k!} ,$$

where λ is the expected value, as determined by the population density and the (real) sampling effort. This source of error is surprisingly large. The variance is equal to the mean λ, and so the CV is $1/\sqrt{\lambda}$.

Statistical and Numerical Methods

We have an explicit model, and we want to estimate the values of the parameters of the model from field data. Given the explicit error model, a maximum likelihood *estimator* can be used.

If each group of individuals taken from the field is counted for several instars at once, as is the usual case, then the numbers of individuals recorded in the various instars share the same sampling effort and success error, but are independently Poisson distributed with different means. (More complicated situations are possible, in which different instars are sampled together but with different sampling efficiencies. We limit ourselves here to the simpler case.) For one sample date, let Y_i be the count of individuals of instar i (a random variable), and y_i the actual number observed. Let z_i be the real expected values of the counts. If the samples from different dates (possibly replicate samples taken on the same data) are independent, then the log likelihood function is (Braner, 1988):

$$\log(L) = n(\alpha \log \alpha - \log \Gamma(\alpha)) - G + \sum_t \left[\sum_i y_{ti} \log z_{ti} - (\alpha + w_t) \log(\alpha + \sum_i z_{ti}) + \log \Gamma(\alpha + w_t) \right] ,$$

where n is the number of sample dates, $w_t = \Sigma_i y_{ti}$ (a constant, independent of the values of the parameters), and $G = \Sigma_t \Sigma_i \log(y_{ti}!)$, which is also a constant. Finding the maximum likelihood estimator is equivalent to finding the parameter values that maximize the log–likelihood. The statistical problem has now been reduced to a numerical problem: find the values of the parameters that maximize $\log(L)$, given the data. This is an explicit but difficult problem.

Since our merit function (the log–likelihood) is a very nonlinear function of the parameters, we have adapted Powell's algorithm (see Press et al., 1986, p. 297) to our needs. More details about the numerical scheme, source code, and detailed instructions, are given by Braner (1988).

Validation of the Method

In order to test the utility of the method we have developed for cohort analysis, we tried it out in cases where the answers are known. Such data are created by simulation. A detailed test of this sort is given by Braner (1988). It shows that the method works correctly in the face of significant random noise.

In these Proceedings Manly gives a set of simulated data, that he used for a similar validation test of various proposed methods. Those numbers were generated using his "method 3" simulation model, with Poisson type sampling errors. Our method fitted Manly's data set to our model, and recovered his original parameters quite well (Table 1). We see that our method, applied to noisy data from another model, yielded good estimates of the known parameter values.

Table 1. Manly's simulation parameters, and the values fitted by our method to his simulated data. (The CV of the Gamma error term was set at 10%.)

Stage	Number entering		Stage duration		Daily survival	
	est.	true	est.	true	est.	true
1	241	200	1.2	1.5	0.70	0.70
2	157	117	1.8	2.0	0.63	0.75
3	69	66	1.8	2.0	0.81	0.80
4	47	42	1.6	2.0	0.62	0.70
5	22	21	—	—	0.52	0.60

Analysis of Field Data

We have started to apply the method of cohort analysis described above to real data. Populations of small (1mm long) copepods (*Diaptomus sanguineus*) in several ponds in Rhode Island have been sampled regularly over several years. Here we present the results of carrying out our method of analysis on two cohorts that developed during the same year in the same pond. The first cohort developed in winter 1984–1985, the second in spring 1985, with a small amount of overlap between the two.

Duplicate copepod samples of known water volumes were collected weekly or biweekly using a Clarke–Bumpus quantitative sampler fitted with a $75\mu m$ mesh net, preserved with Formalin, and subsampled and counted in the laboratory at 25X using a stereomicroscope. Population size estimates were corrected for seasonal variations in pond volume and expressed as number of individuals per pond.

In addition to collection of these stage–frequency data, development times of the instars were measured independently during the development of the second cohort in the pond. Forty

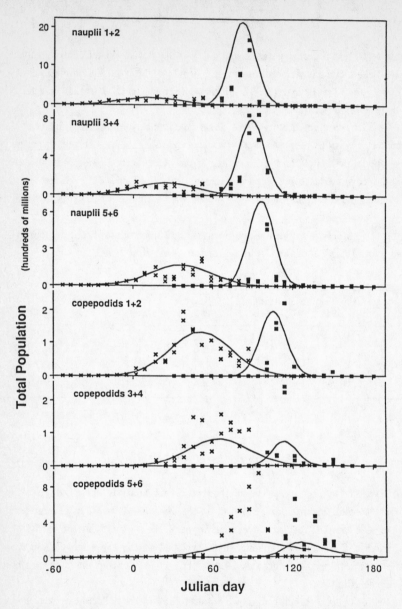

Fig. 2. Abundance of *Diaptomus sanguineus* in Bullhead Pond, RI in late 1984 and early 1985 as a function of time (days from January 1, 1985). Depicted are field data (points) and fitted model predictions (curves). The two cohorts were analyzed separately. For the analysis, we grouped the 12 copepod instars two by two into 6 stages. Note that the vertical scale is not the same in all the panels.

newly hatched copepods were placed individually into 70ml vials with $53\mu m$ mesh windows, suspended in the pond at 1.0m, and scored for developmental stage daily. A total of 29 copepods survived to maturity. These *in situ* estimates of development time serve as a real–life validation for the cohort analysis.

The data points and the fitted model curves for both cohorts are presented in Fig. 2. The curve for some stages seems somewhat offset from the data points, but the method searches for the best overall fit, and the model equations force some tradeoffs between consecutive stages. A better fit could possibly be achieved by changing the model itself, e.g., by using physiological time rather than calendar time. That is a goal for future work. But already several interesting patterns emerge from the data.

Table 2. Fitted estimates of mortality rates (μ) and development time (τ) for the two cohorts of *Diaptomus sanguineus* illustrated in Fig. 2. Field measurements of development times for the second cohort are given for comparison.

Instar		Winter cohort		Spring cohort		
		μ(day^{-1})	τ(days)	μ(day^{-1})	fitted τ(days)	measured τ(days)
Nauplii	1+2	0.001	13.0	0.199	6.3	4.2
	3+4	0.001	10.4	0.002	7.8	10.5
	5+6	0.004	11.9	0.238	7.1	6.4
Copepodid	1+2	0.054	14.8	0.021	11.2	7.1
	3+4	0.001	14.7	0.002	5.0	5.4
	5+6	0.005	n.a.	0.056	n.a.	n.a.
Totals		0.013	64.8	0.085	37.4	33.6

It is apparent from inspection of Fig. 2 that the development time of the winter cohort is substantially longer than that of the spring cohort. The model best fit gives total development times of 64.8 days in the winter and 37.4 days in the spring (Table 2). There is a well established physiological relationship between temperature and copepod developmental rate with a Q_{10} on the order of 2.2 to 2.8 (development rate increases by this factor over a 10 degree temperature rise, e.g., Geiling & Campbell 1972, Landry 1975). The mean pond temperature during the development of the winter cohort was 2.6^0C and during the spring was 10.4^0C, for a temperature difference of 7.8^0C. The ratio of development times over this temperature difference represents a Q_{10} of 2.02, close to the range of established values.

Monitoring of development time in the *in situ* vial experiment began on 14 March and the mean maturation date was 17 April. The model fitted a mean starting date for the second cohort of 21 March with maturation to the last two instars on 27 April. Thus development time was estimated by the two methods over essentially the same period. Although there are differences between the development times in the various copepod instars as fitted by the model and as determined *in situ*, the values are quantitatively close, and the total development times only differ by about 10 percent.

Mortality rates are 6.5 times higher in the spring cohort than in the winter cohort. We have no independent means of establishing mortality of *Diaptomus sanguineus*, but the trend is consistent with what we would expect. The activity of the major predator of the copepods (bluegill sunfish) is greatly reduced at winter as compared with spring temperatures (Hairston et al., 1983).

Taken together, these results give us reason to be encouraged about the quality of the information we can extract from field collected cohort data. The method could still be refined, as is evident from Fig. 2, but the estimated parameters are useful as they are.

Conclusion

This new method for the analysis of one–cohort stage–frequency data rests on solid ground both logically and statistically. The models are explicitly stated and the assumptions are consistent within themselves and with biological knowledge. The "best fit" criteria are chosen to comply with statistical principles and a reasonable model of the errors involved in this kind of sampling. The numerical procedures have been carefully optimized to perform reliably and efficiently in the nontrivial task of simultaneously fitting many parameters of a nonlinear model to data. The resulting method has been shown to work correctly in cases where the real answers are known, and to yield biologically reasonable estimates when applied to real data.

The method we have developed does have some disadvantages. In particular, computing a single set of estimates takes about 30 minutes on a typical laboratory microcomputer. That is not an unreasonable time for the analysis of data that took months to gather, and in a few years computer technology will probably cut the computing time by an order of magnitude. Unfortunately, Tukey's "Jackknife" technique (see Manly, 1977) for the precise determination of the confidence ranges of the estimated parameters — which requires many repetitions of the procedure — will most likely be out of reach for quite some time, not to mention the even more computationally intensive bootstrap methods.

The proposed method of analysis does not provide the researcher with very accurate estimates of all the biologically interesting parameters, in particular stage–specific mortality. These parameters are hidden inside typical data sets by large amounts of random noise. There is only a limited amount of information in such a data set, and no method can hope to squeeze out of the data more than that limited amount. It is hoped that the method proposed here approaches that theoretical limit.

We have modeled time as a uniform stream of development. But in nature, developmental rates depend on environmental factors, in particular on temperature. Given the right laboratory data, it would be interesting to correct the calendar time to fit the physiological clock (e.g., by using "degree days") and then do the cohort analysis. That is a topic for future research.

Acknowledgements

We thank E. J. Olds and T. A. Dillon for counting zooplankton samples, C. E. McCulloch for statistical advice, and S. A. Levin for mathematical advice and financial support. The field research was funded by NSF grants BSR–8307350 and BSR–8516724 to N. G. Hairston, Jr. The theoretical work was funded in part by NSF and NYS Hatch grants to S. A. Levin.

References

Aksnes & Høisaeter. 1987. Obtaining life table data from stage–frequency distributional statistics. *Limnol. Oceanogr.* 32: 514–517.

Ashford, J. R., K. L. Q. Read, & G. G. Vickers. 1970. A system of stochastic models applicable to studies of animal population dynamics. *J. Anim. Ecol.* 39: 29–50.

Braner, M. 1988. Dormancy, dispersal and staged development: Ecological and evolutionary aspects of structured populations in random environments. Ph.D. Thesis, Cornell University.

Geiling, W. T. & R. S. Campbell. 1972. The effect of temperature on the development rate of the major life stages of *Diaptomus pallidus* Herrick. *Limnol. Oceanogr.* 17: 304–306.

Hairston, N. G., Jr. W. E. Walton, & K. T. Li. 1983. The causes and consequences of sex–specific mortality in a freshwater copepod. *Limnol. Oceanogr.* 28: 935–947.

Hairston, N. G., Jr. & S. Twombly. 1985. Obtaining life table data from cohort analyses: A critique of current methods. *Limnol. Oceanogr.* 30: 886–893.

Hairston, N. G., Jr., M. Braner & S. Twombly. 1987. Perspective on prospective methods for obtaining life table data. *Limnol. Oceanogr.* 32: 517–520.

Kempton, R. A. 1979. Statistical analysis of frequency data obtained from sampling an insect population grouped by stages. Pp. 401–418 in *Statistical Distributions in Ecological Work*, J. K. Ord et al., (eds.), International Cooperative Publishing House, Fairland, Maryland.

Landry, M. R. 1975. The relationship between temperature and the development of life stages of the marine copepod *Acartia clausi* Giesbr. *Limnol. Oceanogr.* 20: 854–857.

Manly, B. F. J. 1974a. Estimation of stage–specific survival rates and other parameters for insect populations developing through several stages. *Oecologia* 15: 277–285.

Manly, B. F. J. 1974b. A comparison of methods for the analysis of insect stage–frequency data. *Oecologia* 17: 335–348.

Manly, B. F. J. 1976. Extensions to Kiritani and Nakasuji's method for analyzing insect stage–frequency data. *Res. Pop. Ecol.* 17: 191–199.

Manly, B. F. J. 1977. A further note on Kiritani and Nakasuji's model for stage–frequency data including comments on the use of Tukey's Jackknife technique for estimating variance. *Res. Pop. Ecol.* 18: 177–186.

Manly, B. F. J. 1985. Further improvements to a method for analyzing stage–frequency data. *Res. Pop. Ecol.* 27: 325–332.

Mills, N. J. 1981. The estimation of mean duration from stage frequency data. *Oecologia* 51: 206–211.

Mood, A. M., F. A. Graybill, & D. C. Boes. 1974. Introduction to the Theory of Statistics, third edition. McGraw–Hill, New York. 564pp.

Press, W. H., B. P. Flannery, S. A. Teukolski, & W. T. Vetterling. 1986. Numerical Recipes: the Art of Scientific Computing. Cambridge University Press. 818pp.

Read, K. L. Q. & J. R. Ashford. 1968. A system of models for the life cycle of a biological organism. *Biometrika* 55: 211–221.

Rigler, F. H. & J. M. Cooley. 1974. The use of field data to derive population statistics of multivoltine copepods. *Limnol. Oceanogr.* 19: 636–655.

Woodroofe, M. 1975. Probability with Applications. McGraw–Hill. 372pp.

Estimation of Stage–Specific Demographic Parameters for Zooplankton Populations: Methods Based on Stage–Classified Matrix Projection Models

Hal Caswell[1] and Saran Twombly[2]

ABSTRACT Our understanding of zooplankton dynamics is constrained by the difficulty of estimating stage–specific vital rates from field data. In this paper, we approach demographic estimation as an inverse problem, using stage–classified matrix projection models. We allow the vital rates to vary between stages and over time, and do not assume stable age distributions or that discrete cohorts can be identified. We obtain least–squares estimates of survival and growth probabilities, which can be obtained from as few as three consecutive censuses. However, the estimates are ill–conditioned in the presence of sampling noise. Two regularization methods, truncated singular value decomposition and Tikhonov regularization (ridge regression) are examined as possible solutions.

Estimates of age– or stage–specific vital rates (survival, growth, fertility) provide valuable information about the distribution and abundance of organisms. Estimation of these rates is particularly difficult for organisms such as insects and zooplankton, individuals of which cannot be marked and followed over time. In these cases, vital rates must be inferred from observed changes in population structure and abundance. In this paper we report an estimation procedure for such populations. We have attempted to make only minimal assumptions. In particular, we do not assume that the population is at a stable age distribution, we do not assume that the enviroment is constant, and we do not assume that discrete cohorts can be identified.

Two classes of methods are presently used to estimate vital rates from zooplankton abundance data. When populations exhibit exponential growth and stable age distributions (e.g., short–lived organisms with rapid and continuous reproduction, such as rotifers and cladocerans), "egg ratio" techniques are used to estimate birth rates from the abundance of eggs and adults in the population (Edmondson 1960, Caswell 1972, Paloheimo 1974, Taylor & Slatkin 1981). Population growth rate is estimated from the observed change in population size, and death rate is estimated from the difference between birth and population growth rates.

Egg–ratio methods, however, provide only crude birth and death rates, with no information about how those rates differ among age or size classes. Unstable population structure (common in field populations) can produce large errors in estimates obtained from egg ratio methods (Threlkeld 1979, Gabriel et al. 1987). In addition, these methods are inappropriate for longer–lived organisms with complex life cycles, such as copepods (Taylor & Slatkin 1981, Lynch 1983).

---

[1]Biology Department, Woods Hole Oceanographic Institution, Woods Hole, MA. 02543.
[2]Department of Zoology, University of Rhode Island, Kingston, RI. 02281.

Rigler & Cooley (1974) and Gehrs & Robertson (1975) introduced a second class of methods for copepod populations exhibiting discrete cohorts. These methods are related to estimation procedures for insect populations (see review by Manly, this symposium). Rigler & Cooley's (1974) method is seriously flawed, being valid only in the biologically unrealistic case where survivorship does not vary between successive stages (Hairston & Twombly 1985, Hairston et al. 1987). In all other cases, this method can generate erroneous estimates of stage duration and survivorship. Some of these problems have been resolved by Aksnes (1986; Aksnes & Hoisaeter 1987) and Braner and Hairston (this symposium).

Other cohort—based estimation methods (Manly, this symposium) are superior to Rigler & Cooley's, but all of them assume either that survival is constant over time or constant across stages. They thus cannot provide any insight into temporal patterns of stage—specific vital rates. In addition, they require the identification of cohorts, which is impossible when generations overlap widely.

Finally, both egg—ratio and cohort methods lack an explicit demographic basis, which is needed if the estimated parameters are to be interpreted in terms of their contributions to population growth rate, population structure, reproductive value, sensitivity, and other standard demographic methods.

Matrix Projection Models

The estimation method we present here is based on population projection matrix models for complex life cycles (e.g., Lefkovitch 1965, Caswell 1978, 1982, 1983, 1985, 1986). These models incorporate the structure of the life cycle in relation to the sampling interval, identify the necessary demographic parameters, and provide a powerful set of analytical methods for evaluating these parameters once they are estimated.

We suppose that the population dynamics at time t are described by the matrix projection equation

$$n(t+1) = A_t n(t), \tag{1}$$

where $n(t)$ is a vector whose entries $n_i(t)$, $i = 1,2, ..., s$ give the abundance of each of s stages at time t, and A_t is a $s \times s$ non—negative matrix whose entries a_{ij} incorporate the demographic transitions (easily specified by use of a life cycle graph (Caswell 1982)) of the individuals in the population.

The demographic properties of the population that can be derived directly from A include the asymptotic rate of increase λ, given by the dominant eigenvalue of A, the stable stage distribution w and the reproductive value distribution v, given by the corresponding right and left eigenvectors of A, and the sensitivities (or selection gradients (Lande 1982)) on the demographic parameters, given by

$$\frac{\partial \lambda}{\partial \alpha_{ij}} = \frac{v_i w_j}{<\mathbf{w},\mathbf{v}>} , \tag{2}$$

where $<\cdot,\cdot>$ denotes the scalar product.

The Standard Size–Classified Model

We focus our attention on a standard size–classified model, appropriate when the population is classified by size classes or instars, and when the time interval $[t,t+1]$ is shorter than the duration of the shortest stage. The first row of the projection matrix contains stage–specific fertilities F_i. The subdiagonal contains the probabilities G_i of surviving and growing from stage i to $i+1$. The diagonal contains the probabilities P_i of surviving and remaining in the same stage. All other entries are zero.

The probabilities G_i and P_i can be expressed in terms of more basic survival and growth probabilities as

$$G_i = \sigma_i \gamma_i \tag{3}$$

$$P_i = \sigma_i(1-\gamma_i) \tag{4}$$

where σ_i is the survival probability of stage i and γ_i the probability that a surviving individual grows from stage i to $i+1$.

Like any matrix projection model, this model assumes that demography is truly stage–dependent. The age of an individual is irrelevant to its survival, growth, and reproduction. Since the age distribution of individuals within a stage is ignored, the stage duration is assumed to be geometrically distributed. In fact, the distribution of instar duration is typically unimodal, often skewed to the left, but not geometric (e.g., Sharpe et al. 1977; Mills 1981). Caswell (1983, 1987) and Law (1983) have considered models which include ages within each stage, permitting specification of arbitrary stage duration distributions. Such models are unlikely to be useful for parameter estimation for zooplankton, because their detail generates a profusion of additional parameters, and because zooplankton cannot usually be aged from field samples (see Threlkeld 1979). Given the difficulty of measuring ages within stages, the standard size–classified model is the most reasonable basis on which to develop methods of demographic estimation of zooplankton.

Demographic Inverse Problems

Keller (1976) defines two problems as inverses of one another if the formulation of each involves the solution of the other. One problem is typically more well understood that the other, and is referred to as the direct problem. The direct problem in demography is to project the dynamics of the population stage structure $n(t)$, given the vital rates contained in the sequence of matrices A_t. The inverse problem is to estimate the vital rates, given a sequence of stage distribution vectors $n(t)$ generated by those vital rates.

We present here a solution to the problem of estimating the P_i and G_i from a sequence of census vectors $n(t)$. A subsequent paper will consider the estimation of the F_i and the comparison of observed and predicted sequences of population vectors (Caswell and Twombly in prep.).

Estimating P_i and G_i

Consider a sequence of $m+1$ census vectors $n(t),...,n(t+m)$. For any stage i ($i>1$; the contribution of reproduction to n_i may complicate things) we can write the projection equations:

$$
\begin{aligned}
n_i(t+1) &= G_{i-1}n_{i-1}(t) + P_i n_i(t) \\
n_i(t+2) &= G_{i-1}n_{i-1}(t+1) + P_i n_i(t+1) \\
n_i(t+3) &= G_{i-1}n_{i-1}(t+2) + P_i n_i(t+2)
\end{aligned}
$$

$$
\vdots
$$

$$
n_i(t+m) = G_{i-1}n_{i-1}(t+m-1) + P_i n_i(t+m-1)
$$

Rearranging terms yields a system of inverse equations in the unknown parameters G_{i-1} and P_i:

$$
\begin{bmatrix}
n_{i-1}(t) & n_i(t) \\
n_{i-1}(t+1) & n_i(t+1) \\
\vdots & \vdots \\
n_{i-1}(t+m-1) & n_i(t+m-1)
\end{bmatrix}
\begin{bmatrix}
G_{i-1} \\
P_i
\end{bmatrix}
=
\begin{bmatrix}
n(t+1) \\
n_i(t+2) \\
\vdots \\
n_i(t+1)
\end{bmatrix}
\tag{5}
$$

or, in matrix form,

$$
N\beta = d \tag{6}
$$

If $m > 2$, this system of equations is overdetermined, and will not in general possess a solution. However, least–squares solutions $\hat{G}_{i-1}$ and $\hat{P}_i$ can be found by calculating the generalized inverse (the Moore–Penrose pseudo–inverse[1]) N^+ of the matrix N and writing

$$
\beta = N^+ d. \tag{7}
$$

The generalized inverse of a matrix N satisfies

$$NN^+N = N$$
$$(NN^+)' = NN^+$$
$$(N^+N)' = N^+N$$

It can be written down explicitly as

$$N^+ = \lim_{\delta \to 0}(N'N + \delta^2 I)^{-1}N'$$

(Albert 1972).

This is equivalent to treating the problem as a linear regression and solving for β

$$\beta = (N'N)^{-1}N'd. \tag{8}$$

Lefkovitch (1964) used multiple regression to estimate parameters in a less structured stage–classified model, but did not attempt to follow temporal variation in the resulting matrices.

Note that if the population is at a stable stage distribution, each row of the matrix N is proportional to the first, and it is not possible to solve (6).

Ill–Conditioning in the Presence of Noise

This method works exactly when applied to noise–free artificial data. A size–classified matrix is used to generate a series of census vectors $n(t)$ from some initial vector. When the resulting data is plugged into (7), the matrix entries are recovered exactly.

Real data, however, are always corrupted by sampling noise, the presence of which will obviously reduce the accuracy of the estimation procedure. Ideally, however, the errors in the estimates should be small when the error in the data is small. Methods for which this is not true are said to be *ill–conditioned*. Many linear inverse problems are ill–conditioned, including ours.

Fig. 1 shows the results of estimating P_i and G_i from noisy data generated from a 6 x 6 matrix A, with all $P_i = 0.5$ and all $G_i = 0.4$. A 20–day sequence of census vectors was generated by iterating an initial population vector whose entries were uniformly distributed random numbers between 0 and 1. A 25% sampling error was simulated by multiplying each data point by a random number log–uniformly distributed between $1/1.25$ and 1.25. Estimates of P_i, $i = 2,...,6$ and G_i, $i = 1,...,5$ were calculated for values of $m = 2,...,19$.

The results show considerable variation. The parameter estimates fluctuate widely and independently, with no sign of convergence to their actual values as the number of census vectors is increased.

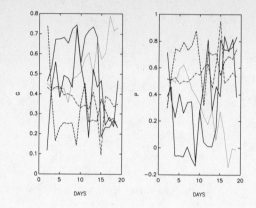

Fig. 1. Results of applying the generalized inverse to artificial data with 25% sampling noise. Estimates of G_i (actual values 0.4) and P_i (actual values 0.5) shown as a function of the number m of data points used.

Regularization Methods

Ill–conditioning is a frequently encountered difficulty in inverse problems in a variety of fields (e.g., Bertero 1986). Tikhonov (1965, see review in Tikhonov & Arsenin 1977) introduced the approach of *regularization* to deal with the problem. Regularization introduces a family of approximations to the ill–conditioned inverse operator, depending on a parameter α. Tikhonov & Arsenin (1977) discuss the criteria that an operator must satisfy to qualify as a regularization algorithm. Basically, as the parameter $\alpha \to 0$, the regularized solution must converge to the least square solution.

Two commonly used regularizing algorithms for the generalized inverse are *truncated singular value decomposition* and *Tikhonov regularization* (also known as ridge regression).

Truncated Singular Value Decomposition

Any r x c matrix N has a singular value decomposition

$$N = USV'$$

(9)

where U and V are unitary matrices of dimension r x r and c x c, respectively, and S is an r x c matrix with the singular values s_j of N on the diagonal. The singular values are the square roots of the eigenvalues of $N'N$. The generalized inverse N^+ of N can be written in terms of the singular values and vectors as

$$\mathbf{N^+} = \sum_j s_j^{-1} \mathbf{v}._j \mathbf{u}'._j \; , \tag{10}$$

where $\mathbf{v}._j$ and $\mathbf{u}._j$ denote the jth columns of $\mathbf{V}$ and $\mathbf{U}$.

Ill conditioning of the generalized inverse arises when one or more singular values are close to zero, in which case the s_j^{-1} term in (10) becomes very large. Regularization by truncated singular value decomposition consists of carrying our the summation in (10) only for those terms in which s_j is greater than some threshold value.

The resulting least squares solution is biased, but less subject to drastic fluctuations due to errors in the data.

Tikhonov Regularization

The least squares solution of the inverse problem can be written

$$\beta = (\mathbf{N'N})^{-1} \mathbf{N'd}. \tag{11}$$

Ill–conditioning of the generalized inverse corresponds to ill–conditioning of the matrix $\mathbf{N'N}$. This arises in ordinary multiple linear regression problems when there are dependencies among the would–be independent variables. Hoerl & Kennard (1970 a, b) introduced the method of "ridge regression" to stabilize these estimates; the same technique had been independently proposed earlier by Tikhonov. It consists of replacing the ill–conditioned inversion problem by estimating β by

$$\beta = (\mathbf{N'N} + \alpha m \mathbf{I})^{-1} \mathbf{N'd} \; , \tag{12}$$

where α is the regularization parameter.

The results of this method depend strongly on the choice of α. There are a number of methods in the literature, but one of the best and most intuitively appealing is that of *generalization cross–validation* (Wahba 1973, Golub et al. 1979). In cross–validation, each data point is deleted from the data set, and a corresponding estimate of β produced. If the value of α is good, the resulting estimate should do a good job of predicting the missing data point. The eventual value of α chosen is that which minimizes the sum of squared prediction errors.

Golub et al (1979) show that this is equivalent to finding the value of α which minimizes the function

$$V(a) = \frac{\frac{1}{m}\|(\mathbf{I} - \mathbf{C}(\alpha))\mathbf{d}\|^2}{\left[\frac{1}{m}\text{tr}(\mathbf{I} - \mathbf{C}(\alpha))\right]^2} \; , \tag{13}$$

where

$$C(\alpha) = N(N'N + m\alpha I)^{-1}N'.$$

Results

Artificial Data

We begin by applying truncated singular value decomposition and Tikhonov regularization to the artifical data set used in Fig. 1. The results are shown in Figs. 3 and 2.

In the model used here, the matrix N is $m \times 2$, and thus has only 2 non–zero singular values. The only truncation possible is to retain only the larger of the two in calculating the generalized inverse. This clearly stabilizes the estimates of P_i and G_i, once m is greater than 4 or 5. The bias in the estimates is also apparent, with G_i converging to approximately 0.5 instead of 0.4, and P_i converging to about 0.3 instead of 0.5.

Tikhonov regularization produces less stabilization, but also results in a smaller bias.

The 25% sampling noise level in these data is certainly not excessive for field samples of zooplankton. We are not encouraged by the amount of variation in the estimates of P_i and G_i remaining after applying these methods, although they are clear improvements over un–regularized least squares solutions.

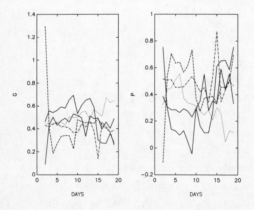

Fig. 2. Results of applying Tikhonov regularization to artificial data with 25% sampling noise. Estimates of G_i (actual values 0.4) and P_i (actual values 0.5) shown as a function of the number m of data points used.

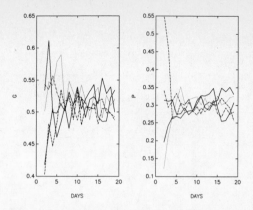

Fig. 3. Results of applying truncated singular value decomposition to artificial data with 25% sampling noise. Estimates of G_i (actual values 0.4) and P_i (actual values 0.5) shown as a function of the number m data points used.

Real Data

We applied these methods to data on the copepod *Diaptomus sanguineus*. The population was sampled daily for 26 days, starting on 2 July 1982, in a small temporary pond in Rhode Island (Hairston & Twombly 1985). Samples were collected in triplicate, from 5 different locations in the pond, using a 27.5 1 Schindler trap (65 um). 12 developmental stages (six naupliar, five copepodite and the adult stage) were identified. The data are shown in Fig. 4. The resulting 12 x 26 array of data was analyzed using ordinary least squares, truncated singular value decomposition, and Tikhonov regularization.

Fig. 5 shows the resulting values of G_i and P_i as functions of time for the three methods. In all cases, $m = 5$ days.

The generalized inverse method produces extremely variable results. P_i and G_i (and also σ_i and γ_i, which are not shown here) are frequently negative or greater than 1, which is impossible for probabilities.

The Tikhonov regularization method produces estimates which are less variable than those produced by the generalized inverse, but many of the values are still outside their possible range. Neither of these methods is improved significantly by increasing m from 5 to 10 days.

The results of the truncated singular value decomposition method appear much more satisfactory. The resulting estimates almost all fall between 0 and 1. The patterns for $m = 5$ and $m = 10$ are very similar.

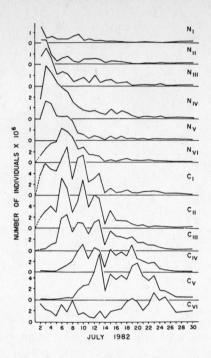

Fig. 4. Abundance data for the copepod *Diaptomus sanguineus* (Hairston & Twombly 1985). Figures are shown for six naupliar stages and six copepodite stages; the last copepodite stage is the adult.

Fig. 6 shows the values of γ_i and σ_i calculated from the P_i and G_i. Encouragingly, most of these values also fall between 0 and 1. In addition, there is encouraging evidence that differences between the vital rates of different stages are consistent over time. Daily survival probabilities σ_i are high, averaging about 0.8 over all stages, while growth probabilities γ_i lie between 0.4 and 0.5.

These results are shown in more detail, as a function of time, in Figs. 7 and 8.

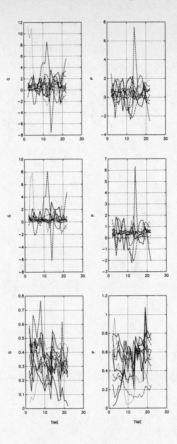

Fig. 5. Estimates of G_i and P_i for *Diaptomus sanguineus*, as a function of time, using the unregularized generalized inverse (top), Tikhonov regularization (middle), and truncated singular value decomposition (bottom). In all cases, a data window of $m = 5$ days was used.

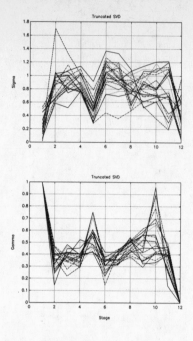

Fig. 6. Estimates of σ_i and γ_i for *Diaptomus sanguineus*, as a function of stage i, using truncated singular value decomposition with a data window of $m = 5$ days. Each line represents a different sample date.

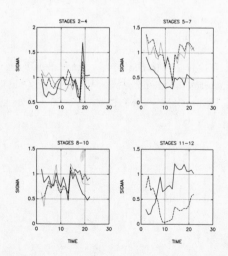

Fig. 7. Estimates of stage–specific survival probabilities σ_i for *Diaptomus sanguineus*, as a function of time, using truncated singular value decomposition with a data window of $m = 5$ days.

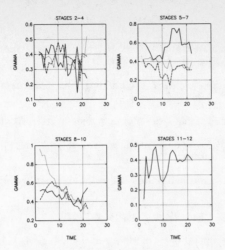

Fig. 8. Estimates of stage–specific growth probabilities γ_i for *Diaptomus sanguineus*, as a function of time, using truncated singular value decomposition with a data window of $m = 5$ days.

Conclusions

The solution of the demographic inverse problem posed in this paper is simple in concept, but limited in practice by ill–conditioning in the presence of noise. Of the methods examined here, truncated singular value decomposition seems to provide consistent and reasonable estimates of stage–specific vital rates.

Several unsolved problems remain, however. First, the results with noisy artificial data clearly show that truncated singular value decomposition produces biased estimates. Without any way to evaluate this bias, it is hard to know how much faith to place in the resulting estimates, although comparison of actual and predicted population structure suggests that in this case the estimates are good (Caswell and Twombly in prep.). Second, confidence intervals must be associated with the estimates. Without such intervals, it is impossible to know, for instance, whether the estimates which fall outside of the acceptable [0,1] range are "really" outside that range, or if the precision in their estimation is too poor to permit such a conclusion. We are working on this problem, and will present results in a subsequent paper.

Acknowledgements

This research was supported by NSF Grant BSR86–9395. Our thanks to Jules Jaffe for advice and to Dan Smith for programming. Woods Hole Oceanographic Institution Contribution 6718.

References Cited

Aksnes, D. L. 1986. Resource allocation in the study of copepod population dynamics. Ph.D. Thesis, University of Bergen, Bergen, Norway.

Aksnes, D. L. & T. J. Hoisaeter. 1987. Obtaining life table data from stage–frequency distributional statistics. *Limnol. Oceanogr.* 32: 514–517.

Albert, A. 1972. Regression and the Moore–Penrose pseudoinverse. Academic Press, New York.

Bertero, M. 1986. Regularization methods for linear inverse problems. pp. 52–112 in G. Talenti [ed.] Inverse Problems. Springer–Verlag, New York.

Caswell, H. 1972. On instantaneous and finite birth rates. *Limnol. Oceanogr.* 17: 787–791.

Caswell, H. 1978. A general formula for the sensitivity of population growth rate to changes in life history parameters. *Theor. Pop. Biol.* 14: 215–230.

Caswell, H. 1982. Stable population structure and reproductive value for poplations with complex life cycles. *Ecology* 63: 1223–1231.

Caswell, H. 1983. Phenotypic plasticity in life–history traits: demographic effects and evolutionary consequences. *Amer. Zool.* 23: 35–46.

Caswell, H. 1985. The evolutionary demography of clonal organisms. In L. Buss, J. B. C. Jackson and R. Cook [eds.]. The Population Biology of Clonal Organisms. Yale Univ. Press, New Haven.

Caswell, H. 1986. Life cycle models for plants. *Lect. Math. Life Sci.* 18: 171–223.

Caswell, H. 1987. Approaching size and age in matrix population models. Proc. Symposium on Dynamics of Size–Classified Populations (in press).

Edmondson, W. T. 1960. Reproductive rates of rotifers in natural populations. *Mem. Ist. Ital. Idrobiol.* 12: 21–77.

Gabriel, W., B. E. Taylor & S. Kirsch–Prokosch. 1987. Cladoceran birth and death rates estimates: experimental comparisons of egg–ratio methods. *Freshwat. Biol.* 18: 361–372.

Gehrs, C. W. & A. Robertson. 1975. Use of life tables in analyzing the dynamics of copepod populations. *Ecology* 56: 665–672.

Golub, G. H., M. Heath, & G. Wahba. 1979. Generalized cross–validation as a method for choosing a good ridge parameter. *Technometrics* 21: 215–223.

Hairston, N. G., Jr. & S. Twombly. 1985. Obtaining life table data from cohort analyses: a critique of current methods. *Limnol. Oceanogr.* 30: 886–893.

Hairston, N. G., Jr., M. Braner & S. Twombly. 1987. Perspective on prospective methods for obtaining life table data. *Limnol. Oceanogr.* 32: 517–520.

Hoerl, A. E. & R. W. Kennard. 1970a. Ridge regression: biased estimation for nonorthogonal problems. *Technometrics* 12: 55–67.

Hoerl, A. E. & R. W. Kennard. 1970b. Ridge regression: applications to nonorthogonal problems. *Technometrics* 12: 69–82.

Keller, J. B. 1976. Inverse problems. *Amer. Math. Monthly* 83: 107–118.

Lande, R. 1982. A quantitative genetic theory of life history evolution. *Ecol.* 63: 607–615.

Law, R. 1983. A model for the dynamics of plant populations containing individuals classified by age and size. *Ecology* 64: 224–230.

Lefkovitch, L. P. 1964. Estimating the Malthusian parameter from census data. *Nature* 204: 810.

Lefkovitch, L. P. 1965. The study of population growth in organisms grouped by stages. *Biometrics* 21: 1–18.

Paloheimo, J. E. 1974. Calculation of instantaneous birth rate. *Limnol. Oceanogr.* 19: 692–694.

Rigler, F. H. & J. M. Cooley. 1974. The use of field data to derive population statistics of multivoltine copepods. *Limnol. Oceanogr.* 19: 636–655.

Sharpe, P. J. H., G. L. Curry, D. W. DeMichele & C. L. Cole. 1977. Distribution model of organism development times. *J. Theor. Biol.* 66: 21–38.

Taylor, B. E. & M. Slatkin. 1981. Estimating birth and death rates of zooplankton. *Limnol. Oceanogr.* 26: 143–158.

Threlkeld, S. T. 1979a. Estimating cladoceran birth rates: the importance of egg mortality and the egg age distribution. *Limnol. Oceanogr.* 24: 601–612.

Tikhonov, A. N. 1965. Ill–posed problems of linear algebra and a stable method of solving them. Doklady Akad. Nauk USSR 163: 6.

Tikhonov, A. N. & V. Y. Arsenin. 1977. Solutions of ill–posed problems. V. H. Winston and Sons, Washington, D.C.

Wahba, G. 1977. The approximate solution of linear operator equations when the data are noisy. *SIAM J. Num. Anal.* 14: 651–667.

Compartmental Models in the Analysis of Populations

Malcolm J. Faddy[1]

ABSTRACT Compartmental models can apply to populations where individuals in the population develop through a series of specific stages (compartments), and can thus form a basis for the analysis of stage–frequency data. A review of the theory of compartmental systems is given.   Both deterministic and stochastic theory are covered with mention of phase–type compartmental residence time distributions (covering exponential, gamma and mixtures of gamma), and interactions between individuals within the compartments.

Compartmental models can apply to populations where individual members are classified into discrete compartments, with changes in classification occurring over time. For example, in an insect population a natural system of compartments would be the clearly identifiable stages of development: egg stage, larval stages and adult stage, with an orderly progression through these successive stages as each insect develops (Read & Ashford 1968; Kempton 1979); of interest are the dynamics of this progression, and of death at the different stages. Compartmental modelling provides a theoretical framework for such a study of populations, and it is the purpose of this paper to give a short personal review of the general theory and its applicability.

Consider then a system of compartments numbered 1, 2, ..., m, and suppose that all individuals in the population are classified into these compartments. Over time the individuals change compartments by developing in some way, or dying. It is convenient to think of these changes dynamically in terms of individuals moving out of one compartment and into another according to their development, or moving out of the system altogether by death. Although the theory will be developed for quite general movement of individuals between the compartments, models for population development are likely to be progressive with sequential movement of individuals through the compartments, resulting in many of the parameters in the general specification being zero.

Deterministic Formulation

Here it is assumed that the population is large enough for the numbers of individual in the compartments to behave as continuous variables according to the assumptions:

(D1)   individuals move out of compartment i and into compartment j $(\neq i)$ at fractional rate γ_{ij};

---

[1]Department of Mathematics and Statistics, University of Otago, Box 56, Dunedin, New Zealand.

(D2)   individuals move out of compartment i (by death) at fractional rate μ_i;

and   (D3)   new individuals move into compartment i (from outside) at rate λ_i.

According to this formulation, the numbers of individuals in the m compartments at time t, $x_1(t), x_2(t), ..., x_m(t)$, will satisfy the system of differential equations (j = 1, 2, ..., m):

$$\frac{dx_j}{dt} = \lambda_j + \sum_{\substack{i=1 \\ i \neq j}}^{m} x_i \gamma_{ij} - x_j \left[\sum_{\substack{k=1 \\ k \neq j}}^{m} \gamma_{jk} + \mu_j \right]$$

or

$$\frac{dx_j}{dt} = \lambda_j + \sum_{i=1}^{m} x_i \gamma_{ij}$$

where

$$\gamma_{jj} = - \left[\sum_{\substack{k=1 \\ k \neq j}}^{m} \gamma_{jk} + \mu_j \right] . \tag{1}$$

In a natural matrix notation, these differential equations may be written:

$$\frac{d\mathbf{x}}{dt} = \lambda + \mathbf{x}\mathbf{V} . \tag{2}$$

where the matrix $\mathbf{V}$ has non–negative off–diagonal elements, negative diagonal elements and row sums (i = 1, 2, ..., m):

$$\sum_{\substack{j=1 \\ j \neq i}}^{m} \gamma_{ij} + \gamma_{ii} = -\mu_i \leq 0 . \tag{3}$$

Much has been written about the solution of differential equations (2): see, for example, Jacquez 1985, Chapter 2. Briefly, the solution may be written in the following forms:

(a) with no movement of new individuals into system ($\lambda = 0$)

$$\mathbf{x}(t) = \mathbf{x}(0) \exp(\mathbf{V}t)$$

$$= \mathbf{x}(0) \sum_{n=0}^{\infty} \mathbf{V}^n t^n / n! \qquad (\mathbf{V}^0 = \mathbf{I}) . \tag{4}$$

(b) with zero numbers of individuals initially in the system ($\mathbf{x}(0) = \mathbf{0}$)

$$\mathbf{x}(t) = \int_0^t \lambda \exp(\mathbf{V}s)ds$$

$$= \lambda \, \mathbf{V}^{-1} \{\exp(\mathbf{V}t) - \mathbf{I}\} \; . \tag{5}$$

Notice that the series representation of the matrix exponential function on the right hand side of equation (4) is convergent, since the number (m) of compartments in the system is finite.

Stochastic Formulation

For many populations a stochastic formulation will be preferred, and such a formulation need not result in too many complications. Assume that the individuals in the population behave independently according to a Markov process described by the infinitesimal probabilities:

(S1) $\quad P\left\{\begin{array}{l} \text{individual moves out of compartment } \; i \\ \text{into compartment } \; j \; \text{ in time } (t, \; t + dt) \end{array}\right\} = \gamma_{ij}dt;$

(S2) $\quad P\left\{\begin{array}{l} \text{individual moves out of compartment} \\ i \; \text{ (by death) in time } (t, \; t + dt) \end{array}\right\} = \mu_i dt;$

and (S3) $\quad P\left\{\begin{array}{l} \text{new individual moves into compartment} \\ i \; \text{ (from outside) in time } (t, \; t + \; dt) \end{array}\right\} = \lambda_i dt \; .$

According to the standard theory of Markov processes (see, for example, Cox & Miller 1965, Chapter 4) the transition probabilities for each individual in the population:

$$p_{ij}(t) = P\left\{\begin{array}{l} \text{in compartment} \\ j \; \text{ at time } \; t \end{array} \; \middle| \; \begin{array}{l} \text{in compartment} \\ i \; \text{ at time } \; 0 \end{array}\right\}$$

($i, j = 1, 2, ..., m$) will satisfy the differential equations:

$$\frac{dp_{ij}}{dt} = \sum_{\substack{k=1 \\ k \neq j}}^{m} p_{ik}\gamma_{kj} - p_{ij}(t)\left[\sum_{\substack{k=1 \\ k \neq j}}^{m} \gamma_{jk} + \mu_j\right] \; .$$

Or in matrix notation, using (1):

$$\frac{d\mathbf{P}}{dt} = \mathbf{P}\,\mathbf{V} \; , \tag{6}$$

cf. equation (2). The solution of (6) subject to the natural initial condition $P(0) = I$ may be written:

$$P(t) = \exp(Vt) \ . \tag{7}$$

cf. equation (4). The solution for arbitrary initial numbers of individuals now follows from the assumption of independence of their behavior.

(a) $N_i(0)$ individuals initially (time 0) present in compartment i will at time t be multinomially distributed across the m compartments according to probabilities:

$$N_i(0)! \ \prod_{j=1}^{m} \ \frac{\{p_{ij}(t)\}^{x_j}}{x_j!} \ \frac{\left\{1 - \sum_{j=1}^{m} p_{ij}(t)\right\}^{N_i(0) \ - \sum_{j=1}^{m} x_j}}{\left\{N_i(0) - \sum_{j=1}^{m} x_j\right\}!} \ ; \tag{8}$$

$$0 \le x_1, x_2, ..., x_m, \ \sum_{j=1}^{m} x_j \le N_i(0) \ .$$

(b) New individuals will move into compartment i according to a Poisson process of rate λ_i (assumption S3), so that the number of new individuals entering the system via compartment i in time interval $(0,t)$ will be Poisson distributed with mean $\lambda_i t$. Also, conditional on some number entering compartment i during such a time interval, their times of entry will be independent random variables uniformly distributed on the interval $(0,t)$ (Cox & Lewis 1966, Chapter 2); each individual will then move about the system, being in compartment j at time t with probability:

$$q_{ij}(t) = \int_0^t \frac{1}{t} p_{ij}(t-s) \ ds$$

independently of one—another. So unconditionally the numbers of new individuals in the compartments at time t that entered via compartment i will be independently Poisson distributed with means $(j = 1, 2, ..., m)$:

$$\lambda_i t q_{ij}(t) = \lambda_i \int_0^t p_{ij}(t-s) ds \ , \tag{9}$$

cf. equation (5).

Generally, the distribution of numbers of individuals in the compartments will be given by a convolution of *independent* multinomial distributions like (8) (for each $N_i(0) \neq 0$) and Poisson distributions (for each $\lambda_i \neq 0$ and each compartment).

This direct approach to the solution making use of the independent behaviour of all individuals within the compartmental system obviates the need for more complicated arguments involving generating functions and partial differential equations (Matis & Hartley 1971; Capasso & Paveri–Fontana 1981). Also, note from comparison of equations (8) with (4) and (9) with (5), using (7), that the mean numbers of individuals in the compartments from the stochastic formulation of this section are the same as the deterministic values from the previous section.

The matrix exponential (7) is a convenient representation of the matrix $P(t)$ of compartmental transition probabilities. However, it is not at all convenient for numerical computation, because the negative values appearing on the diagonal of the matrix V lead to alternating series for the individual $p_{ij}(t)$. Kohlas (1982, Section 3.5) gives a satisfactory numerical procedure by writing:

$$Q = I + \frac{1}{c}V,$$

where $c = \max\{-\gamma_{ii}\}$, $1 \leq i \leq m$, so that the matrix Q has all its elements non–negative and row sums ≤ 1, cf. (3). Now

$$P(t) = \exp(Vt) = \exp(Qct)e^{-ct}$$

$$= \sum_{n=0}^{\infty} Q^n \frac{(ct)^n}{n!} e^{-ct}, \tag{10}$$

and truncating the infinite power series expansion (10) for $P(t)$ will result in errors bounded by the tail of a Poisson distribution with mean ct.

Generalizations

In most applications it is unlikely that the basic model will be entirely satisfactory (Matis and Wehrly 1979), prompting the need for generalizations. Some possibilities are as follows.

(I) Time dependence in the parameters γ_{ij}, μ_i and λ_i may be incorporated with algebraic modifications (Faddy 1977); e.g. the Poisson mean (9) replaced by:

$$\int_0^t \lambda_i(s)p_{ij}(s,t)ds,$$

with the individual compartmental transition probabilities

$$p_{ij}(s,t) = P\left\{ \begin{array}{c} \text{in compartment} \\ \text{j at time t} \end{array} \left| \begin{array}{c} \text{in compartment} \\ \text{i at time s} \end{array} \right. \right\}$$

satisfying (in matrix notation):

$$\frac{\partial P(s,t)}{\partial t} = P(s,t)\, V(t)\, .$$

Unfortunately there is no convenient matrix exponential representation of $P(s,t)$ in general (Godfrey 1983, Chapter 10) unless the matrices $V(t)$ and $\int_0^t V(s)ds$ commute; however, step–function forms for $\gamma_{ij}(t)$ and $\mu_i(t)$ can be handled conveniently (Faddy 1976).

   (II)   Non–Poisson input of new individuals into the system of compartments cannot be accommodated simply, as it would induce some dependence among the individuals' conditional entry times and so the argument leading to (9) would break down.

   (III)   Perhaps the most restrictive consequence of the Markov assumption is that of exponentially distributed residence–times: random time spent in compartment i before moving to another compartment is exponentially distributed with mean

$$1 \Big/ \sum_{\substack{j=1 \\ j \neq i}}^{m} \gamma_{ij}$$

and random time spent in compartment i before dying is exponentially distributed with mean $1/\mu_i$, with probability of survival in compartment i

$$\sum_{\substack{j=1 \\ j \neq i}}^{m} \gamma_{ij} \Bigg/ \left[\sum_{\substack{j=1 \\ j \neq i}}^{m} \gamma_{ij} + \mu_i \right] \, .$$

Weiner & Purdue (1977) consider a semi–Markov formulation using a general approach; many (compartmental residence–time) distributions can be derived in terms of exponential distributions, and can be incorporated within the framework developed in the previous section.

   A gamma (k, γ) distributed random variable with probability density function $\gamma^k t^{k-1} e^{-\gamma t}/(k-1)!$, $t > 0$, may be written as a sum of k independent exponentially distributed random variables each with mean $1/\gamma$. Hence, a compartment consisting of k sub–compartments with movement of individuals sequentially through these sub–compartments according to infinitesimal probability rate γ (cf. assumption (S1)) will result in a residence–time in the main compartment distributed as gamma (k,γ), cf. Matis (1972).

The gamma (k,γ) distribution has a coefficient of variation of $1/\sqrt{k}$ (< 1 if $k \geq 2$); distributions formed by mixtures of two or more gamma distributions can have coefficients of variation > 1 and exhibit long–tailed behaviour. A compartment consisting of several sets of sub–compartments with movement of individuals sequentially through each set of sub–compartments and random starting of individuals in the first sub–compartment of each set will result in residence–times in the main compartment distributed according to a mixture of several gamma distributions.

As a simple example, consider two compartments both with sub–compartments and movement of individuals between sub–compartments according to the infinitesimal probability rate matrix:

$$\mathbf{V} = \left[\begin{array}{cc|cc} -\gamma_1 & \gamma_1 & 0 & 0 \\ 0 & -\gamma_1 & p\gamma_1 & (1-p)\gamma_1 \\ \hline 0 & 0 & -\gamma_2 & 0 \\ 0 & 0 & 0 & -\gamma_3 \end{array} \right].$$

Here main compartment 1 consists of two sub–compartments with all individuals starting in the first sub–compartment, so that the residence–times of individuals in this main compartment before movement into compartment 2 will be gamma distributed with probability density function $\gamma_1^2 t e^{-\gamma_1 t}$ ($t > 0$); main compartment 2 also consists of two sub–compartments, but individuals start randomly in one or other of these sub–compartments, so that the residence–times of individuals in this main compartment before movement out will be hyper–exponentially distributed with (mixed) probability density function $p\gamma_2 e^{-\gamma_2 t} + (1 - p)^{\gamma_3} e^{-\gamma_3 t}$ ($t > 0$).

Other residence–time distributions for any number of main compartments can be constructed by making use of sub–compartments within each main compartment, with movement of individuals between sub–compartments and random starting sub–compartments; such residence–time distributions belong to a class of distributions known as phase–type (Neuts 1981). Note that gamma distributed compartmental residence times so constructed are not subject to any restrictions like those of Read & Ashford (1968) or Kempton (1979). Also note that random death of individuals within main compartments according to assumption (S2) can simply be incorporated by adding $-\mu_i$ to all those diagonal elements of the infinitesimal probability rate matrix, $\mathbf{V}$, corresponding to the sub–compartments of main compartment i; a similar change just to the diagonal elements of $\mathbf{V}$ corresponding to those sub–compartments from which individuals can leave a main compartment will describe death of individuals occurring only at the time of leaving.

(IV) Interaction between individual causes the assumption of independent individuals' behaviour to be violated. This results in the distributions of numbers of individuals in the compartments being quite different from the multinomial–Poisson distributions of the previous section; in particular, the pattern of random variation is changed: for example, positive interaction between the individuals, where each individual becomes relatively less likely to move

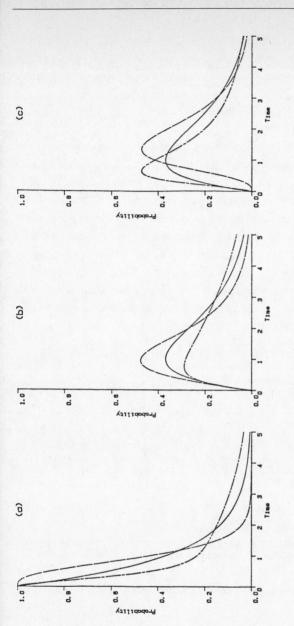

Fig. 1. Compartmental probabilities

$$P_{ij}(t) = P \left\{ \begin{array}{l} \text{individual in compartment} \\ j \text{ at time } t \end{array} \middle| \begin{array}{l} \text{in compartment} \\ i \text{ at time } s \end{array} \right\}$$

as functions of elapsed time t: (a) $P_{11}(t)$ for differing compartment 1 residence–time distributions; (b) $P_{12}(t)$ for differing compartment 2 residence–time distributions with fixed (exponential) compartment 1 residence–time distribution; (c) $P_{12}(t)$ for differing compartment 1 residence–time distributions with fixed (exponential) compartment 2 residence–time distribution; Key: ———— exponential, with coefficient of variation 1; — — — gamma, with coefficient of variation 0.5; —··—··— mixture of 2 gammas, with coefficient of variation 1.5.

from a compartment as the number of individuals in the compartment increases, results in greater random variation (Faddy 1985).

Some Numerical Results

To illustrate the effects of different residence–time distributions on the compartmental probabilities for a progressive system of compartments, three distributions (all with mean 1) were used: (i) exponential with coefficient of variation 1, (ii) gamma with coefficient of variation 0.5, and (iii) a mixture of two gammas with coefficient of variation 1.5. Fig. 1 shows some compartmental probabilities for an individual starting in compartment 1: (a) probability of still being in compartment 1 as a function of elapsed time, for differing compartment 1 residence–time distributions; (b) and (c) probability of being in compartment 2 as a function of elapsed time, for (b) differing compartment 2 residence–time distributions with fixed compartment 1 residence–time distribution, and (c) differing compartment 1 residence–time distributions with fixed compartment 2 residence–time distribution. Quite different compartmental probabilities may be noted, even though the residence–time distributions all have the same mean.

References Cited

Capasso, V. & S. L. Paveri–Fontana. 1981. Some results on linear stochastic multicompartment systems. *Math. Biosci.* 55: 7–26.

Cox, D. R. & H. D. Miller. 1965. The Theory of Stochastic Processes. Methuen, London.

Cox, D. R. & P. A. W. Lewis. 1966. The Statistical Analysis of Series of Events. Methuen, London.

Faddy, M. J. 1976. A note on the general time–dependent stochastic compartmental model. *Biometrics* 32: 443–448.

Faddy, M. J. 1977. Stochastic compartmental models as approximations to more general stochastic systems with the general stochastic epidemic as an example. *Adv. Appl. Prob.* 9: 448–461.

Faddy, M. J. 1985. Non–Linear stochastic compartmental models. *IMA J. Math. Appl. Med. Biol.* 2: 287–297.

Godfrey, K. 1983. Compartmental Models and Their Application. Academic Press, London.

Jacquez, J. A. 1985. Compartmental Analysis in Biology and Medicine. University of Michigan, Ann Arbor.

Kempton, R. A. 1979. Statistical analysis of frequency data obtained from sampling an insect population grouped by stages. In Statistical distributions in ecological work, J. K. Ord, G. P. Patel & C. Taillie [eds.]: pp 401–418. International Co–operative Publishing House, Fairland, Maryland.

Kohlas, J. 1982. Stochastic Methods of Operations Research. Cambridge University Press.

Matis, J. H. & H. O. Hartley. 1971. Stochastic compartmental analysis: model and least squares estimation from time–series data. *Biometrics* 27: 77–97

Matis, J. H. 1972. Gamma time–dependency in Blaxter's compartmental model. *Biometrics* 28: 597–602.

Matis, J. H. & T. E. Wehrly. 1979. Stochastic models of compartmental systems. *Biometrics* 35: 199–220.

Neuts, M. F. 1981. Matrix Geometric Solutions in Stochastic Models. Johns Hopkins University, Baltimore.

Read, K. L. Q. & J. R. Ashford. 1968. A system of models for the life cycle of a biological organism. *Biometrika* 55: 211–221.

Weiner, D. & P. Purdue. 1977. A semi–Markov approach to stochastic compartmental models. *Commun. Stats.* A6: 1231–1243.

Modeling Grasshopper Phenology with Diffusion Processes

William P. Kemp[1], Brian Dennis[2], Patricia L. Munholland[3]

ABSTRACT Understanding the development of pest insects is central to any integrated management system. Development in insects is characterized by a progression through a variable number of discrete stages. The compelexity and number of stages depend in part on whether they are hemi – or holometabolous. With rangeland grasshoppers, for example, most species pass through at least five nymphal stages prior to adulthood.

A number of modeling approaches have been used to describe the development of grasshoppers and other insects. Recent applications of Brownian motion diffusion models to insect development problems have shown promise in insect phenology modeling. Under diffusion processes, the amount of development accumulated by an insect at any time is considered to be the sum of many independent increments. Applications of diffusion processes to describe development in insects has worked well for the western spruce budworm and rangeland grasshoppers.

This paper discusses the application of diffusion process models to rangeland grasshoppers. Details on underlying assumptions and application considerations useful to insect pest management are presented. Previous applications of this approach have been limited to describing stage proportions in insect populations through time. We extend the approach to incorporate mortality for application to grasshopper stage frequency data. We propose a product Poisson–multinomial likelihood function for modeling the combined plot and sweep net sampling scheme used frequently by land management agencies for monitoring field grasshopper populations.

Grasshoppers represent the most important threat to rangeland forage production in the western United States. Recent estimates suggest that 21–23% of the annual production on western rangelands is lost to grasshoppers (Hewitt & Onsager 1983).

Central to the effort to construct an integrated pest management system for rangelands in the western United States is the development of phenology models for grasshoppers. Recent studies in phenology modeling make use of a combination of temperature dependent development and inherent stochastic variation (Osawa et al. 1983, Stedinger et al. 1985, Dennis et al. 1986, Kemp et al. 1986, Kemp 1987). In these research efforts, the cumulative distribution function (cdf) of a time dependent normal or logistic probability distribution was used to obtain expected proportions of the grasshopper population in each development stage (instar 1 – 5, adult), given the number of accumulated heat units. This diffusion type modeling approach, where

---

[1]USDA/ARS, Rangeland Insect Laboratory, Montana State University, Bozeman, MT, SA, 59717 – 0001.

[2]Department of Forest Resources, College of Forestry, Wildlife and Range Sciences, University of Idaho, Moscow, ID, USA 83843.

[3]Department of Statistics and Actuarial Science, University of Waterloo, Waterloo, Ontario, Canada N2L 3GL

development at a given date is the sum of many small independent development increments, has been applied successfully to rangeland grasshoppers (Dennis & Kemp 1988, Kemp & Onsager 1986, Kemp 1987).

Though useful for describing proportions of an insect population in each development stage through time, these models have the obvious drawback of not accounting for mortality in the different stages. Furthermore, a grasshopper population monitoring scheme used by land management agencies collects joint samples for both overall density and stage proportions. The maximum likelihood parameter estimates described by Dennis et al. (1986) are based on a product multinomial likelihood not applicable to this situation. The objectives of this paper are to discuss the application of diffusion process models for describing rangeland grasshopper development, and to extend the models to include mortality and the agency sampling scheme. As a numerical example, parameter estimates for the extended model are calculated using a grasshopper data set from a site in Montana.

Methods

Data Collection

Weekly samples of rangeland grasshopper density, phenology, and species composition were collected at Broadus, MT in 1986. Sampling began on May 20, 1986 and continued through September 22 of that year. Density estimates were determined using 36 0.10 m² ring samples. Sweep net collections in the location of the density rings were used to provide information on grasshopper development stages present and species composition throughout the summer. Data for these analyses were not separated into species and, therefore, model estimates for densities are valid only in describing the density and development stages of the overall community on the site. The reason for this approach is that ranchers and land managers do not generally consider individual species complexes. Rather, total density and general development of grasshoppers are considered when making management decisions. Daily maximum and minimum temperatures for the purposes of calculating thermal heat units (degree days, $17.6°C$ threshold) were obtained from the local airport facility ca. 4.8 km from the sample site (Kemp & Onsager 1986).

Development and Brownian Motion — The Diffusion Model

Let $X(t)$ be the amount of development accumulated by an insect at time t. If $X(t)$ is the sum of many small independent development increments, the Central Limit Theorem ensures that $X(t)$ will converge in distribution to a normal random variable with mean and variance proportional to t. Since $X(t)$ is not observed directly in field samples, then the units of $X(t)$ can be taken such that

$$E[X(t)|X(0) = 0] = t \tag{1}$$

without loss of generality (any proportionality constant becomes absorbed in the a_j values defined below). Thus $X(t) \sim$ normal (t, vt), where v is a positive constant.

Osawa et al. (1983) proposed this development model and applied it to bud development in balsam fir. Like insects, the buds go through discrete, identifiable stages. Osawa et al. used the model to describe proportions of buds in each development stage as follows. Let $0 < a_1 < a_2 < ...$ $< a_{r-1} < \infty$, where a_j denotes the development level necessary for an individual to enter the (j + 1)th stage, where $j = 1, 2, ..., r$. The a_j values are signposts such that the individual is in the jth stage if $a_{j-1} < X(t) \le a_j$ (with $a_0 = -\infty$, $a_r = +\infty$). This definition of stage membership is not adequate for modeling individual grasshoppers, since the process $X(t)$ can decrease as well as increase, attaching nonzero probabilities to individuals re − entering earlier stages. However, it is quite satisfactory as a statistical description of structure at the population level since the overall forward drift of many realizations of $X(t)$ produce population level advancement through the development stages. See Munholland et al. (1988) for a diffusion process modeling approach suitable for describing an individual's life history.

The data considered by Osawa et al. consist of the counts $\{x_{ij}\}$, $i = 1, ..., q$, $j = 1, ..., r$, where x_{ij} is the number of individuals in stage j observed in a sample taken at time t_i. Conditional on the ith sample size given by $x_{i1} + x_{i2} + ... + x_{ir} = n_i$, Osawa et al. used a multinomial distribution for the counts. The multinomial probabilities arise from the underlying normal distribution of $X(t)$:

$$p_j(t_i) = \Phi((a_j - t_i)/\sqrt{vt_i}) - \Phi((a_{j-1} - t_i)/\sqrt{vt_i}). \tag{2}$$

Here $p_j(t_i)$ is the proportion of the population in stage j at sampling time t_i; and Φ is the cdf of a standard normal distribution. The likelihood function is product multinomial in form:

$$L(a_1, ..., a_{r-1}, v) = C \prod_{i=1}^{q} \prod_{j=1}^{r} [p_j(t_i)]^{x_{ij}}, \tag{3}$$

and is a function of the r parameters $a_1, a_2, ..., a_{r-1}, r$; C is a combinatorial constant that does not depend on the unknown parameters. The model was quite successful in describing balsam fir bud development (Osawa et al. 1983).

Dennis et al. (1986) recast this model in the context of insects. Dennis et al. used the more convenient cdf of a logistic distribution in place of normal cdf in (2):

$$p_j(t_i) = 1/\{1 + \exp[-(a_j - t_i)/\sqrt{vt_i}]\}$$

$$- 1/\{1 + \exp[-(a_{j-1} - t_i)/\sqrt{vt_i}]\}. \tag{4}$$

The logistic distribution for X(t) has a mean t and variance $(\pi^2/3)vt$. For most data sets it is virtually indistinguishable from the normal, reminiscent of the similarities between logit and probit analyses for binary data. Dennis et al. observed that the calculations to maximize the likelihood function (3) could be performed using iteratively reweighted least squares; thus any entomologist with access to a nonlinear regression computer package can easily obtain maximum likelihood parameter estimates. The model has worked well with data on western spruce budworm (Kemp et al. 1986) and rangeland grasshoppers (Kemp & Onsager 1986, Dennis & Kemp 1988).

Stedinger et al. (1985) extended the model to provide for the possibility of spatially variable development rates. They allow the multinomial proportions, $p_j(t_i)$, to vary according to a Dirichlet distribution, with mean values essentially following the Brownian motion model (2). This extension could be extremely useful for describing populations in heterogeneous regions.

Incorporating Stage — dependent Mortality

The diffusion process models described above, though successful in describing proportions in each development stage through time, have the obvious drawback of not including mortality. Recently, Dennis & Munholland (unpublished) incorporated mortality into the Brownian motion model using the mathematical theory of diffusion processes with killing. Only the details pertinent to the development of such models for rangeland grasshoppers will be repeated here.

First, we denote the cdf of the logistic distribution by

$$F(x,t) = 1/\{1 + \exp[-(x-t)/\sqrt{vt}\,]\} \,. \tag{5}$$

Next, the grasshopper life cycle can be divided into seven stages (Fig. 1) instead of six as described by Kemp & Onsager (1986). In this reformulation, the egg stage (pre — emergence) is unobserved and, since at time zero grasshoppers have not emerged, we assume they are all in stage 0. The value a_j now separates stage $j-1$ from stage j. It follows from Fig. 1 and (5), that the probability of emerging by time t is

$$1 - F(a_1,t) \,, \tag{6}$$

and the probability, given emergence and survival, of being in stage j (j = 1, ..., 6) at time t is

$$p_j(t) = \{F(a_{j+1},t) - F(a_j,t)\}/\{1 - F(a_1,t)\} \,. \tag{7}$$

Earlier formulations of the model (4) use only the numerator portion of equation (7).

Let n be the mean density (number per unit sample) at time zero and let $\exp(-ct)$ be the probability of surviving to time t. The constant survival rate c is a reasonable assumption for rangeland grasshoppers in Montana for most of the growing season (Onsager & Hewitt 1982). Then the mean density of emerged grasshoppers at time t is

$$m(t) = n[1-F(a_1,t)]\exp(-ct) .\tag{8}$$

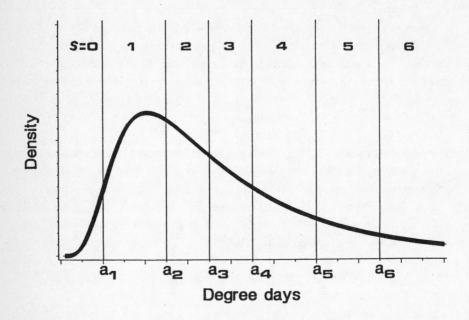

Fig. 1. Hypothetical rangeland grasshopper density and development as a function of accumulated heat (S = stage of development, parameter $a_7 = +\infty$). Eggs (S = 0) at the beginning of the year are unobserved.

The grasshopper data collected at each sampling time t_i consist of the observed density n_i from $0.10m^2$ ring samples (Onsager & Henry 1977), and observed stage frequencies ($x_{i1}...x_{ir}$, where r is the number of observed development stages present) from sweep net collections. For each sampling time t_i the stage frequencies have a multinomial distribution conditional on $x_{1+} = \sum_{j=1}^{r} x_{ij}$, the total number of grasshoppers collected in a sweep net sample at time t_i, so that

$$[x_{i1}, x_{ij}, ..., x_{ir}] \sim \text{multinomial } (x_{i+}, p_1(t_i), ... p_r(t_i)) .\tag{9}$$

Furthermore, it has been shown that grasshopper counts (n_i) derived from ring samples have a Poisson distribution (Onsager 1977), so that

$$n_i \sim \text{Poisson } (m(t)). \tag{10}$$

Estimating Parameters

Given (9) and (10) it follows that the likelihood function is a product of Poisson and multinomial terms. Let

$$\underline{\theta} = [a_1, ..., a_r, v, c, n]' \tag{11}$$

be the vector of unknown parameters. The likelihood function is

$$L(\underline{\theta}) = \left\{ \prod_{i=1}^{q} (\exp(-m(t_i))[m(t_i)]^{n_i}/n_i!) \right\} \left\{ C \prod_{i=1}^{q} \prod_{j=1}^{r} [p_j(t_i)]^{x_{ij}} \right\}, \tag{12}$$

where C is the product of multinomial combinatorial constants. The maximum likelihood estimates ($\hat{\underline{\theta}}$) of the parameter vector ($\underline{\theta}$) are those that maximize $\ln[L(\underline{\theta})]$. Programs written in GAUSS (Edlefsen & Jones 1986) by Dennis (unpublished) were used to compute parameter estimates.

Results and Discussion

The population at Broadus, MT in 1986 was composed primarily (86%) of 3 rangeland grasshopper species *Ageneotettix deorum* (Scudder) (37%) *Aulocara elliotti* (Thomas) (12%) and *Melanoplus sanguinipes* (F.) (37%). All three of these grasshoppers are spring emerging species that are often found together on typical northern mixed grass prairie sites. All three grasshopper species are frequently found in densities that cause economic damage to forage production. As noted, densities and development stages were considered characteristic of the total population, since management decisions are based on total populations rather than on individual species.

Maximum likelihood parameter estimates ($\hat{\underline{\theta}}$) were computed from the methods described previously and are contained in Table 1. Parameters a_1 through a_6 have units of degree days and v is roughly equivalent to a variance term. The parameter n represents the number of individual eggs present in the 3.6 m² area (36 rings X 0.10 m² area/ring) prior to emergence. The parameter c is the mortality rate per degree day for the population at the Broadus site in 1986.

Table 1.   Parameter estimates for the grasshopper diffusion process phenology model, Broadus, MT, 1986.

Parameter	Maximum likelihood estimate
a_1	94.6
a_2	128.4
a_3	163.4
a_4	206.9
a_5	274.4
a_6	374.3
v	3.9
n	194.7
c	1.94×10^{-3}

The relationship between the proportion of grasshoppers in instar 1 or greater and total density can be seen in Fig. 2. As degree days accumulate in the spring, the proportion of emerged individuals increases (Fig. 2A) as does density (with mortality, Fig. 2B).   At the time when emergence is nearly complete (recruitment declines), the exponential portion of equation (8) predicts populations will decline throughout the remaining portion of the summer.   Comparisons of observed data versus model estimates for overall densities as well as proportions emerged show good concurrence.   The model, at present, over — estimates densities late in the field season. However, this is not unexpected, since the variable occurrence of killing frosts during the fall on rangelands causes abrupt drops in grasshopper populations.

Fig. 3 shows good correspondence between observed versus expected proportions of the population in each stage given that t degree days have accumulated.

Previous work on grasshopper modeling (Kemp & Onsager 1986) with the Dennis et al. (1986) model have yielded similar results, but did not provide for the estimation of population densities (Fig. 2B) nor for expected stage frequencies.

Work continues at present to evaluate model assumptions and streamline computational procedures.   Additionally, a system of sentinel sites developed for rangeland grasshopper collections is in its second year of existence. Twelve sites representing a wide range of ecological conditions across the rangeland of Montana will be used to provide data for parameterizing the model described herein.   The possibility of spatially variable development rates will also be explored, as in Stedinger et al. (1985).   Ultimately, we look toward the development of a phenological prediction system for rangeland grasshoppers that can be used by ranchers and land managers as part of an overall pest management system for grasshoppers on rangeland.

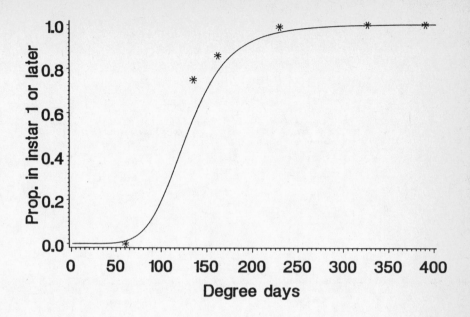

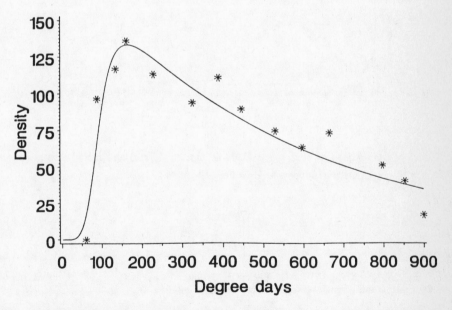

Fig. 2. Comparison of observed field data (Plotted points) and model results (line) for a population of rangeland grasshopeprs at Broadus, MT, 1986.

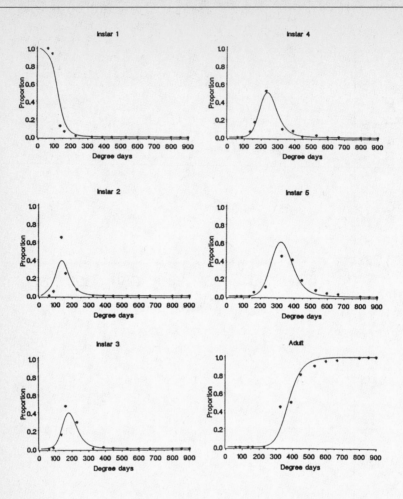

Fig. 3. Comparison of observed field data (plotted points) and model results (line) for a population of rangeland grasshoppers at Broadus, MT, 1986.

References Cited

Dennis, B., W. P. Kemp & R. C. Beckwith. 1986. Stochastic model of insect phenology: Estimation and testing. *Environ. Entomol.* 15: 540–546.

Dennis, B. & W. P. Kemp. 1988. Further statistical inference methods for a stochastic model of insect phenology. *Environ. Entomol.* (in press).

Edlefsen, L. E. & S. D. Jones. 1986. GAUSS programming language manual Ver. 1.00. Aptech Systems, Inc., Kent, WA.

Hewitt, G. B. & J. A. Onsager. 1983. Control of grasshoppers on rangeland in the United States — A perspective. *J. Range Mgmt.* 36: 202–207.

Kemp, W. P. 1987. Predictive phenology modeling in rangeland pest management, pp. 351–368, In: Capinera, J. L. [ed.], Integrated pest management on rangeland: A shortgrass prairie perspective. Westview Press, Boulder, CO.

Kemp, W. P., B. Dennis & R. C. Beckwith. 1986. A stochastic phenology model for the western spruce budworm (Lepidoptera:Tortricidae). *Environ. Entomol.* 15: 547–554.

Kemp, W. P. & J. A. Onsager. 1986. Rangeland grasshoppers (*Orthoptera: Acrididae*): Modeling phenology of natural populations of six species. *Environ. Entomol.* 15: 924–930.

Munholland, P. L., J. D. Kalbfleisch & B. Dennis. 1988. A stochastic model for insect life history data. (These proceedings.)

Onsager, J. A. 1977. A note on the Poisson distribution in integrated pest management. *Bull. Entomol. Soc. Amer.* 27: 119–120.

Onsager, J. A. & J. E. Henry. 1977. A method for estimating the density of rangeland grasshoppers (*Orthoptera: Acrididae*) in experimental plots. Acrida 6: 231–237.

Onsager, J. A. & G. B. Hewitt. 1982. Rangeland grasshoppers: average longevity and daily mortality among six species in nature. *Environ. Entomol.* 11: 127–133.

Osawa, A., C. A. Shoemaker & J. R. Stedinger. 1983. A stochastic model of balsam fir bud phenology utilizing maximum likelihood parameter estimation. *Forest Sci.* 29: 478–490.

Stedinger, J. R., C. A. Shoemaker & R. F. Tenga. 1985. A stochastic model of insect phenology for a population with spatially variable development rates. *Biometrics* 41: 691–701.

Estimation of Relative Trappabilities by Age and Development Delays of Released Blowflies

Richard Morton[1] and W. G. Vogt[2]

ABSTRACT   Marked blowflies *Lucilia cuprina* (Wiedmann) were released as newly–emerged adults and trapped continuously over several consecutive days. Females were classified into reproductive age classes according to their cycle and stage of ovarian devleopment. The experiment was aimed at determining the relative trappability of females in each age class and durations of developmental delays associated with limited availability of protein and oviposition sites in the field. A reasonably natural model led to a complicated multinomial likelihood function involving 15 unknown parameters. Considerable difficulties were experienced in maximizing this likelihood because of the number of parameters and some inequalities between them. Computer algorithms often failed to converge or to report a negative definite Hessian for the log–likelihood. Some estimates obtained strayed very far from realistic values. We conclude that some parameters are poorly identifiable, so that to obtain estimates with sufficiently small standard errors would require impractically large sample sizes.

Females of the Australian sheep blowfly, *Lucilia cuprina* (Wiedmann), complete a sequence of distinct ovarian cycles in which all ovarioles develop synchronously (Vogt et al. 1974). Second and subsequent cycles are indistinguishable but are distinct from the first cycle. Within cycles females can be assigned to age classes on the basis of ovarian devleopment; 10 in the first cycle and 8 in subsequent cycles (Vogt et al. 1985b). The last stage in each cycle is the gravid stage when eggs are ready to be deposited.

In female *L. cuprina*, ovarian development rates are temperature dependent so that female age is measured in 'day degrees' (DD) rather than in real time, i.e., accumulated temperature above 8°C (Vogt et al. 1985b). Development rates are also influenced by the availability of protein–rich material in the field. Females that obtain insufficient protein to mature their full egg complements either fail to mature any eggs (all cycles) or resorb some and mature the remainder. In the latter case, maturation is prolonged because resorption slows the rate of ovarian development (Vogt et al. 1985b). Protein–induced delays are confined to age classes 3, 4 and 5. Further delays can occur in the gravid stage, when females must locate a suitable oviposition site; either susceptible live sheep or carrion. An estimate of the total delay in DD for classes 3–5 can be obtained by estimating the number of eggs resorbed multiplied by 0.3.

Field populations of *L. cuprina* are sampled with carrion–baited traps (Vogt & Havenstein 1974) to obtain relative estimates of population size (Vogt et al. 1983, 1985a). However, changes in population age–structure are a potential source of bias for these estimates, since not all females are equally responsive to traps. Age–distributions of trapped females indicate that responsiveness

---

[1]CSIRO Division of Mathematics and Statistics, Canberra, Australia.
[2]CSIRO Division of Entomology, Canberra, Australia.

to traps (trappability) varies in other classes, even after allowance for differences in relative durations of the age classes. Gravid females appear to be more trappable than other classes.

Mark–recapture experiments were conducted aimed at estimating the relative trappability of females in each age class and the duration of the oviposition delay experienced by gravid females. A secondary objective was to estimate protein–induced developmental delays in the field directly for comparison with estimates obtained from data on egg resorption.

The experiment consisted of 9 releases of blowflies in age class 1 marked with a coloured dust. Catches weere over 3–hour periods 6–9, 9–12, 12–15, 15–18, but since no blowflies were caught after 18, the traps were left open all night. Period 6–9 was abandoned after a few days because of low catches. Table 1 shows the best data set (Red 3). The total of classes 1–18 is sometimes less than the total females caught since when catches were large only a subsample of about 50 blowflies were dissected and classified. The rate refers to the numbers of blowflies expected to be caught from a constant population. It is calculated from environmental variables and has an arbitrary scale factor (Vogt et al. 1983).

The progress through the age classes is clear, though some class 9's appeared before the first class 8. Flies classified as 11–13 near the end could be in class 19–21 since it is not possible to distinguish second and third cycles. We ignored classes 11 onwards because of small numbers. For purposes of analysis, d for the mid–period was subsequently taken to the final d minus 1.

From Vogt et al. (1985b) the 50% transition times in laboratory trials were 23.7, 32.0, 36.1, 41.5, 44.1, 47.1, 51.7, 53.9, 56.9 for entry into classes 2, 3, ..., 10, under conditions where there was no egg resorption. These are lower than would appear from Table 1, but a shift in origin of about 6DD would fit very reasonably. For the Red 3 trial the delays were estimated to be only about 2DD, indicating that protein was readily available in the field.

A Parametric Model

First we model the proportions of blowflies in the field in each age class. Denote by $p_k(d)$ the proportion of blowflies in class k or beyond at d DD, and assume that its form is a logistic in $\ell n(d)$:

$$p_k(d) = \exp\{\beta(\ell n(d) - L_k)\}/[1 + \exp\{\beta(\ell n(d) - L_k)\}], \tag{1}$$

where β is a common 'slope' parameter not varying with age class, and L_k is $\ell n(LT_{50})$ for blowflies in the field entering class k. The reason for using $\ell n(d)$ instead of d on the natural scale is to make the slope parameter constant and correct the skewness. Kemp et al. (1986) also use logistic functions but cope with the increasing spread over time by using $\beta(d-LT_{50})/\sqrt{d}$ in the exponent instead of taking logs. Their model does not allow different trappabilities. If at the j–th trapping time the number of day degrees is d_j, then the proportion in class k, but not beyond, is

Table 1. Data Set Red 3. A total of 38859 females were released on 23 February 1982. Values shown are DD, the degree days; RINV, the reciprocal rate; FEM, the total number of females trapped; and the numbers trapped in the age classes 1—18.

38859 RED FEMALES RELEASED 23/2/82

DD	RINV	FEM	1	2	3	4	5	6	7	8	9	10	11	12	13	14	15	16	17	18
14.1	19.8	1	1																	
15.7	4.7	9	9																	
17.7	2.0	196	40																	
19.7	2.7	60	24																	
24.4	12.9	2	2																	
26.2	2.5	74	39	11																
28.5	.9	203	21	9																
31.0	.9	139	14	24	11															
37.5	11.2	5	1	1	1	2														
39.4	2.4	47	1	15	20	9	2													
41.9	1.1	132		3	24	21	2													
44.3	1.2	114		3	20	23	3	1												
50.2	7.2	20				1	9	4	5	1										
52.0	2.7	65			10	28	6	4	2											
53.7	4.4	24				10	7	3	3		1									
58.1	3.9	35			1	8	6	8	5	1	2	4								
60.3	1.2	78			2	8	11	9	12	2	2	10								
62.6	1.1	33			2	2	3	4	7	5	1	8								
69.1	3.4	32						3	4		1	22								
71.3	2.0	31						1	1		1	27	1							
73.4	1.8	21										15	4	1			1			
79.9	6.5	26						1				21			1	1				
81.7	3.2	61										59			1	2				
83.5	3.7	8										3	1							
89.7	2.5	42										35			5	1	1			
91.8	1.3	35										13	11		3	2	1	1		
93.8	3.1	5										3	1		1					
111.6	3.1	21										17			2	1				
113.8	1.1	31										6	2		6	1	6	2		4
116.2	1.1	37										6	10	16	1	1	1			2
124.6	3.4	10											1	1		1				6
127.0	1.1	12											2		1		3			6
129.0	1.7	9											2		1	1	2			3
137.8	2.0	9												2	1			1		4
140.4	1.7	10											5	3	1				1	1
143.1	1.4	10											3	1	1		2			2
155.2	1.9	6												2			1			3
158.3	2.1	1																		1
161.0	2.3	2																		1

$$p_{jk} = p_k(d_j) - p_{k+1}(d_j) \qquad (2)$$

Since we start in class 1, $p_1(d_j) = 1(L_1 = -\infty)$ is understood. If at the jth time the trapping rate is r_j and the trappability of class k is λ_k, then the expected number of class k caught would be $\mu_{jk} = r_j \lambda_k p_{jk}$. The expected proportion of class k in the traps at time j is then $p^*_{jk} = \mu_{jk}/\mu_{j\cdot}$, where the dot notation denotes summation. There is an arbitrary scale factor in r_j and so we can only determine the trappabilities $\{\lambda_k\}$ up to unknown scale. We therefore change to <u>relative</u> trappabilities $\rho_k = \lambda_k/\lambda_1$, ($\rho_1 = 1$ fixed) which can be determined. Using class 1 as the reference is arbitrary; if we desire, others can be computed e.g. $\lambda_k/\lambda_5 = \rho_k/\rho_5$. Thus

$$p^*_{jk} = \rho_k p_{jk} / \sum_{k=1}^{k} \rho_k p_{jk} , \tag{3}$$

with p_{jk} given by (1) and (2).

The total numbers caught per trapping time are not very informative. Their correlation with the rate is not extremely high, because of natural variation, mortality and escape from the trapping area. We shall therefore remove the row totals by using the conditional distribution. Let X_{jk} be the number of females classified as age k in row j. Then we assume that each row has a multinomial distribution with total X_j. and probabilities p^*_{jk}. That is, the log–likelihood is

$$\ell n \, L = \sum_{j=1}^{N} \sum_{k=1}^{K} X_{jk} \ell n \, p^*_{jk} , \tag{4}$$

where N is the number of trapping occasions. The parameters to be estimated by maximum likelihood are β, L_2, L_3, ..., L_K, ρ_2, ρ_3, ..., ρ_K since $L_1 = -\infty$ and $\rho_1 = 1$ is fixed.

If, as appears with classes 8–9, the higher class number was observed earlier, the estimates obtained by maximizing the log–likelihood lead to $L_8 > L_9$, which is physically impossible. One solution is to constrain the maximization so as to avoid this, but the theory of the statistical behavior of constrained maximum likelihood estimates is very awkward. Also, if $L_k = L_{k+1}$ then ρ = ∞ is absurd. Instead we merge classes 7–9. We recognize that if trappability varies between classes to be merged, the combined trappability would not represent each one separately.

Poor Identifiability of Parameters

When K = 2, $p_{j1} = 1 - p_2(d_j)$, $p_{j2} = p_2(d_j)$ and then $p^*_{j1} = 1 - p^*_{j2} = 1/\{1 + \exp{(\beta d_j + \gamma)}\}$, where $\gamma = \ell n(\rho_2) - \beta L_2$. Since p^*_{jk} depends only on β and γ, only those can be estimated. That is, given any value of L_2, however ridiculous, provided that we set $\rho_2 = \exp(\gamma + \beta L_2)$ we get exactly the same fit. Thus (ρ_2, L_2) are not identifiable.

For K $\geq$ 3, suppose that the third class hardly overlaps the first. Then (5) holds to close approximation for all j until the first class 3 appears, and by that time we have no more information on the overlap between classes 1 and 2. Thus the situation has hardly altered from the case K = 2 considered above, and so (ρ_2, L_2) would be very poorly identifiable. This would appear as a high correlation between their estimates. The same argument holds whenever the overlap between classes (k–1,k) is isolated from the others, and this indicates that the pairs $\{\ell n(\rho_k/\rho_{k-1}), L_k\}$ are liable to be highly correlated for any k = 2, ..., K, and therefore some other parameters are likely to be poorly identifiable. In particular, the last classes (K–1,K) are in the same situation as classes (1,2) if time is viewed as running backwards. Essentially the idea is that

L_2 can vary between $-\infty$ and an upper bound (due to the appearance of class 3), and likewise L_K between a bound (due to the last appearance of class K–2) and $+\infty$. Under these variations we can get almost identically good fits by adjusting the ρ's. For intermediate classes, there are bounds on both sides and the situation is less drastic.

A consequence of this explanation is that by merging classes to the extent that the overlapping regions are well separated in time, intermediate classes are also poorly estimated. That is, it helps to have many classes observed simultaneously.

Analysis of Red 3 Data

The constraints $L_2 \leq L_3 \leq \ldots \leq L_{10}$ affect the method of estimation. If the maximum of the likelihood occurs on one or more boundaries, then estimation based on derivatives breaks down. Thus we used the optimization program MINIM which employs the simplex method of Nelder & Mead (1965). A penalty function was included in the likelihood to enforce the constraints. With 15 parameters convergence was slow; even after 10,000 iterations we were not satisfied that we had reached a good answer.

Next we merged classes 7–9 to remove the need for constraints. Using 17 starting points with 3 stopping criteria, MINIM always reported convergence. But the 51 solutions had a wide range of estimates (Table 2). a particularly extreme solution gave parameter estimates 3.1, 2.0, 7.7, 0.9, 22.0, 0.3, 232.7, 3.42, 3.55, 3.78, 3.83, 3.96, 3.97, 4.60, but the value $\ell n L = -670.2$ showed that it fitted as well as any.

Table 2. Ranges of parameter estimates and $\ell n L$ for Red 3 data from 17 starting values x 3 stopping criteria.

Class	2	3	4	5	6	7–9	10
ρ:	0.9–3.4	1.0–2.2	1.2–7.7	0.3–0.9	0.5–22.0	0.1–0.3	1.3–232.7
L:	3.33–3.42	3.55–3.56	3.72–3.78	3.83–3.85	3.94–3.96	3.96–3.99	4.27–4.60

β: 13.1–14.1 $\qquad$ $-\ell n L$: 670.2–671.7

Simulations with K = 3

To study the statistical properties of the model, we constructed several sets of simulated data. For simplicity, we restricted the problem to 3 classes. The 'true' parameter values we chose were comparable to those estimated for Red 3. Simulated data set 1 was based on 11 sampling times similar to Table 1 but assumed all row totals 50. Sample sizes were increased by inserting intermediate sampling times (to make 21) and by increasing the row totals. Data set 3 had row totals 1000 and was 'ideal' in that expected numbers were used; this was done partly to check that we would recover the known parameter values.

Table 3 summarizes the results. Sets 1 and 2 showed high association between ρ_2, L_2. The IMSL subroutine ZXMIN utilizes first and second derivatives. It proved to be more efficient than MINIM in general. However, ZXMIN failed to converge for simulation 1 after 900 iterations. Although it found a higher likelihood than did MINIM, it sent ρ_3 to ridiculous values.

Table 3. Range of estimates from MINIM with 17 starting points and 3 stopping criteria. Estimates and standard errors from ZXMIN. Simulation 3 used expected instead of random values.

Algorithm	MINIM		ZXMIN		
Simulation	1	2	1	2	3
Row Totals	50	100	50	100	1000
No. occasions	11	21	11	21	21
True values	Estimate Range		Estimate (s.e.)		
ρ_2 = 1.5	0.08–1.99	0.94–2.48	0.49	1.30(0.94)	1.46(0.32)
ρ_3 = 2.3	0.93–20.4	1.08–2.78	185	1.72(1.14)	2.30(0.51)
L = 3.37	3.15–3.39	3.32–3.39	3.3	3.35(0.05)	3.37(0.01)
L = 3.58	3.56–3.85	3.54–3.58	4.0	3.57(0.50)	3.59(0.02)
β = 13.0	12.5–14.1	12.8–13.5	12.9	13.0(0.75)	13.0(0.25)
ℓnL	187.7–189.1	799.4–799.9	187.6	794.4	–

Discussion

Despite the appearance of the data, which suggests that parameters could be estimated fairly precisely, all indications are that some parameters are poorly identifiable. The explanation given above, though not proved formally, does seem to be consistent with what we have found by computation. The extreme solution cited above shows how $(\ell n(\rho_{10}/\rho_{7-9}), L_{10})$ are highly associated in such a way that L_{10} and $\ell n(\rho_{10}/\rho_{7-9})$ can get unrealistically large. The simulations with K = 3 show how $(\ell n(\rho_2/\rho_1), L_2)$ are similarly highly associated, leading to low $\ell n(\rho_2/\rho_1)$ or high $\ell n(\rho_3/\rho_2)$. Table 3 shows that some intermediate ρ's can also be poorly estimated.

The sample sizes required to get good estimates of relative trappability by this analysis would be at least 500 for each period, and that might well be insufficient. If there is extra–multinomial variation, larger sample sizes would be needed. These conclusions are not due to a quirk in the particular data set Red 3, but appear to hold quite generally. Similar problems have occurred in analyzing stationary blowfly populations (Crellin 1987).

Some information has been discarded due to conditioning on row totals, no doubt. But this information would be very hard to use because the row totals are greatly affected by unaccounted variations in the catching rate and by emigration and mortality. A method of estimating mortality is proposed by Manly (1987), but it assumes constant trappability by age and period and no migration.

Points to consider when dealing with maximizing a difficult likelihood function with so many parameters:

(i)   It is reasonable to try to fit the data straight away. If you run into problems, try solving the simplest problem with artificial data, and see if that works.

(ii)   Explicit derivatives are highly desirable. ZXMIN (using derivatives) worked much better than MINIM.

(iii)   The derivatives computed from formulae should be checked by computing first and second differences. This may be less trouble than trying to get an obstinate optimization package to print out the answer.

(iv)   Explore the model using simulated data and by computing large numbers of function evaluations.

Acknowledgements

The authors are grateful to K. W. J. Malafant, B. A. Ellem and especially S. Evans for doing almost all the computing.

References Cited

Crellin, N. 1987. Modelling the age distribution of wild female sheep blowflies. Report ACT 87/24 CSIRO Div. Maths. and Stats., Canberra.

Kemp, W. P., B. Dennis & R. C. Beckwith. 1986. Stochastic phenology model for the western spruce budworm (Lepidoptera: Tortricidae) *Environ. Entomol.* 15: 547–554.

Manly, B. F. J. 1987. A multiple regression method for analyzing stage frequency data. *Res. Popul. Ecol.* 29: 119–127.

Nelder, J. A. and R. Mead. 1965. A simplex method for function minimization. *Comp. J.* 7: 308–315.

Vogt, W. G. & D. E. Havenstein. 1974. A standardized bait trap for blowfly studies. *J. Aust. Ent. Soc.* 13: 249–253.

Vogt. W. G., T. L. Woodburn, R. Morton & B. A. Ellem. 1983. The analysis and standardization of trap catches of *Lucilia cuprina* (Wiedmann) (Diptera: Calliphoridae). *Bull. Ent. Res.* 73: 609–617.

Vogt, W. G., T. L. Woodburn, R. Morton & B. A. Ellem. 1985a. The influence of weather and time of day on trap catches of males and females of *Lucilia cuprina* (Wiedmann) (Diptera: calliphoridae). *Bull. Ent. Res.* 75: 315–319.

Vogt. W. G., T. L. Woodburn & M. Tyndale–Biscoe. 1974. A method of age–determination in *Lucilia cuprina* (Wied.) (Diptera: Calliphoridae) using cyclic changes in the female reproductive system. *Bull. Ent. Res.* 64: 365–370.

Vogt, W. G., T. L. Woodburn & A. C. M. van Gerwen. 1985b. The influence of oocyte resorption on ovarian development rates in the Australian sheep blowfly, *Lucilia cuprina*. *Entomol. Exp. Appl.* 39: 85–90.

A Stochastic Model for Insect Life History Data

P.L. Munholland[1], J.D. Kalbfleisch[2] and B. Dennis[3]

ABSTRACT A stochastic model is developed for an insect's life history: time may be measured on a chronological time scale or on some operational time scale such as degree–days. We assume a Brownian motion process, with drift, for development, which is similar to the development model underlying Stedinger et al. (1985), but its use in this paper is different and more theoretically appealing. Independent of development, the mortality process is defined by a two–state Markov process with nonhomogeneous mortality rate; biological considerations may suggest appropriate forms of the mortality rate. The life history process is obtained by superimposing the two stochastic processes and subsequently aggregating the state space; the aggregation is necessary because a development stage is observable, rather than the exact level of development. Stage occupancy is determined by the stage recruitment times, which are inverse Gaussian random variables. Defining stage occupancy in this manner allows for an interpretation at the microscopic level, and leads to a semi–Markov model for life history processes with inverse Gaussian stage transition rates if the mortality rate is constant. The proposed model is extended to incorporate recruitment to the first development stage. Stage–specific recruitment, sojourn times and mortality rates are expressed as functions of the model parameters. This model provides a bridge between the macroscopic models suggested by various authors (e.g. Manly 1974 and Stedinger et al. 1985) and the microscopic models developed by others (e.g. Read & Ashford 1968).

The model is fitted to longitudinal data on the prevalence of insects in the development stages assuming a product–Poisson likelihood for the counts. Maximum likelihood estimates of the parameters and their asymptotic standard errors are obtained via a Gauss–Newton algorithm for iteratively reweighted least squares. An example is provided and the results are compared to those obtained by Read & Ashford (1968), and Kempton (1979).

Models describing an individual's life cycle are useful to researchers studying the microdynamics of an insect population and may assist in the development of an integrated pest–management program. Read & Ashford (1968) developed one of the first stochastic models at the individual level, in which they assume sojourn time in each development stage has a gamma distribution with constant shape parameter. Kempton (1979) assumes that stage sojourn times are either normal, gamma or inverse Gaussian variables. Bellows & Birley (1981) and Bellows et al. (1982) describe stage development times with a reciprocal normal distribution. There are some drawbacks to the above methods. While each study describes a statistical estimation procedure, in applications difficulties may arise in fitting the models. For example, Kempton's

---

[1]Department of Mathematical Sciences, Montana State University, Bozeman, Montana 59715.

[2]Department of Statistics and Actuarial Science, University of Waterloo, Waterloo, Ontario, Canada N2L 3G1.

[3]College of Forestry, Wildlife and Range Sciences, University of Idaho, Moscow, ID 93943.

procedure does not allow for simultaneous estimation of the model parameters, so that some important correlations are ignored. Also none of these approaches incorporates the effect of temperature on an insect's life history, although each could be generalized easily by measuring time in degree—days, for example.

Once a model for an individual's life cycle is specified, the macroscopic properties of the population can be deduced. An alternative approach, however is to model the macroscopic properties directly. Generally, the macroscopic models are marginal in that they stipulate the structure of the population at any point in time without specifying a complete model for an individual's life history. Manly (1974) describes a model of this type: he considers each development stage separately and assumes a distributional form for stage recruitment, with a constant mortality rate. More recently, Stedinger et al. (1985) and Dennis et al. (1986) specify a parametric density function for an individual's level of development at each point in time, the probability of residency in each stage being determined as a definite integral of the assumed density. Both models incorporate the effect of temperature.

Model Description

We begin by assuming individuals accumulate small increments of development over time; in the absence of mortality, the development level, $X(t)$ is defined as the amount of development an insect has accumulated up to that time. The development process $\{X(t)\}$ is assumed to be a Brownian motion process with positive drift coefficient v and variance parameter σ^2, which commences at the time origin so that $X(0) = 0$. Time t may be measured in days or as a function of days (e.g. degree—days).

At any time $t \geq 0$ a discrete development stage, rather than the exact level of development, $X(t)$ is observed. Suppose k development stages can be recognized, which may include the adult stage and any immature stages (e.g. eggs, nymphs or larvae and pupae). Let $\alpha_1, \alpha_2, ..., \alpha_k$ denote the development thresholds for stages $1, 2, ..., k$ respectively, where $\alpha_1 = 0$, and $\alpha_{k+1} = \infty$, so that a development level of α_i is necessary for entry into stage i. It follows that T_i, the recruitment or entry time to stage i, is the first time that the development process $\{X(t)\}$ hits α_i, $i = 1, 2, ..., k$. Since T_i is a hitting time for a Brownian motion process, its probability density function, $g_i(\cdot)$ is the inverse Gaussian density with parameters $\mu_i = \alpha_i/v$ and $\lambda_i = \alpha_i^2/\sigma^2$

$$g_i(t) = (\alpha_i/\sigma\sqrt{t^3})\phi[x_{it}] \quad G_i(t) = \phi[x_{it}] + e^{2v\alpha_i/\sigma^2}\Phi[-y_{it}] \tag{1}$$

where $G_i(\cdot)$ denotes the distribution function with $x_{it} = (vt - \alpha_i)/\sigma\sqrt{t}$, $y_{it} = (vt + \alpha_i)/\sigma\sqrt{t}$ and $\phi(\cdot)$ and $\Phi(\cdot)$ are the standard normal density and distribution functions, respectively. An intuitive derivation of (1) is given by Whitmore & Seshadri (1987). For stage 1, $T_1 = 0$ since $P\{X(0) = 0\} = 1$. The density (1) is unimodal and positively skewed with mean μ_i and variance

λ_i / μ_i^3. The sojourn or residence time in stage i, $X_i = T_{i+1} - T_i$, is also a hitting time and is an inverse Gaussian random variable with parameters $\mu_i^* = \beta_i / v$ and $\lambda_i^* = \beta_i^2 / \sigma^2$ where $\beta_i = (\alpha_{i+1} - \alpha_i)$.

An individual is said to be in development stage i, $i = 1, 2, ..., k$ at time t if $T_i \leq t < T_{i+1}$, that is, in the absence of mortality, stage occupancy is determined by the stage recruitment times. This definition of stage occupancy differs from the criteria specified by both Stedinger et al. (1985) and Dennis et al. (1986). The former assumes the development level at time t, $X(t)$ is normally distributed while the latter specifies a logistic distribution for $X(t)$ and both stipulate residence in stage i at time $t \geq 0$ if $\alpha_i \leq X(t) < \alpha_{i+1}$ $i = 1, 2, ..., k$. Their criteria attach a nonzero probability to the event that an individual in stage i at time t may reside in an earlier stage j a short time later. By determining stage residency on the basis of stage recruitment times, this problem is circumvented, since if $\alpha_1 < \alpha_2 < ... < \alpha_k$, then $T_1 < T_2 < ... < T_k$.

Independent of the development process $\{X(t)\}$, the mortality process $\{M(t), t \geq 0\}$ is assumed to be a two–state nonhomogeneous Markov process with instantaneous transition intensity

$$q_{01}(t) = \theta(t) = \lim_{\delta t \to 0} P\{M(t + \delta t) = 1 \mid M(t) = 0\} / \delta t, \quad q_{10}(t) = 0, \tag{2}$$

where an individual is alive at time $t > 0$ if $M(t) = 0$ and dead if $M(t) = 1$. This definition of mortality implies the death rate is age– rather than stage–dependent. Several models for the rate have been suggested; a constant mortality rate, $\theta(t) = \theta$ is widely assumed in the literature (e.g. Manly 1974, Ashford et al. 1970). A variety of monotonic failure rates has also been proposed (e.g. the gamma hazard function, Kempton 1979). Munholland (1988) discusses the specification of $\theta(\cdot)$ if a non–monotonic rate is more appropriate, and reviews the general area. Biological considerations and the data themselves may suggest reasonable models.

We describe the life history for an individual by the stochastic process $\{Z(t), t \geq 0\}$ with state space $S = \{-2, 1, 2, ..., k\}$, where

$$Z(t) = \begin{cases} i & \text{if } T_i \leq t < T_{i+1}, \ M(t) = 0 \\ -2 & \text{if } M(t) = 1, \end{cases} \tag{3}$$

states $1, 2, ..., k$ correspond to the development stages and state -2 denotes death. Model (3) results from superimposing the mortality process $\{M(t)\}$ on the development accumulation process $\{X(t)\}$ and subsequently aggregating the continuous state space to conform to the discrete development stages. Thus the process $\{Z(t)\}$ is a nonhomogeneous semi–Markov process with finite state space S and the initial condition that the individual enters development stage 1 at time $t = 0$. The instantaneous intensity of transitions from stage i to $i + 1$ depends only on the time spent in stage i and corresponds to the inverse Gaussian failure rate for the random variable X_i. Model (3) also specifies the intensity of mortality in stage i, $i = 1, 2, ..., k$ to be a function of time, measured from the origin $t = 0$, given by (2). A time–homogeneous semi–Markov model for an individual's life history results when the mortality rate is assumed constant.

Let $p_i(t)$ denote the probability that an individual is alive and in stage i at time $t \geq 0$, that is $p_i(t) = P\{Z(t) = i\}$. It follows from (3) that

$$p_i(t) = [G_i(t) - G_{i+1}(t)] \exp\left[-\int_0^t \theta(u)\,du\right], \tag{4}$$

for $i = 1, 2, ..., k$ and $t \geq 0$; $G_i(\cdot)$ is given by (1). Given a model for the mortality rate (2), the $p_i(\cdot)$ describe the structure of the population at an arbitrary time $t > 0$.

Kempton (1979) obtains the same form for the $p_i(\cdot)$, $i = 1, 2, ..., k$ although his reasoning appears to be different for he requires the densities of the independent random variables X_i to possess a property of additivity so that the distribution of a sum of them can be obtained explicitly. The inverse gaussian distribution is mentioned as one with the additivity property, but throughout most of his paper the X_i are assumed to be gamma variables with common rate parameter.

In developing the model (3), we assume that all individuals enter development stage 1 at time $t = 0$, but in many situations, individuals are recruited to the first stage over time. Recruitment can be incorporated by allowing the development threshold for stage 1, α_1 to be positive. In this case, T_1 also has density (1) implying population recruitment times, in the absence of mortality, are inverse Gaussian random variables.

Estimation

Consider a population whose individuals act independently and are randomly distributed over a confined homogeneous area and suppose that individuals are recruited to the first development stage over time, the procedure being easily modified for populations with members in stage 1 at the time origin. Suppose that population is sampled at times $t_1 < t_2 <, ..., < t_L$, where, at time t_1, a small proportion of the area, c is randomly selected and the observed stage–frequencies are recorded. Let n_{il} be the number of insects in stage i in the lth sample, where $i = 1, 2, ..., k$ and $l = 1, 2, ..., L$. For c small, the n_{il} are Poisson random variables (approximately) with means $\eta p_i(t_l)$, where η denotes the expected mean abundance over the sampling area and the $p_i(\cdot)$ are given by (4). The log–likelihood function for these data is

$$\log L = \sum_{l=1}^{L} \sum_{i=1}^{k} n_{il}\log[\eta p_i(t_l)] - \eta p_i(t_l). \tag{5}$$

In obtaining (5), time and development time are assumed to have coincidental origins at $t = 0$. In practice, the latter origin is usually unknown. Let t denote time measured from an arbitrary origin and define t_0 to be the time of first recruitment. Then the model may be expressed as a function of development time by writing $t_1 - t_0$ in place of t_1 in (5) and including t_0 as a model parameter. An alternative description for $t_1 - t_0$ is that the stage recruitment times have inverse Gaussian distributions with guaranteed holding–time parameter, t_0.

Some of the parameters associated with (5) may be nearly non–identifiable. The parameters v, α_i and σ of the distribution function of T_i, (1), share a common scale parameter so that the $G_i(\cdot)$ are unaffected by a change of scale. Thus, without loss of generality we assume $\sigma = 1$. Inclusion of the parameter t_0 specifies a three–parameter distribution for T_1, in which location and origin are interdependent. Read & Ashford (1968), and Kempton (1979), whose respective models are similar to (3), allude to parallel problems with model overparameterization. Kempton assumes t_0 is known, whereas Read & Ashford fix the gamma shape parameter. We treat t_0 as a known constant. Thus the likelihood may be represented as a function of the independent parameters η, v, β_0, β_1, ..., β_{k-1} and the r parameters of $\theta(\cdot)$. Expressing (5) as a function of the β_i is computationally preferable since it circumvents the order restriction $\alpha_1 < \alpha_2 < ... < \alpha_k$.

Once a value of t_0 is specified (usually by an educated guess), maximum likelihood estimates of the remaining $k + r + 2$ parameters, collectively denoted by ρ, and their asymptotic standard errors are available via a Gauss–Newton algorithm for iteratively reweighted least squares (Jennrich & Moore 1975). The algorithm requires the first partial derivatives of the $p_i(\cdot)$ and initial estimates of the parameters. Previous knowledge of the population and the data themselves will usually suggest reasonable starting values, subsequently refitting the model with various t_0 should indicate an appropriate value for this unknown. Nonlinear regression packages such as the SAS procedure NLIN (1985) are useful in fitting the model and estimates of the asymptotic standard errors and correlations are usually included in the output so that an estimate of the covariance matrix, $\Sigma(\rho)$ may be readily obtained. A SAS program for maximizing (5) is available upon request from the first author.

Asymptotic theory, suitably applied, indicates the estimator $\hat{\rho}$ of ρ has an approximate multivariate normal distribution with mean ρ and covariance matrix $\Sigma(\rho)$. Thus, further inferences for the model (3) are possible. For example, if the mortality rate is constant, $\theta(t) = \theta$ the expected mean number (η_i) entering stage i, the expected sojourn time (τ_i) in stage i and the stage–specific mortality rate ($p_{i-2} = P\{$death in $i|$ enters $i\}$) are

$$\eta_i = \eta \exp\{\alpha_i \gamma\}, \quad p_{i-2} = 1-\exp\{\beta_i \gamma\} \quad \text{and} \quad \tau_i = p_{i-2}/\theta,$$

where $\gamma = v - \sqrt{v^2 + 2\theta}$. Substituting $\hat{\rho}$ in these yields maximum likelihood estimates and approximate standard errors are easily obtained.

The adequacy of the model may be assessed using the likelihood ratio statistic. Let $\log L(\hat{\rho}, t_0)$ denote the maximized value of (5) and define $\log L(n)$ to be the maximized log likelihood obtained by allowing a free Poisson mean for n_{j1}. The likelihood ratio statistic $-2\log R$ $= -2(\log L(\hat{p}, t_0) - \log L(n))$ has an asymptotic χ^2 distribution with $Lxk - (k+3+r)$ degrees of freedom when the L samples are independent. Alternatively, the Pearson χ^2 statistic, based on the differences between the observed and expected frequencies may be used. In either case, small expected frequencies must be pooled for the distributional approximation to be valid.

An Example

The model is used to analyze the stage–frequency data for the grasshopper *Chorthippus parallelus* in Devon (England). The data, from Ashford et al. (1970), consist of the number of insects in 5 stages collected in 3 to 4 day intervals, between 20 May and 23 September 1964. We assume a constant mortality rate and time is measured from 1 April while the development time origin is fixed at Ashford et al.'s estimate of 9.2 May. While this time origin facilitates the following comparisons among models, our studies indicate several values of t_0 are equally appropriate. The results are listed in Table 1, which includes Ashford et al.'s values for comparison. The maximum likelihood procedure discussed above is used to fit Kempton's (1979) gamma model to the data, thus estimating the model parameters simultaneously, so that uncertainty in the mortality and development rates is incorporated in the estimation of the development thresholds and their standard errors. The results obtained from this procedure and those reported by Kempton are also given in Table 1, with the origin for development time fixed at 9.2 May.

Table 1. Estimates of population structure under three models. The data appear in Ashford et al. (1970). All models, other than model (3), incorporate a β parameter for stage 5.

Quantity	Model (3)	Ashford et al. (1970)	Gamma	Kempton (1979)
θ	0.020 ± 0.002	0.015 ± 0.008	0.016 ± 0.004	0.014
η_1	23.7 ± 2.7	20.5 ± 6.1[a]	21.5 ± 3.0	21.0
τ_0	14.3 ± 1.0	15.8 ± 3.4	14.2 ± 1.2	16.1
τ_1	14.7 ± 1.4	17.4 ± 4.3	15.7 ± 1.9	18.0
τ_2	13.5 ± 1.2	12.9 ± 3.2	13.8 ± 1.4	16.6
τ_3	8.7 ± 1.3	8.8 ± 2.8	8.9 ± 1.4	10.2
τ_4	10.8 ± 1.5	15.6 ± 3.5	11.0 ± 1.6	12.9
log L	112.57 (137 df)	83.7 (136 df)	112.99 (136 df)	
χ^2	27.43 (18 df)	54.94 (17 df)	24.89 (17 df)	

[a]estimated mean population recruitment total

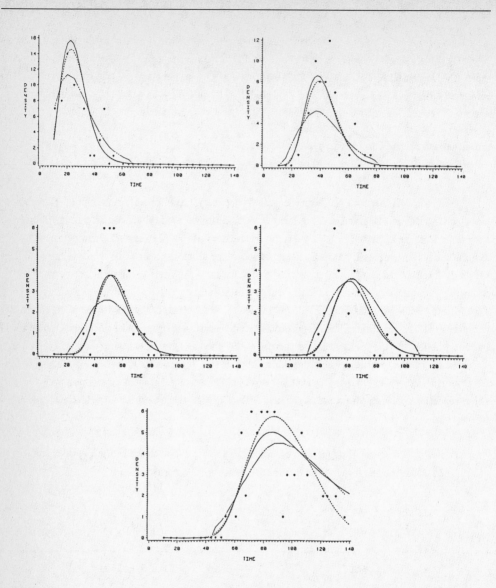

Fig. 1. Observed and expected frequencies, under three models: model (3) (————);
Kempton's gamma model (– – – –); Ashford et al. (——– – –). (a) – (e) correspond to stages 1 to
5.

The estimates of the stage–specific sojourn times and their respective standard errors obtained from fitting the inverse Gaussian model, (3), and the gamma model are very similar. While the values reported by Ashford et al. are comparable, the standard errors for their estimates are consistently larger. As a consequence of his method of estimation, asymptotic standard errors for Kempton's estimates cannot be calculated.

Both the inverse Gaussian and gamma models appear to provide a reasonable fit to the data and either is preferred to the Read & Ashford (1968) approach (see Fig. 1). the goodness–of–fit statistics also indicate the former models adequately describe the data, while the latter performs poorly. For each model, logL($\mathbf{n}$) has value 160.2. The χ^2 statistics are based on groups for which the expected frequency is at least five and, when necessary, samples are pooled across time and within stage, with an equivalent pooling structure applied to all three sets of predicted values.

Conclusions

The stochastic model developed in this paper appears to provide a reasonable description of an insect's life history and gives results which are comparable to those obtained from fitting a gamma model. This similarity is expected, since both approaches assume a unimodal and positively skewed distribution for the stage development times in the absence of mortality. The inverse Gaussian model is more appealing from a biological viewpoint and is computationally simpler than the other methods considered here. Clearly the estimation procedure discussed above is preferable to that suggested by Kempton, since standard errors can be calculated. The method used by Ashford et al. to analyze the data, used to illustrate our model, performs relatively poorly. In their aproach the shape parameter, rather than the development time origin is assumed known. While stage 1 data provide most of the information on the origin, all the data contain information on the β parameter. For these reasons, it seems preferable to fix the development time origin.

Acknowledgements

This research was funded in part by the Natural Sciences and Engineering Research Council of Canada. We are grateful to D.L. McLeish and G.W. Bennett (Department of Statistics and Actuarial Science, University of Waterloo) for their manuscript reviews. We also thank Dr. B.F.J. Manly (Department of Mathematics and Statistics, University of Otago) for his editorial comments.

References

Ashford, R. A., K. L. Q Read & G. G. Vickers. 1970. A system of stochastic models applicable to studies of animal population dynamics. *J. Anim. Ecol.* 37: 29–50.

Bellows, T. S. & M. H. Birley. 1981. Estimating development and mortality rates and stage recruitment from insect stage–frequency data. *Res. Pop. Ecol.* 23: 232–244.

Bellows, T. S, M. Ortiz, J. C. Owens & E. W. Huddleston. 1982. A model for analyzing insect stage–frequency data when mortality varies with time. *Res. Pop. Ecol.* 24: 142–156.

Dennis, B., W. P. Kemp, & R. C. Beckwith. 1986. Stochastic model of insect phenology: estimation and testing. *Environ. Entomol.* 15: 540–546.

Jennrich, R. I. & R. H. Moore. 1975. Maximum likelihood estimation by means of nonlinear least squares. In Proceedings, Statistical Computing Section, American Statistical Association, pp. 57–65.

Kempton, R. A. 1979. Statistical analysis of frequency data obtained from sampling an insect population grouped by stages. In Statistical Distributions in Ecological Work. J. K. Ord, G. P. Patil and C. Taille [eds.]: 401–418. Fairland, Maryland: International Cooperative Publishing House.

Manly, B. F. J. 1974. Estimation of stage–specific survival rates and other parameters for insect populations developing through several stages. *Oecologia* 15: 277–285.

Munholland, P. L. 1988. Statistical Aspects of Field Studies on Insect Populations. Ph.D. thesis (in preparation), University of Waterloo, Waterloo.

Read, K. L. Q. & J. R. Ashford. 1968. A system of models for the life cycle of a biological organism. *Biometrika* 55: 211–221.

SAS Institute. 1985. SAS User's Guide: Statistics. SAS Institute, Cary, N.C.

Stedinger, J. R., C. A. Shoemaker & R. F. Tenga. 1985. A stochastic model of insect phenology for a population with spatially variable development rates. *Biometrics* 41: 691–701.

Whitmore, G. A. & V. Seshadri. 1987. A heuristic derivation of the inverse Gaussian distribution. *Amer. Statist.* 41: 280–281.

Nonparametric Estimation of Insect Stage Transition Times

Jeffrey S. Pontius[1], John E. Boyer, Jr.[2]
and Michael L. Deaton[3]

ABSTRACT Consider an experiment on an insect where the stages (or stadia) can only be observed by sacrificing the insect or its habitat. Choose a fixed sequence of samples in time. At each sample time, sacrifice a subset of a cohort and record the number of insects in each stage. We review nonparametric point estimators of the time to each stage and evaluate the estimators by computer simulation. The simulation results, under five survival distributions, indicate that overall the estimators provide reasonable estimates of parametric values.

Consider the following experiment on an insect where the stages (or stadia), 0, 1, ..., A, can only be observed by sacrificing the insect or its habitat (example: an insect parasitoid). A cohort of these insects begins in stage 0 and subsets of the cohort are sampled, by sacrifice, periodically in time until the cohort is in stage A. At each sample time, the number of insects in each stage is recorded. Note that the samples are independent in time.

Review of Point Estimators

Let t_i, $i = 0, ..., F$, be an increasing sequence of fixed samples in time such that the cohort of insects begins in stage $s = 0$ at $t_0 = 0$ and ends in stage A at t_F. A sample of n_i insects is selected at each t_i and the stage of each recorded. Let $n_{i,s}$ be the number of insects observed in stage s at t_i. Note that $n_i = \Sigma\, n_{i,s}$, summed over stage s.

Let the random variable $T_s \in [0,\infty)$ be the time to stage s and have the cumulative distribution function F_s. The statistical survival function is $P(T_s > t) = 1 - F_s(t) = G_s(t)$. For each t_i, let $p_{i,s} = P(T_s > t_i) = G_s(t_i)$ and estimate $p_{i,s}$ by

$$\hat{p}_{i,s} = (1/n_i) \sum_{j=0}^{s-1} n_{i,j}$$

(i.e.; the proportion of the cohort in the sample <u>not</u> yet in stage s by t_i). The quantity $n_i \hat{p}_{i,s}$ has a binomial $(n_i, p_{i,s})$ distribution.

---

[1]1669 Saratoga, Pocatello, ID 83201.

[2]Department of Statistics, Kansas State Univ., Manhattan, KS 66506.

[3]E. I. duPont deNemours and Co., Waynesboro, VA. 22980.

We know that

$$E(T_s) = \int_0^\infty G_s(t)dt$$

so we can estimate $E(T_s)$ by averaging upper and lower Riemann sums with the partition $t_0, \ldots,$ t_F to obtain an approximate expression for the mean time to stage s as

$$E(T_s) \doteq (1/2) \sum_{i=0}^{F-1} (G_s(t_i) + G_s(t_{i+1}))(t_{i+1} - t_i).$$

Substituting $\hat{p}_{i,s}$ for $G_s(t_i)$, $\hat{p}_{i+1}$ for $G_s(t_{i+1})$ and noting that $t_0 = 0$, $\hat{p}_{0,s} = 1$ and $\hat{p}_{F,s} = 0$ we obtain

$$\hat{E}(T_s) = (1/2)t_1 + (1/2) \sum_{i=1}^{F-1} \hat{p}_{i,s}(t_{i+1} - t_{i-1})$$

as an estimator for $E(T_s)$. The absolute bias of $\hat{E}(T_s)$ is then less than or equal to $(1/2)\max(t_{i+1} - t_i)$. The $\hat{p}_{i,s}$ are independent for different t_i (assuming that the population size is infinitely large) so an estimator of the variance of $\hat{E}(T_s)$ is

$$\hat{Var}(\hat{E}(T_s)) = (1/4) \sum_{i=1}^{F-1} (1/n_i)(\hat{p}_{i,s}(1 - \hat{p}_{i,s}))(t_{i+1} - t_{i-1})^2.$$

An estimator of the mean time to stage s from stage $s^* < s$ is

$$\hat{E}(T_s - T_s^*) = (1/2) \sum_{i=1}^{F-1} (\hat{p}_{i,s} - \hat{p}_{i,s}^*)(t_{i+1} - t_{i-1}).$$

When $s^* = s-1$, $\hat{E}(T_s - T_s^*)$ is the mean duration time for stage s.

We now review two estimators, $Var_T(T_s)$ and $Var_L(T_s)$, of the variance of the time to stage s, $Var(T_s)$. The estimators are determined from $Var(T_s) = E(T_s^2) - [E(T_s)]^2$ after deriving an estimator of the second moment

$$E(T_s^2) = 2 \int_0^\infty tG_s(t)dt .$$

For $Var_T(T_s)$, $E_T(T_s^2)$ is estimated by averaging Reimann sums over each $[t_i, t_{i+1}]$ to obtain

$$\hat{E}_T(T_s^2) = (1/2) \sum_{i=0}^{F-1} (\hat{p}_{i,s} + \hat{p}_{i+1,s})(t_{i+1}^2 - t_i^2).$$

$E_L(T_s^2)$ is estimated by assuming $G_s(t)$ is a linear function on $[t_i, t_{i+1}]$, integrating, and simplifying to obtain

$$\hat{E}_L(T_s^2) = (1/3) \sum_{i=0}^{F-1} [\hat{p}_{i,s}(2t_i + t_{i+1})] + \hat{p}_{i+1,s}(t_i + 2t_{i+1})](t_{i+1} - t_i).$$

We have shown that $\text{Var}_T(T_s) > \text{Var}_L(T_s)$ and $|\text{Var}_T(T_s) - \text{Var}_L(T_s)| \to 0$ as $\Delta t \to 0$ where $\Delta t = [t_{i+1} - t_i]$, under the assumption of nonincreasing monotonicity of $G_s(t)$. Note that the $\hat{p}_{i,s}$ may not be monotonic because of sampling variation, but we have always observed these relations to occur with actual data.

Evaluation of Estimators

We next compare the performance of $\hat{E}(T_s)$, $\hat{\text{Var}}_T(T_s)$, $\hat{\text{Var}}_L(T_s)$, and $\hat{\text{Var}}(\hat{E}(T_s))$ to their respective expected values $E(T_s)$ or $\text{Var}(T_s)$ by computer simulation. To evaluate the robustness of the estimators we selected five survival distributions $\{G_s(t)\}$ (Table 1) that provided a wide variety of shapes. We assumed the 'insect' had two stages, $s = \{0, 1\}$. The distributions are from standard densities except that the beta distribution was chosen to have a concave shape. There was a constant time interval (Δt) between t_i and the numbers sampled (n_i) were variables. To construct a similar sampling regimen over the distributions, $\Delta t = c\sqrt{\text{Var}(T_s)}$, $c = 1, 1/2, 1/4$ and $1/8$, was chosen. For each c and t_i, $n_i = n = 5, 10, 20$ or 40. Fifty 'cohorts' were simulated for each pair $(c;n)$.

The statistics we used to evaluate the estimators were the mean and root mean square error (RMS) of $\hat{E}(T_s)$, $\hat{\text{Var}}_T(T_s)$, $\hat{\text{Var}}_L(T_s)$ and the mean of $\hat{\text{Var}}(\hat{E}(T_s))$, over the fifty cohort simulations. We also calculated 95% confidence intervals for the mean of $\hat{E}(T_s)$ by $\text{mean}(\hat{E}(T_s)) \pm 1.96 \, [\text{mean}(\hat{\text{Var}}(\hat{E}(T_s)))/50]^{1/2}$.

We used PROC MATRIX of SAS, version 5.16, (SAS Institute, 1985) for simulation because our computer algorithms for calculating estimates were already coded in SAS (see Pontius 1987).

The simulation algorithm is as follows. Instructions apply to each survival distribution. First, initialize $E(T_s)$, $\text{Var}(T_s)$, c and n. Next generate $t_0 = 0$ and $t_{i+1} = t_i + \Delta t$, $i = 0, ..., F-2$. To ensure that $p_{F,s} = 0$, calculate $t_F = t_{F-1} + \Delta t$ where $t_{F-1} = \max\{t_i \, \epsilon \, [0,1]\}$ for the uniform and beta distributions and $t_{F-1} = \max\{t_i : P(T_s < t_i) \leq 0.9999\}$ for the exponential, normal and gamma distributions.

For each t_i, calculate a corresponding $p_{i,0}$. Set $p_{0,0} = 1$. Calculate $p_{i,0}$, $i = 1, ..., F-1$, for uniform and exponential distributions (see Table 1), and $P_{i,s} = 1 - P(T_s < t_i)$ for the beta,

Table 1. Survival distribution used in simulations.

survival distribution	parameter value(s)	$E(T_s)$	$Var(T_s)$
uniform: $G_s(t) = (1-t)I_{[0,1]}(t)$	–	0.5	0.083
exponential: $G_s(t) = e^{-rt}I_{[0,\infty)}(t)$	$r = 1$	1.0	1.0
beta:	$a = 2.0$	0.8	0.046
$G_s(t) = 1 - \int_0^t \frac{1}{B(a,b)} x^{a-1}(1-x)^{b-1}I_{[0,1]}(t)$	$b = 0.5$		
normal:	$u = 3.5$	3.5	1.0
$G_s(t) = 1 - \int_0^t \frac{1}{\sqrt{2\pi}s} \exp\left[-\frac{1}{2}\left(\frac{x-u}{s}\right)^2\right]I_{(0,\infty)}(t)$	$s = 1.0$		
gamma:	$a = 2.0$	2.0	2.0
$G_s(t) = 1 - \int_0^t \frac{1}{\Gamma(a)} x^{a-1}e^{-x}I_{[0,\infty)}(t)$			

normal and gamma distributions using the SAS probability generators PROBBETA, PROBNORM and PROBGAM, respectively. Set $p_{F,0} = 0$.

Now determine the number observed in stage 0. For each of the fifty cohort simulations do the following. Initialize $n_{i,0} = 0$, $i = 1, ..., F$. Set $n_{0,0} = n$. For each t_i, $i = 1, ..., F$, choose $n_{i,0}$ using the binomial random number generator RANBIN with parameters n and $p_{i,0}$. Stop sampling when the first $n_{i,0} = 0$, $i > 0$. Calculate $n_{i,1} = n - n_{i,0}$, $i = 1, ..., F$. Calculate $\hat{E}(T_s)$, $\hat{Var}_T(T_s)$, $\hat{Var}_L(T_s)$ and $\hat{Var}(\hat{E}(T_s))$ and store the estimates. After the fifty cohorts have been simulated, calculate and print the evaluation statistics. This ends the algorithm.

Results and Discussion

It will be seen that $\hat{E}(T_s)$, $\hat{Var}_T(T_s)$, and $\hat{Var}_L(T_s)$ performed similarly in relation to the shapes of the distributions. Based on means and RMS of the estimates, these estimators best estimated their respective expected values under the uniform and beta distributions and performed worst under the gamma distribution. Note that $\hat{\text{R}}$MS, and hence deviations of means of estimates from expected values, can be expressed as a function of c and 1/n (Pontius, 1987).

Our overall evaluation of $\hat{\text{E}}(\text{T}_\text{s})$ is based on trends in means and $\hat{\text{R}}$MS (Table 2) and 95% confidence intervals for $\text{E}(\text{T}_\text{s})$(Table 3). If $\text{E}(\text{T}_\text{s})$ is contained in the confidence interval then we consider $\hat{\text{E}}(\text{T}_\text{s})$ to be a good estimator of $\text{E}(\text{T}_\text{s})$ under the particular survival distribution. Note that the confidence intervals may be used to test H_0: $\hat{\text{E}}(\text{T}_\text{s})$ is an unbiased estimate of $\text{E}(\text{T}_\text{s})$ against H_a: $\hat{\text{E}}(\text{T}_\text{s})$ is biased. If H_0 is rejected, bias can be estimated by bias = mean($\hat{\text{E}}(\text{T}_\text{s})$) − $\text{E}(\text{T}_\text{s})$.

$\hat{\text{E}}(\text{T}_\text{s})$ is a good estimator for the uniform and beta distributions and for most (c;n) under the normal distribution but appears biased and to underestimate $\text{E}(\text{T}_\text{s})$ for the exponential and gamma distributions, particularly for smaller Δt (c = 1/4 or 1/8). Possibly the bias when c = 0.25 or c = 0.125 is because some t_i (in the right part of the distribution) are consistently not being sampled. Overall, $\hat{\text{R}}$MS decreased as Δt decreased and n increased. As expected (see Boyer & Deaton 1984), means of $\hat{\text{Var}}(\hat{\text{E}}(\text{T}_\text{s}))$ decreased as Δt decreased and n increased (Table 4).

Overall, $\hat{\text{Var}}_\text{T}(\text{T}_\text{s})$ better estimated Var(T_s) for smaller c and larger n under the uniform, beta and normal distributions; and for smaller c under the exponential and gamma distributions (Table 5). Means of $\hat{\text{Var}}_\text{T}(\text{T}_\text{s})$ were closer overall to Var(T_s) of the beta distribution and farther from Var(T_s) of the exponential and gamma distributions. Trends in $\hat{\text{R}}$MS were variable and appeared to depend upon each distribution.

Mean of $\hat{\text{Var}}_\text{L}(\text{T}_\text{s})$ (Table 6) were less than the means of $\hat{\text{Var}}_\text{T}(\text{T}_\text{s})$ for all distributions and (c;n) (equality is the result of rounding of means). Also, as c decreased and n increased, $|\hat{\text{Var}}_\text{T}(\text{T}_\text{s}) - \hat{\text{Var}}_\text{L}(\text{T}_\text{s})| \to 0$. In general, statements about $\hat{\text{Var}}_\text{T}(\text{T}_\text{s})$ pertain to $\hat{\text{Var}}_\text{L}(\text{T}_\text{s})$. Noting the similarities in $\hat{\text{R}}$MS, the variance estimators give similar ranges of bias as Δt decreases and n increases.

Overall, $\hat{\text{E}}(\text{T}_\text{s})$, $\hat{\text{Var}}_\text{T}(\text{T}_\text{s})$ and $\hat{\text{Var}}_\text{L}(\text{T}_\text{s})$ best estimated their respective expected values under the uniform, concave beta and normal distributions. Because in applications the shape of the survival distribution is rarely known (and when the distribution is assumed known, parametric estimators may be preferred), possibly graphing the $\hat{p}_{i,s}$ and observing the shape of the graph may aid in evaluating how 'good' the nonparametric estimates are.

$\hat{\text{E}}(\text{T}_\text{s})$ better estimates $\text{E}(\text{T}_\text{s})$ as the number of t_i in $[t_0, t_\text{F}]$ and/or n_i, $1 \leq i \leq$ F−1, are increased. $\hat{\text{Var}}(\hat{\text{E}}(\text{T}_\text{s}))$ becomes smaller as the number of t_i in $[t_0, t_\text{F}]$ and/or n_i are increased. In particular, better estimates of $\text{E}(\text{T}_\text{s})$ and smaller $\hat{\text{Var}}(\hat{\text{E}}(\text{T}_\text{s}))$ are obtained by increasing the number of t_i in $[t_0, t_\text{F}]$ and n_i when stage transitions occur.

$\hat{\text{Var}}_\text{T}(\text{T}_\text{s})$ and $\hat{\text{Var}}_\text{L}(\text{T}_\text{s})$ better estimate Var(T_s) for larger n_i, but underestimate Var(T_s) as the number of t_i in $[t_0, t_\text{F}]$ increases, especially for smaller n_i. $\hat{\text{Var}}_\text{T}(\text{T}_\text{s})$ would probably be preferred for graphs (see graphing of $\hat{p}_{i,s}$ above) similar to normal, exponential or gamma

Table 2. $\hat{E}(T_s)$ results from simulations. Mean of $\hat{E}(T_s)$ and (root mean square error) are listed for each c and n combination under each survival distribution.

DISTRIBUTION:		uniform	exponential	beta	normal	gamma
$E(T_s)$		.5	1.0	.8	3.5	2.0
c	n					
1	5	.50	1.06	.80	3.45	1.88
		(.10)	(.35)	(.07)	(.34)	(.46)
	10	.53	1.02	.79	3.53	1.96
		(.09)	(.21)	(.05)	(.28)	(.29)
	20	.51	1.06	.80	3.53	2.05
		(.05)	(.16)	(.04)	(.17)	(.24)
	40	.51	1.08	.80	3.53	1.98
		(.04)	(.14)	(.04)	(.11)	(.18)
.5	5	.48	.83	.80	3.41	1.81
		(.08)	(.31)	(.05)	(.28)	(.39)
	10	.50	.95	.81	3.43	1.90
		(.06)	(.18)	(.04)	(.21)	(.25)
	20	.50	.99	.80	3.51	1.98
		(.03)	(.09)	(.03)	(1.2)	(.19)
	40	.50	1.00	.80	3.50	1.99
		(.02)	(.06)	(.02)	(.10)	(.11)
.25	5	.50	.81	.79	3.45	1.78
		(.06)	(.28)	(.03)	(.22)	(.33)
	10	.50	.90	.80	3.47	1.90
		(.04)	(.16)	(.02)	(.10)	(.19)
	20	.50	.92	.80	3.50	1.94
		(.02)	(.12)	(.02)	(.07)	(.13)
	40	.50	.99	.80	3.49	1.97
		(.02)	(.07)	(.01)	(.07)	(.10)
.125	5	.44	.76	.79	3.34	1.65
		(.09)	(.29)	(.03)	(.22)	(.42)
	10	.49	.84	.80	3.43	1.78
		(.04)	(.20)	(.02)	(.12)	(.25)
	20	.50	.93	.80	3.49	1.94
		(.02)	(.10)	(.01)	(.05)	(.12)
	40	.50	.96	.80	3.49	1.96
		(.01)	(.05)	(.01)	(.04)	(.09)

Table 3. 95% confidence intervals for $E(T_s)$. Confidence intervals are listed for each c and n combination under each survival distribution.

DISTRIBUTION:		uniform	exponential	beta	normal	gamma
$E(T_s)$		.5	1.0	.8	3.5	2.0
c	n					
1	5	(.48,.53)	(.99,1.13)	(.78,.82)	(3.36,3.54)	(1.77,1.99)*
	10	(.51,.54)*	(.97,1.07)	(.78,.80)	(3.47,3.59)	(1.88,2.04)
	20	(.50,.52)	(1.02,1.10)*	(.79,.81)	(3.48,3.58)	(1.99,2.11)
	40	(.50,.52)	(1.05,1.11)*	(.79,.81)	(3.50,3.56)	(1.94,2.02)
.5	5	(.46,.50)	(.79, .87)*	(.79,.81)	(3.35,3.47)*	(1.74,1.88)*
	10	(.49,.51)	(.91, .99)*	(.80,.81)	(3.39,3.47)*	(1.84,1.96)*
	20	(.49,.51)	(.96,1.02)	(.79,.81)	(3.48,3.54)	(1.94,2.02)
	40	(.49,.51)	(.98,1.02)	(.80,.81)	(3.48,3.52)	(1.96,2.02)
.25	5	(.49,.51)	(.78, .84)*	(.79,.81)	(3.41,3.49)*	(1.73,1.83)*
	10	(.49,.51)	(.87, .93)*	(.80,.81)	(3.44,3.50)	(1.86,1.94)*
	20	(.49,.51)	(.90, .94)*	(.80,.81)	(3.48,3.52)	(1.91,1.97)*
	40	(.50,.51)	(.98,1.01)	(.80,.80)	(3.47,3.51)	(1.95,1.99)*
.125	5	(.43,.45)*	(.73, .78)*	(.78,.80)	(3.31,3.37)*	(1.62,1.68)*
	10	(.48,.50)	(.82, .86)*	(.80,.81)	(3.41,3.45)*	(1.75,1.81)*
	20	(.50,.51)	(.92, .94)*	(.80,.80)	(3.47,3.51)	(1.92,1.96)*
	40	(.50,.50)	(.95, .97)*	(.80,.80)	(3.48,3.50)	(1.95,1.98)*

* confidence intervals that do not contain $E(T_s)$.

distributions used in simulations. $\hat{\text{Var}}_L(T_s)$ would probably be preferred when the number of t_i in $[t_0, t_F]$ is small for distributions similar to the uniform or concave beta distributions. Because $\text{Var}_T(T_s) > \text{Var}_L(T_s)$, one could choose $\hat{\text{Var}}_T(T_s)$ as a conservative estimate when the graph of $\hat{p}_{i,s}$ yields little information on distributional shape. However, for $n_i \geq 20$, $1 \leq i \leq F-1$, and a large number of t_i in $[t_0, t_F]$, either variance estimator could be used.

Table 4. $\hat{\text{Var}}(\hat{\text{E}}(T_s))$ results from simulations. Mean of $\hat{\text{Var}}(\hat{\text{E}}(T_s))$ is listed for each c and n under each survival distribution.

DISTRIBUTIONS:		uniform	exponential	beta	normal	gamma
c	n					
1	5	.0080	.0576	.0037	.1024	.1517
	10	.0044	.0327	.0023	.0531	.0879
	20	.0023	.0193	.0011	.0274	.0489
	30	.0011	.0103	.0006	.0138	.0254
.5	5	.0038	.0250	.0020	.0474	.0710
	10	.0022	.0190	.0011	.0239	.0442
	20	.0011	.0108	.0006	.0135	.0244
	40	.0006	.0056	.0003	.0069	.0130
.25	5	.0017	.0134	.0010	.0195	.0311
	10	.0011	.0087	.0005	.0120	.0210
	20	.0006	.0051	.0003	.0066	.0117
	40	.0003	.0029	.0001	.0034	.0062
.125	5	.0008	.0059	.0005	.0087	.0127
	10	.0005	.0040	.0003	.0057	.0093
	20	.0003	.0025	.0001	.0032	.0057
	40	.0001	.0014	.0001	.0017	.0030

Table 5. $\hat{\mathrm{Var}}_T(T_s)$ results from simulations. Mean of $\hat{\mathrm{Var}}_T(T_s)$ and (root mean square error) are listed for each c and n combination under each survival distribution.

DISTRIBUTION:		uniform	exponential	beta	normal	gamma
Var(T_s)		.083	1.0	.046	1.0	2.0
c	n					
1	5	.108	.74	.053	1.39	1.90
		(.056)	(.48)	(.027)	(.83)	(1.02)
	10	.108	.84	.062	1.35	2.17
		(.040)	(.47)	(.024)	(.63)	(.80)
	20	.111	1.01	.060	1.34	2.43
		(.034)	(.35)	(.022)	(.46)	(.81)
	40	.105	1.12	.062	1.33	2.54
		(.025)	(.32)	(.018)	(.40)	(.77)
.5	5	.078	.41	.047	.99	1.31
		(.029)	(.66)	(.023)	(.48)	(1.02)
	10	.089	.68	.049	.92	1.69
		(.023)	(.44)	(.015)	(.23)	(.74)
	20	.086	.83	.051	1.06	1.85
		(.015)	(.28)	(.012)	(.22)	(.53)
	40	.088	.88	.053	1.07	2.09
		(.010)	(.21)	(.010)	(.15)	(.38)
.25	5	.069	.35	.048	.72	.91
		(.032)	(.68)	(.017)	(.41)	(1.18)
	10	.080	.50	.045	.90	1.34
		(.014)	(.54)	(.010)	(.28)	(.77)
	20	.083	.64	.047	.97	1.64
		(.013)	(.40)	(.005)	(.16)	(.54)
	40	.086	.85	.047	.98	1.72
		(.008)	(.23)	(.005)	(.11)	(.38)
.125	5	.057	.27	.043	.62	.64
		(.036)	(.75)	(.012)	(.46)	(1.39)
	10	.076	.42	.047	.80	1.08
		(.019)	(.61)	(.008)	(.26)	(1.01)
	20	.085	.62	.046	.91	1.43
		(.009)	(.40)	(.004)	(.15)	(.63)
	40	.083	.75	.047	.96	1.69
		(.006)	(.28)	(.004)	(.09)	(.40)

Table 6. $\hat{Var}_L(T_s)$ results from simulations. Mean of $\hat{Var}_L(T_s)$ and (root mean square error) are listed for each c and n under each survival distribution.

DISTRIBUTION:		uniform	exponential	beta	normal	gamma
Var(T_s)		.083	1.0	.046	1.0	2.0
c	n					
1	5	.095	.58	.045	1.22	1.57
		(.051)	(.59)	(.026)	(.76)	(1.11)
	10	.094	.67	.054	1.18	1.83
		(.034)	(.55)	(.020)	(.56)	(.80)
	20	.097	.84	.053	1.17	2.10
		(.024)	(.39)	(.018)	(.35)	(.69)
	40	.091	.96	.054	1.16	2.21
		(0.15)	(.30)	(.012)	(.28)	(.59)
.5	5	.075	.37	.045	.95	1.23
		(.030)	(.70)	(.023)	(.48)	(1.08)
	10	.086	.64	.047	.88	1.60
		(.022)	(.47)	(.014)	(.31)	(.78)
	20	.083	.79	.049	1.02	1.76
		(.015)	(.31)	(.010)	(.22)	(.56)
	40	.084	.84	.051	1.02	2.00
		(.009)	(.24)	(.009)	(.13)	(.36)
.25	5	.068	.34	.048	.71	.89
		(.032)	(.69)	(.017)	(.41)	(1.20)
	10	.079	.49	.044	.89	1.32
		(.014)	(.55)	(.010)	(.29)	(.79)
	20	.083	.63	.046	.96	1.62
		(.013)	(.41)	(.005)	(.16)	(.57)
	40	.085	.84	.046	.97	1.70
		(.008)	(.24)	(.005)	(.11)	(.39)
.125	5	.057	.27	.042	.62	.64
		(.036)	(.75)	(.012)	(.46)	(1.40)
	10	.075	.41	.047	.80	1.07
		(.019)	(.61)	(.008)	(.27)	(1.01)
	20	.085	.62	.046	.91	1.44
		(.010)	(.41)	(.004)	(.15)	(.63)
	40	.083	.75	.047	.96	1.69
		(.006)	(.28)	(.004)	(.09)	(.41)

Acknowledgements

We gratefully acknowledge Paul Nelson, Department of Statistics, Kansas State University and Stephen M. Welch, Computer Systems Office, Kansas Cooperative Extension Service, Kansas State University, Manhattan, Kansas for reviewing our manuscript.

References Cited

Boyer, Jr., J. E. & M. L. Deaton. 1984. Estimation of duration from stage frequency data. Tech. Report, Dept. of Statistics, Kansas State University, Manhattan, Kansas.

Pontius, J. S. 1987. Nonparametric estimation of stage transition time from stage frequency data. M. S. report. Kansas State University, Manhattan, Kansas.

SAS Institute. 1985. SAS user's guide: statistics. SAS Institute, Cary, North Carolina.

Problems Associated With Life Cycle Studies
of a Soil–Inhabiting Organism

S. M. Schneider[1]

ABSTRACT Building a simple, distributed–development model of the life cycle of an ectoparasitic nematode requires stage–specific values for stage duration and its standard deviation, survivorship, and fecundity. Direct observation of uniform cohorts of the ectoparasitic nematode, *Paratrichodorus minor* is not possible. Each female contains at most only two mature eggs at a time, making it difficult to obtain an adequate and uniform cohort of first–stage juveniles (J1). Distributed development of the J1 also complicates the task of obtaining uniform cohorts of all subsequent stages. Soil habitats must be destructively sampled and the target organisms extracted from the soil. This procedure results in greater variation than if the same individuals could be observed over time. Plants were inoculated with an approximately uniform cohort of J1 and destructively sampled at time intervals shorter than the expected duration of the shortest stage. The data included numbers of individuals in each stage at each time step, corrected for the stage–specific extraction efficiency. A simulator was constructed based on a conceptual model of one egg, four juvenile, and the adult female stages (males are rare and were not included). Individuals moved from one stage to the next based on the stage duration, its standard deviation, and the stage survivorship. Two data sets were created from the original data, a first generation only data set and a multiple generation data set. A range of possible values was chosen for each parameter to be estimated. Beginning with J1 all combinations of parameter values were used to generate a predicted data set for the first generation. The predicted and observed values were compared and the combination resulting in the smallest weighted least squares was chosen as the initial best estimate. Each subsequent stage was estimated in the same manner, allowing for minor modifications of previous stage estimates. Predicted and observed values were generally in agreement. Using a fecundity value from additional experiments and the best estimates for the life cycle parameters, population data were predicted and compared for multiple generations. The second generation occurred much earlier in the predicted than in the observed data set. Further investigation resulted in the addition of a pre–reproductive adult stage. After this modification, agreement between observed and predicted was acceptable. This simple technique proved adequate for greenhouse and growth chamber data, but functioned poorly for field data. This discrepancy is likely due to additional mortality factors and the influence of population density on the life cycle. Although the technique presented herein can be computationally intensive with large sets of parameter combinations and lengthy simulations, it is conceptually simple and offers much potential for a group of often neglected animals.

Computer simulations of the population dynamics of pests can be useful in crop management models, in testing the understanding of the biology of the pest organism, and in identifying knowledge gaps. Pest models can be linked to plant models and used to evaluate the plant's response to the stress experienced as a result of attack by one or more pest organisms. There are

---

[1]United States Department of Agriculture, ARS, Oxford Research Laboratory, Oxford, NC 27565.

several additional questions that might be addressed. What stages of the pest could be most efficiently managed with available control procedures? Which stages would be most beneficial to control in terms of the marketable yield? In breeding programs, which forms of quantitative resistance would have the greatest impact on final yield? How does host resistance influence the life cycle of the pest? The desirability of pest population models can be demonstrated by the number of research programs involved in this area.

Many of the systems being studied are insects or disease–causing organisms (fungi, bacteria, viruses, mycoplasma, nematodes). Of these, most of the research effort has focused on foliar or above–ground pests. Very little research has been done on modeling soil inhabiting pests in general, and plant parasitic nematodes in particular.

The original goal of this research was to build a simple, distributed development model of the ectoparasitic nematode, *Paratrichodorus minor* (Colbran, 1956) Siddiqi, 1974. The conceptual model included an egg stage, four juvenile stages, and the adult female stage (Fig. 1). Necessary information included the duration of each stage, the survival rate for each stage, and the fecundity rate for the adult. Adult males are rare in this nematode and were not considered in the model.

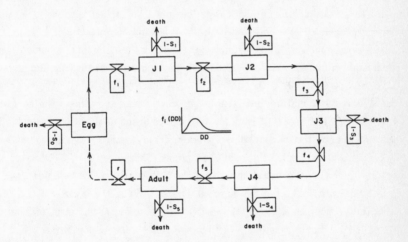

Fig. 1. Conceptual model for the life cycle of *Paratrichodorus minor*. Maturation of an individual from one stage to the next is described by an Erlang distribution, f(DD), defined by a mean and standard deviation for the stage duration. S_i is the survivorship for stage i, $(1-S_i)$ is the mortality for stage i. The rate of egg production is r.

Temperature–Dependent Model

Development of an individual from one stage to the next is governed primarily by time and temperature. Nematodes, like other invertebrates, are poikilothermic organisms. The rate of development depends on the rate of internal biochemical processes which are related to ambient temperature. Cumulative development is a function of the rate of development and elapsed time. The concept of degree–day integrates the impact of temperature and time (Candolle 1855). It assumes a linear relationship between the rate of development and temperature. Provisions are included for basal thresholds, below which development will not occur, and upper thresholds, above which development either continues at a constant rate or stops (Baskerville 1969). The two types of upper thresholds reflect different biological responses. The former response is a result of development having reached its maximum rate and remaining at that rate even though temperature increases. The latter response reflects an upper value above which development will not continue.

Degree–days is a useful concept, but must be used with certain assumptions clearly in mind. The assumption of linearity between development and temperature does not hold for temperatures near the basal threshold or near the optimum (Stinner et al. 1974, 1975, Logan et al. 1976, Curry et al. 1978, Wagner et al. 1984b). Experiments are often conducted at constant temperatures. Biological response to diurnal temperature fluctuations in the field and constant temperatures in a greenhouse or growth chamber might not be the same (Sharpe et al. 1977, Curry et al. 1978). Use of the same degree–day calculations in both situations must be applied cautiously. In addition to upper and lower thresholds, lethal limits must also be taken into account.

Individuals in a population do not all mature at exactly the same rate. Inherent genetic and microenvironmental differences lead to different developmental rates (Ashford et al. 1970, Manetsch 1976, Sharpe et al. 1977, Curry et al. 1978, Wagner et al. 1984a). The basic nematode model assumes individuals mature from one stage to the next according to some distribution determined by a mean and standard deviation. The choice of the distribution was made based on current understanding of the biology of poikilothermic organisms. Developmental rates are governed by enzymes necessary for biochemical reactions (Sharpe et al. 1977, Curry et al. 1978). The distribution of enzyme concentrations within the population can be assumed to be normal, resulting in a normal distribution of developmental rates. When the inverse of this function is taken to calculate developmental times, the distribution becomes asymmetric with an extended right–hand tail. The Erlang function, which is completely defined by a mean and standard deviation, has been suggested as a reasonable choice (Ashford et al. 1970, Manetsch 1976). The function can be offset by a constant to represent the minimum time required for development in a stage. Although an Erlang function was used for the data presented here, the technique used can be applied to any probability function defined by the mean and standard deviation.

Each stage of the life cycle model is divided into substages. At each iteration, according to an appropriately chose time step, the individuals can stay in the same substage, move to the next substage of the same stage, mature to the next stage, or die. If sufficient degree–days have accumulated, the probability function and the survivorship function determine the proportion of the population in the substage that mature, die, or simply move to the next substage.

To build the distributed development model of the nematode life cycle, mean and standard deviation values for the duration of each stage, stage–specific survivorships, and a fecundity value for the adult stage were needed. The ideal system would be one in which individuals could be placed on the host and observed periodically for maturation to the next stage. The result would be a data set of exact stage durations and survivorships for each stage for each individual from one stage to the next. These data would be actual values for the population, not estimates, and could be used to calculate mean population values. For some macroscopic, above–ground pests, these kinds of observations are possible. Another alternative to observing the same individuals at each sampling date, would be to start with a uniform age cohort for a stage and directly observe it periodically through a single generation to determine when the first individual matured to the next stage, the rate of maturation, and the length of time required for the last individual to mature. In other words, the mean and standard deviation for the length of time necessary for a uniform age cohort of stage n to mature to stage n+1.

Sampling Restrictions

For a soil inhabiting organism, direct observation of the same individuals is not possible. Individuals must be extracted from the soil to be observed. This destructive sampling process dictates a large experimental plan to supply sufficient plant/pest units to sample at many sampling dates. There may be a mortality associated with the extraction process, which should be separated from "natural mortality", if a meaningful estimate of mortality is to be calculated from the data. If the extraction process extends over a period of time, there is a possibility of individuals maturing to the next stage during the extraction process, and thus influencing the estimates of numbers in each stage.

It is very difficult to obtain uniform age cohorts of any stage of *P. minor*. The female contains at most two well developed eggs at any point in time. These are laid singly in the soil making recovery of a uniform cohort of eggs almost impossible. The eggs do not hatch synchronously, but according to some distribution giving rise to a distributed cohort of first–stage juveniles (J1). With each successive stage, the population becomes more distributed across life stages, and more difficult to obtain uniform age cohorts (Birley 1979). Egg production continues over a considerable length of time. Egg hatch, signalling the beginning of the second generation, begins before egg production by the first generation ends. There is no easily detected point indicating the end of the first generation. Overlapping stages and overlapping generations are the rule.

Given the obstacles found in the ectoparasitic nematode system, the following data collection method was used. An approximately uniform cohort of J1 was obtained by inoculating plants with gravid females, allowing 24 hours for females to deposit eggs, waiting the average length of time for egg hatch, extracting the nematodes from the soil, and isolating the J1 from the females. The J1 were inoculated onto host plants and destructively sampled at time intervals smaller than the expected value of the shortest stage. The resulting data, numbers of individuals in each stage at each sampling time (Schneider & Ferris 1987), were the values available to estimate the life cycle parameters needed for the model.

The next problem encountered was the appearance in the extracted samples of more second stage juveniles (J2) than original J1. Since the J1 were the only source of J2, this was theoretically impossible. The solution was to adjust for the efficiency of the extraction process. In this study, not only was the extraction process not 100% efficient, the efficiency of extraction varied from one stage to the next. The J2's were more efficiently extracted than the J1 resulting in higher numbers. Extraction efficiency values were determined for each stage by seeding samples with a population of known size and age structure. The seeded samples were processed and the recovery efficiency (number recovered/initial number) calculated. Although this adjustment improved the usefulness of the data, the data were still estimates. The recovery efficiency undoubtedly varied within each stage as well as between stages. The amount of improvement depended on how closely the experimental population age structure matched the population age structure used to determine the recovery efficiency values.

The available data now consisted of adjusted numbers of individuals in each stage at each sampling time (Figs. 2 and 3). The numbers present in each stage are given in Appendix A. From these data, various calculations were made. The times of first and last observation for each stage were noted. Last minus first for the same stage could be used as a maximum limit for the stage duration, but this was a gross overestimation of this value, since the first individual to reach a given stage was probably not the last individual to leave the stage. First observation of stage n+1 minus first observation of stage n was also vague, since the first individual entering stage n+1 might not be the first individual entering stage n (Birley 1979). The peaks of each stage represented the time when the greatest numbers of individuals were in that stage and the point at which the dominant influence changed from being the number entering the stage to the number leaving the stage. If a value had been known for the duration of each stage, the area under the curve could have been used to calculate the number of nematodes (Manly 1976, 1977) and then used for estimating survival from one stage to the next. If the mortality or number of nematodes had been known, the areas under the curves could have been used to estimate stage duration. The rates of increase and decrease observed in each stage were a function of numbers entering and leaving the stage. Many other techniques have been presented to estimate stage durations and survivorships, but most assume information is available for numbers entering a stage, leaving a stage, dying in a stage, or surviving a stage (Richards et al. 1960, Manly 1974a, 1976, 1977, 1985, 1987, Birley 1977, Mills 1981a,b, Bellows et al. 1981, 1982, Fargo 1986). Many techniques are designed to analyze data from a single generation, but some may be adapted to multiple generation data sets if generations are sufficiently discrete (Richards et al. 1960, Ashford et al. 1970, Manly 1974a, b, 1976, 1977, 1985, Birley 1979, Bellows et al. 1981, Mills 1981a,b, Fargo 1986). The interpretation of the data available for this organism is confounded by the fact that the population was distributed over several stages at the same point in time. Multiple generations are encountered leading to more overlap between subsequent stages (Birley 1979) (Fig. 3).

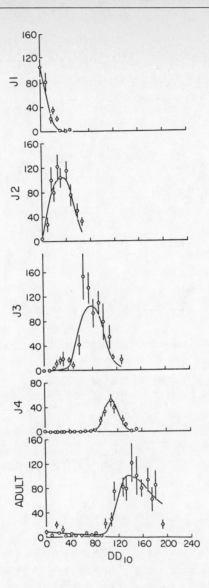

Fig. 2. Observed and predicted data for the first generation of *Paratrichodorus minor*. O are observed data points, mean of 10 replications with standard error bars, —— are predicted values.

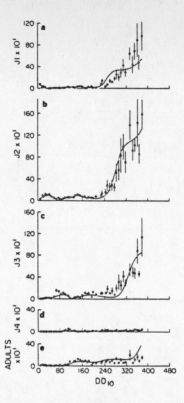

Fig. 3. Observed and predicted data for the multiple generation data set of *Paratrichodorus minor*. O are observed data points, mean of 10 replications with standard error bars, —— are predicted values.

Life Cycle Parameters

A mathematically simple approach for estimating life cycle parameters was adopted (Schneider & Ferris 1986). A population model was written based on the conceptual model described earlier. The model was embedded in a shell which generated predicted population values from all possible combinations of ranges of values for stage duration and its standard deviation, and survivorship for each stage. The shell chose the combination of parameters which minimized the differences between the predicted and observed data.

A two–step approach was used to apply this technique to the *P. minor* data set. The data set was divided into first generation only and the whole data set. Using the first generation data only, and fixing the values for the survivorship and standard deviation of the stage duration, a range of values for the duration of the first stage was tested. For each duration estimate, a set of predicted values for the first stage was generated using the model. The predicted and observed values were compared using a weighted least squares, and the duration estimate resulting in the smallest weighted least square was chosen as the best first estimate. This value was fixed, and a range of values tested for the duration of the second stage in the same manner. The predicted and observed data for the first and second stages were compared, and the best first estimate chosen for the second stage duration. This was continued in a stepwise fashion for each remaining stage. After estimates for all stage durations had been obtained (the coarse fit), the process was repeated leaving the durations fixed and testing ranges for the standard deviations and survivorships. During the fine—tuning step, minor modifications of the stage durations could be made. The life—cycle parameter estimates chosen by this technique predicted population levels that corresponded well with the observed population for the first generation (Fig. 2).

The second step of the two—step process involved using a fecundity estimate derived from other data (Schneider & Ferris 1987) and the duration, standard deviation, and survivorship values estimated in the first step to create a multiple generation predicted data set to compare to the multiple generation observed data set. The second generation appeared much earlier in the predicted data than was observed. The model was modified to include a pre—ovipositional female stage to allow for egg development before egg deposition. After this modification, the predicted and observed values over the whole data set were generally in agreement (Fig. 3).

The estimated life cycle parameters were then used to predict population values for a field data set. The predicted and observed data did not agree. There are many reasons this might occur. The host plant in the field was different than that used in the growth chamber. Different hosts have been shown to influence development, survival, and reproduction (Ferris & Hunt 1979, Ferris et al. 1982, 1984). There are many additional sources of mortality in the field setting: predators, parasites, fluctuations of soil temperature and moisture, and density dependent factors. The growth—chamber experiment had been designed to exclude all of these factors. Another source of variation in the field data is the initial age structure of the population. Although the numbers in each stage can be determined, distribution in the substages is subjective.

Summary

The technique presented here requires some previous knowledge and is computationally intensive. A reasonable estimate of the duration of the shortest stage is needed to ensure the sampling period is less than the shortest duration. If neither survival nor fecundity can be independently determined, the estimated values may be related. Overestimation of fecundity can be compensated by underestimating survival and vice versa. The technique works most efficiently when starting with a reasonably uniform cohort of a single stage, which is not always easy to obtain. If this is not the case, the initialization of the population age structure is largely subjective and can be a source of error.

Despite the potential shortcomings, the technique described in this paper successfully estimated stage durations, standard deviations, and survival using stage frequency data. Many previously published methods required information which was not obtainable for plant parasitic nematodes. This technique offers an alternative for the estimation of life cycle parameters.

References Cited

Ashford, J. R., K. L. Q. Read, G. G. Vickers. 1970. A system of stochastic models applicable to studies of animal population dynamics. *J. Anim. Ecol.* 39: 29–50.

Baskerville, G. L., & P. Emin. 1969. Rapid estimation of heat accumulation from maximum and minimum temperatures. *Ecology* 50: 514–517.

Bellows, T. S. Jr., & M. H. Birley. 1981. Estimating developmental and mortality rates and stage recruitment from insect stage–frequency data. *Res. Pop. Ecol.* 23: 232–244.

Bellows, T. S. Jr., M. Ortiz, J. C. Owens & E. W. Huddleston. 1982. A model for analyzing insect stage frequency data when mortality varies with time. *Res. Pop. Ecol.* 24: 142–156.

Birley, M. 1977. The estimation of insect density and instar survivorship functions from census data. *J. Anim. Ecol.* 46: 497–510.

Birley, M. 1979. The estimation and simulation of variable developmental period, with application to the mosquito *Aedes aegypti* (L.). *Res. Pop. Ecol.* 21: 68–80.

Candolle, A. P. de. 1855. Geographic Botanique Raisonee. Paris: Masson.

Curry, G. L., R. M. Feldman & P. J. H. Sharpe. 1978. Foundations of stochastic development. *J. Theor. Biol.* 74: 397–410.

Fargo, W. S. 1986. Estimation of life stage development times based on cohort data. *Southwest Entomol.* 11: 89–94.

Ferris, H. & W. A. Hunt. 1979. Quantitative aspects of the development of *Meloidogyne arenaria* larvae in grapevine varieties and rootstocks. *J. Nematol.* 11: 168–174.

Ferris, H., S. M. Schneider & M. C. Semenoff. 1984. Distributed egg production functions for *Meloidogyne arenaria* in grapevine varieties and consideration of the mechanistic relationship between plant and parasite. *J. Nematol.* 16: 178–183.

Ferris, H., S. M. Schneider, & M. C. Stuth. 1982. Probability of penetration and infection by root–knot nematode, *Meloidogyne arenaria* in grape cultivars. *Am. J. Enol. and Vitic.* 33: 31–35.

Logan, J. A., D. J. Wollkind, S. C. Hoyt & L. K. Tangoshi. 1976. An analytic model for description of temperature dependent rate phenomena in arthropods. *Environ. Ent.* 5: 1133–1140.

Manetsch, T. J. 1976. Time–varying distributed delays and their use in aggregative models of large systems. IEEE Trans. On Systems, Man Cybernetics, Vol. SMC 6: 547–553.

Manly, B. F. J. 1974a. Estimation of stage–specific survival rates and other parameters for insect populations developing through several stages. *Oecologia* 15: 277–285.

Manly, B. F. J. 1974b. A comparison of methods for the analysis of insect stage–frequency data. *Oecologia* 17: 335–348.

Manly, B. F. J. 1976. Extensions to Kiritani and Nakasuji's method for analyzing insect stage–frequency data. *Res. Pop. Ecol.* 17: 191–199.

Manly, B. F. J. 1977. A further note on Kiritani and Nakasuji's model for stage–frequency data including comments on the use of Tukey's jackknife technique for estimating variance. *Res. Pop. Ecol.* 18: 177–186.

Manly, B. F. J. 1985. Further improvements to a method for analyzing stage–frequency data. *Res. Pop. Ecol.* 27: 325–332.

Manly, B. F. J. 1987. A multiple regression method for analyzing stage–frequency data. *Res. Pop. Ecol.* 29: 119–127.

Mills, N. J. 1981a. The estimation of mean duration from stage frequency data. *Oecologia* 51: 206–211.

Mills, N. J. 1981b. The estimation of recruitment from stage frequency data. *Oecologia* 51: 212–216.

Richards, O. W., N. Waloff, J. P. Spradbery. 1960. The measurement of mortality in an insect population in which recruitment and mortality widely overlap. *Oikos.* 11: 306–310.

Schneider, S. M. & H. Ferris. 1986. Estimation of stage–specific developmental times and survivorship from stage frequency data. *Res. Pop. Ecol.* 28: 267–280.

Schneider, S. M. & H. Ferris. 1987. Stage–specific population development and fecundity of *Paratrichodorus minor*. *J. Nemat.* 19: 395–403.

Sharpe, P. J. H., G. L. Curry, D. W. DeMichael & C. L. Cole. 1977. Distribution model of organism development times. *J. Theor. Biol.* 66: 21–38.

Stinner, R. E., G. D. Butler, Jr., J. S. Bacheler & C. Tuttle. 1975. Simulation of temperature–dependent development in population dynamics models. *Can. Entomol.* 107: 1167–1174.

Stinner, R. E., A. P. Gutierrez & G. D. Butler, Jr. 1974. An algorithm for temperature–dependent growth rate simulations. *Can. Entomol.* 106: 519–524.

Wagner, T. L., H. Wu, P. J. H. Sharpe & R. N. Coulson. 1984a. Modeling distributions of insect development time: a literature review and application of the Weibull function. *Forum: Ann. Entomol. Soc. Amer.* 77: 475–487.

Wagner, T. L., H. Wu, P. J. H. Sharpe, R. M. Schoolfield & R. N. Coulson. 1984b. Modeling insect development rates: a literature review and application of biophysical model. *Forum: Ann. Entomol. Soc. Amer.* 77: 208–225.

Appendix A

Sample counts for developmental stages, corrected for stage–specific extraction efficiencies. Counts are the means of ten replications.

DD_{10}	J1	J2	J3	J4	Adult	Total Population
0.0	105.677	3.70	0.00	1.0471	8.571	119.0
9.4	81.223	27.16	0.00	0.0000	2.857	111.24
17.0	20.961	100.00	4.12	0.0000	20.000	145.08
20.8	34.934	79.56	11.45	0.0000	6.349	132.30
27.8	20.961	120.99	16.49	0.0000	11.429	169.87
31.9	1.747	103.70	18.56	0.0000	1.429	125.44
42.3	0.873	114.81	16.49	0.0000	4.286	136.47
48.9	2.620	75.31	8.25	0.0000	1.429	87.60
59.0	10.674	49.38	41.24	0.0000	1.587	102.88
66.7	6.987	32.10	152.58	1.0471	5.714	198.42
75.3	6.116	51.85	134.02	1.0471	2.857	195.89
83.0	6.987	24.60	92.78	2.0942	5.714	132.27
92.0	11.354	39.51	109.28	17.8010	1.429	179.37
99.3	13.100	58.02	78.35	32.4607	21.429	203.36
109.3	29.694	66.67	53.61	50.2618	28.571	228.80
114.2	20.087	60.49	20.62	38.7435	74.286	214.23
129.2	23.581	62.96	16.49	18.8482	81.429	203.32
133.7	15.721	98.77	32.99	11.5183	78.571	237.57
143.0	35.905	115.23	43.53	1.1635	120.635	316.46
151.4	40.757	100.14	20.62	3.4904	100.000	265.00
159.0	35.905	76.82	73.31	6.9808	79.365	272.38
169.4	6.114	75.31	113.40	21.9895	92.857	309.67
176.0	11.354	46.91	84.54	4.1885	61.429	208.42
181.6	30.568	66.67	101.03	8.3770	84.286	209.93
193.4	33.964	54.87	41.24	11.6347	20.635	162.34
197.6	12.009	50.93	128.87	15.7068	55.357	262.86
209.4	12.227	82.72	82.47	13.6126	48.571	239.60
226.4	82.969	97.53	105.15	17.8010	51.429	354.88
231.3	136.827	138.55	103.09	31.4136	85.714	495.59
244.2	34.934	160.49	111.34	23.0366	71.429	401.23
249.7	70.742	285.19	183.51	32.4607	104.286	676.18
259.4	134.498	261.73	105.15	15.7068	67.143	584.23
267.1	114.508	289.44	151.20	13.9616	98.413	667.52
276.1	177.293	251.85	76.29	17.8010	51.429	574.66
283.0	248.908	515.43	198.45	31.4136	125.000	1119.21
292.1	200.873	563.27	309.28	13.0890	55.357	1141.87
299.8	286.269	827.16	249.71	17.4520	106.349	1486.94
309.5	418.341	986.42	410.31	18.8482	55.714	1889.63
216.4	289.956	707.41	292.78	26.1780	65.714	1382.04
327.2	236.463	796.30	352.58	13.6126	74.286	1523.23
333.1	634.934	1382.72	527.84	23.0366	190.000	2758.52
343.9	431.441	918.52	484.54	18.8482	52.857	1906.20
350.8	682.096	1016.05	465.98	59.7906	108.571	2312.49
360.9	882.096	1425.93	876.29	28.2723	154.286	3366.87
366.8	474.236	856.79	453.61	20.9424	85.714	1891.29
377.6	957.787	1580.25	1134.02	39.5579	146.032	3857.64

SECTION II

MODELLING OF POPULATION DYNAMICS

A Review of Methods for Key Factor Analysis

Bryan F.J. Manly[1]

ABSTRACT Key factor analysis is a term that has become standard for methods for analyzing data on the total numbers entering stages for a series of successive generations of a stage structured population. This paper is a review of these methods, concentrating on ones that have developed from Varley & Gradwell's graphical method based on k values. The winter moth in Wytham Wood is used as an example.

For populations with distinct generations it is often possible to count or estimate the total number of individuals entering different development stages for a series of generations. This is done in order to study the variation in the survival rates in different stages from generation to generation, and hence determine which sources of variation are particularly important for population dynamics. Studies like this have become known as 'key factor analyses' since the object is to find the key factors (i.e., stages) that determine the variation in population numbers.

An example is shown in Table 1. Here the population is the winter moth *Operophtera brumata* in Wytham Wood and estimates of the number of moths at seven distinct points in the life cycle are available for each of the years 1950–66. This is the example that will be used throughout this paper. It has particular interest because of some recent controversy concerning it.

It was Morris (1957) who first pointed out that in considering the analysis of this type of data it is variation in survival rates that is important for population dynamics rather than mean survival rates since if stage–specific survival rates are constant then population numbers become completely predictable. Hence a life table for a single generation reveals little about population dynamics. it cannot be used to indicate the relative importance of the different mortality factors because some of the highest mortalities may be relatively constant while some of the low ones may vary greatly.

A major reason for carrying out a key factor analysis is often the identification of the stages for which the survival rates are so variable as to make it necessary to study them further in order to develop a realistic model for the population. The stage–specific survival rates that vary little can be treated as being constant, at least in the initial stages of model building. The further study of stages with variable survival rates may then involve attempting to relate this variation to environmental variables such as climate, numbers of predators, or the availability of food. In this way it can be hoped that a clearer understanding of the major causes of population changes will emerge.

---

[1]Department of Mathematics and Statistics, University of Otago, P.O. Box 56, Dunedin, New Zealand. On leave in the Department of Statistics, University of Wyoming, Laramie, Wyoming, 82071, during 1988.

Table 1. Life tables for the winter moth in Wytham Wood, Berkshire, as determined from Table F and Fig. 7.4 of Varley et al. (1973).

Year	(1)	(2)	Number of survivors to stage (3)	(4)	(5)	(6)	(7)
1950	4365.0	112.2	87.1	79.4	70.8	14.5	7.41
51	417.0	117.5	114.8	109.6	102.3	17.4	13.8
52	758.6	55.0	54.6	47.5	42.4	7.50	7.03
53	389.0	18.2	17.8	17.0	15.5	7.08	4.90
54	275.4	158.5	157.0	146.6	127.6	23.8	20.2
55	1122.0	77.6	77.1	71.9	65.6	14.7	11.9
56	645.7	95.5	89.1	87.0	83.2	28.2	14.8
57	831.8	275.4	263.0	257.0	229.1	37.2	23.4
58	1288.0	190.5	162.2	154.9	141.3	21.4	14.8
59	812.8	57.5	45.7	41.7	39.8	8.71	6.17
60	346.7	21.4	20.4	17.8	16.6	3.16	1.12
61	61.7	7.59	7.59	6.61	6.03	3.63	3.02
62	166.0	13.5	12.9	11.2	10.5	6.03	5.25
63	288.4	40.7	40.7	37.2	36.3	14.5	11.0
64	602.6	131.8	130.3	127.4	124.5	22.6	16.4
65	891.3	269.2	251.2	245.5	239.9	32.4	24.0
66	1349.0	51.3	46.8	46.2	44.2	6.10	2.85
67	154.9	9.77	9.55	8.91	8.91	2.82	2.82
68	154.9	10.0	10.0	9.77	9.12	3.02	3.02

Stages are: (1) Egg output from the adults in the previous year, which is taken by Varley et al. to be 56.25 eggs per adult; (2) fully grown larvae, (3) after parasitism by *Cyzensis albicans*, (4) after parasitism by non–specific Diptera and Hymenoptera, (5) after parasitism by *Plistophora operophterae*; (6) after pupal predation, (7) adults, after parasitism by *Cratichneumon culex*.

The Varley & Gradwell Graphical Method for Key Factor Analysis

Consider one generation of a population, for which the numbers entering development stages 1 to q are N_1, N_2, ..., N_q. The stage–specific survival rates are then $w_1 = N_2/N_1$, $w_2 = N_3/N_2$, ..., $w_{q-1} = N_q/N_{q-1}$, and the total survival to the adult stage is

$$N_q/N_1 = w_1.w_2...w_{q-1}.$$

Taking logarithms (base 10 is usually used) then gives

$$\log(N_q/N_1) = \log(w_1) + \log(w_2) + ... + \log(w_{q-1}),$$

or

$$K = k_1 + k_2 + ... + k_{q-1},$$

where $K = -\log(N_q/N_1)$ and $k_j = -\log(w_j)$. The changes of sign are simply made to avoid negative values. Varley & Gradwell (1960) suggested plotting K and the k values against the same time axis for a series of generations. The 'key factor' is then the stage for which variation in the k value most closely follows variation in K, as determined by a visual inspection.

The K and k_1 to k_6 values for the winter moth data are shown in Table 2, and plotted on Fig. 1. The key factor seems to be k_1 (over–winter loss) since it is the plot of k_1 that is most similar to the plot of K. The k_5 value (loss of pupae in the soil due to predation) is also quite variable.

Table 2. Calculated K and k values for the winter moth data using base 10 logarithms.

Year	K	k_1	k_2	k_3	k_4	k_5	k_6
1950	2.77	1.59	.11	.04	.05	.69	.29
51	1.48	.55	.01	.02	.03	.77	.10
52	2.03	1.14	.00	.06	.05	.75	.03
53	1.90	1.33	.00	.02	.04	.34	.16
54	1.13	.24	.00	.03	.06	.73	.07
55	1.97	1.16	.00	.03	.04	.65	.09
56	1.64	.83	.03	.01	.02	.47	.28
57	1.55	.48	.02	.01	.05	.79	.20
58	1.94	.83	.07	.02	.04	.82	.16
59	2.12	1.15	.10	.04	.02	.66	.15
60	2.49	1.21	.02	.06	.03	.72	.45
61	1.31	.91	.00	.06	.04	.22	.08
62	1.50	1.09	.02	.06	.03	.24	.06
63	1.42	.85	.00	.04	.01	.40	.12
64	1.57	.66	.00	.00	.01	.74	.14
65	1.57	.52	.03	.00	.01	.87	.13
66	2.68	1.42	.04	.00	.02	.86	.33
67	1.74	1.20	.00	.03	.00	.50	.00
68	1.71	1.19	.00	.01	.03	.48	.00
Mean	1.817	.966	.026	.030	.030	.616	.150
Std. Dev.	.449	.357	.033	.019	.016	.205	.118

Extensions of the Varley & Gradwell Approach

The graphical model of Varley & Gradwell often identifies key factors but unclear cases do arise and this has led a number of authors to propose more objective methods for evaluating the importance of different k values. The correlations between the k values and K have been calculated with the highest value taken to indicate the key factor. Regression coefficients for K on k values have been considered, again with high values indicating key factors. However, Podoler & Rogers (1975) have pointed out that the best approach along these lines is to look at regressions of k values against K and use the regression slopes to compare the importance of different k values in determining K.

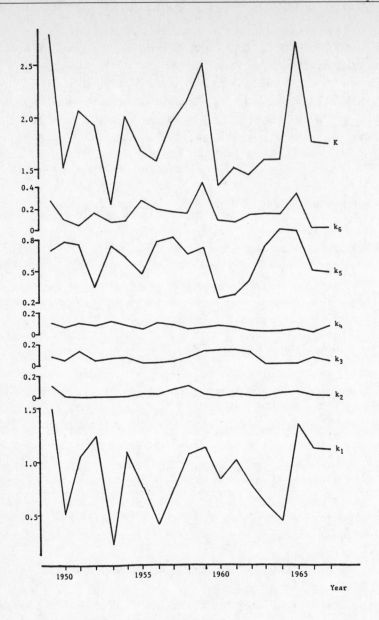

Fig. 1. K and k value for the winter moth data.

The regression slope for k_j regressed on K is given by the usual equation

$$b_j = \sum_{i=1}^{G} (k_{ij} - \overline{k}_j)(K_i - \overline{K}) / \sum_{i=1}^{G} (K_i - \overline{K})^2 ,$$

where k_{ij} is the value of k for the survival in stage j in generation i, $\overline{k}_j$ is the mean k value for survival in stage j, K_i is the K value in generation i, $\overline{K}$ is the mean of K for all generations, and there are G generations of data. It follows immediately from this equation that the b_j values for stages 1 to q−1 add to 1. Thus the b_j values have the immediate interpretation of indicating the fraction of the total variation of K, as measured by $\Sigma(K_i-\overline{K})^2$, that is accounted for by k_j.

Although the idea of using the regression coefficients of k values on K to distinguish key factors seems to have been first put forward by Podoler & Rogers, what amounts to the same thing was proposed earlier by Smith (1973). He suggested measuring the relative importance of k_j by the covariance of k_j with K, which is just b_j multiplied by a constant. Smith also proposed a further analysis after the key factor has been found with the aim being to ascertain the relative importance of the remaining k values. Thus suppose that k_p is the key factor. The residual killing power in stages other than p is then $K'_i = K_i - k_{ip}$ in the ith generation, and the contribution of stage j to this residual can be measured by the covariance of K' and k_j or, equivalently, by the regression coefficient of k_j on K'. This allows the second most important k value to be determined. Continuing in this way, removing the k values one at a time, allows them to be ranked in terms of importance.

For the winter moth data the b_j values are stated by Podoler & Rogers to be as follows: $b_1 = 0.743$, $b_2 = 0.038$, $b_3 = 0.012$, $b_4 = 0.004$, $b_5 = 0.053$, and $b_6 = 0.150$. The stage 1 survival is confirmed as the clear key factor. Removing k_1 values from K and calculating regression coefficients for the remaining k values on their total, along the lines suggested by Smith, indicates that the second most important factor is k_5. Removing this from the total and calculating regression coefficients indicates that the third most important factor is k_6, etc. Table 3 summarizes the results of these calculations.

Table 3. Key factor determination for the winter moth data by the method of Podoler, Rogers and Smith.

Step	factor	b_1	b_2	b_3	b_4	b_5	b_6
				Key			
1	k_1	0.62	0.05	−0.03	0.00	0.18	0.19
2	k_5	−	0.05	0.00	0.00	0.66	0.27
3	k_6	−	0.13	0.07	0.03	−	0.77
4	k_3	−	0.21	0.63	0.16	−	−
5	k_2	−	0.80	−	0.20	−	−

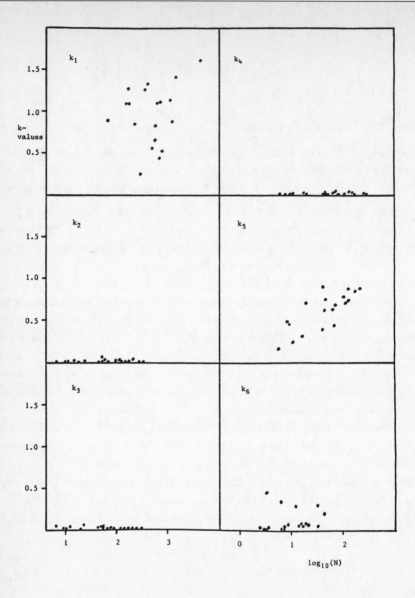

Fig. 2. The relationship between k values and the population densities on which they act.

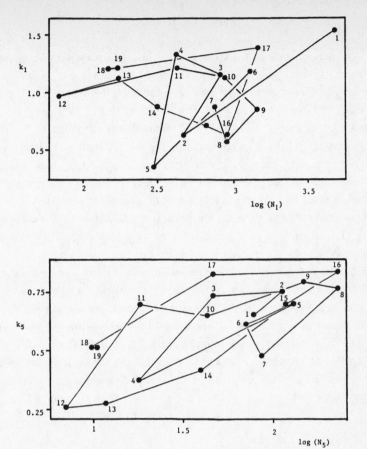

Fig. 3. Plots of k_1 and k_5 indicating an element of spiralness with the winter moth data.

Detecting Density–Dependent k Values

Following the establishment of key factors the usual practice has been to graph k values against the logarithms of the population densities on which they act in order to see if there is any evidence for density–dependent survival. A positive relationship between k_j and log(density) is what indicates this. Plots for the winter moth data are shown in Fig. 2. There does seem to be density dependence for k_5 (pupal predation), with a fitted regression line $k_5 = 0.10 + 0.31$ log(density). Unfortunately, testing regression lines like this for statistical 'significance' is not straightforward since sampling errors in k values may well induce regression relationships of precisely this type. This matter is considered further below. Here we can merely note that ordinary regression methods for testing significance are not generally appropriate.

A further possibility that is often considered is that there is delayed density–dependent mortality arising, for example, from a parasite acting as proposed by Nicholson (1933) and Nicholson & Bailey (1935). In this case the relationship between k values and density is positive when the host and parasite population are both increasing, and negative when the host population is increasing and the parasite population decreasing, or vice versa. This may show up in an anti–clockwise spiral graph when k values for successive generations are joined (Varley et al., 1973, p. 64, 124). Patterns of this type are suggested for k_1 and k_5 with the winter moth data (Fig. 3) so it seems that this may be an example where this occurs. The question of how one measures 'spiralness', and determines whether an observed amount is significant is considered below. Of course, if estimates of parasite densities are available then it will be more direct to plot k values against these.

The Manly (1977) Model for Key Factor Analysis

As before, let $k_j = -\log(N_{j+1}/N_j)$ using base 10 logarithms, where N_j is the number entering stage j of the life cycle in one generation. Also, let $R_j = \log(N_j)$, so that $k_j = R_j - R_{j+1}$. Furthermore, assume that the k_j values are dependent on the number entering stage j by a regression relationship

$$k_j = \tau_j + \delta_j R_j + \epsilon_j, \tag{1}$$

where τ_j and δ_j are constants, and ϵ_j is a random disturbance with mean zero and variance $Var(\epsilon_j)$. This model then permits mortality rates to be density–dependent ($\delta_j > 0$), density independent ($\delta_j = 0$), or inverse density–dependent ($\delta_j < 0$).

It follows from equation (1) that the expected value of R_q is

$$E(R_q) = \Theta_0 E(R_1) - \sum_{j=1}^{q-1} \Theta_j \tau_j \tag{2}$$

and the variance is

$$\mathrm{Var}(R_q) = \Theta_0{}^2\mathrm{Var}(R_1) + \sum_{k=1}^{q-1} \Theta_j{}^2\mathrm{Var}(\epsilon_j), \tag{3}$$

where

$$\Theta_j = \begin{bmatrix} (1-\delta_{j+2})(1-\delta_{j+2})...(1-\delta_{q-1}), & j=0,1,...,q-2, \\ \\ 1, & j=q-1. \end{bmatrix}$$

These results are given by Manly (1977) with $\alpha_j = -\tau_j$ and $\beta_j = \delta_j$.

For some sets of data the results for successive generations are directly related in the sense that the number entering stage 1 in one generation can be calculated from the number in the last stage in the previous generation by multiplying by a potential fecundity rate. This is the situation with the winter moth data where the fecundity rate is taken as 56.25. In other cases the relationship between the numbers in successive generations is more obscure since the individuals entering stage 1 in each generation may be produced largely or partly by immigrant adults, or there may be contributions from several previous generations. In these cases it may be realistic to regard the number entering stage 1 in a generation as a random variable that is independent of the number in the last stage in the previous generation. This distinction between data with related and unrelated generations is important as far as the use of equations (1) to (3) is concerned and the two situations must therefore be considered separately.

With unrelated generations the right hand side of equation (2) gives the expected value of R_q as a function of the Θ and τ values, and of the mean of $R_1 = \log(N_1)$. Similarly, the right hand side of equation (3) shows how the variance of R_q depends on the variance of R_1, the variances of ϵ_j values, and the Θ parameters. Defining

$$A_u = \begin{bmatrix} \Theta_0{}^2\mathrm{Var}(R_1), & u = 0, \\ \\ \Theta_u{}^2\mathrm{Var}(\epsilon_u), & u = 1,2,...,q-1, \end{bmatrix} \tag{4}$$

it can be seen that A_0 is the contribution of the variance of R_1, and A_j $(j>0)$ is the contribution of the variance of ϵ_j, to the variance of R_q. Hence the A_j values indicate the relative importance of random variation in different stages in determining the amount of variation in stage q. They are conveniently expressed as percentages of the total sum ΣA_j.

If most of the variation in the survival in a stage is a density–dependent response to the number entering the stage then the A value for the stage may be quite small. However, the importance of the stage will show up if a calculation is made to see how $\text{Var}(R_q)$ is affected if the survival in the stage were made constant. This amounts to fixing the corresponding k value at its mean or, in other words, making the δ and ϵ values equal zero for the stage. From equation (3) it can be shown that if this is done for stage u then the value of $\text{Var}(R_q)$ is expected to change to

$$B_u = \sum_{j=0}^{u-1} A_j/(1-\delta_j)^2 + \sum_{j=u+1}^{q-1} A_j \ . \tag{5}$$

Clearly, if B_u is very different from the observed variance of R_q then variation in the survival in stage u seems to be important for population dynamics. It is convenient to express B_u as a percentage of the sum ΣA_j.

Consider now the situation where results are related from one generation to the next so that the number entering stage 1 in generation i is determined by multiplying the number entering the last stage in generation i–1 by a constant C. Thus $N_{i1} = CN_{i-1q}$, so that

$$R_{i1} = \log(N_{i1}) = \log(N_{i-1q}) + \log(C) = R_{i-1q} + \log(C) \ .$$

From equation (2) this means that the relationship between the mean values of R_q in generations i–1 and i is

$$E(R_{iq}) = \Theta_0\{E(R_{i-1q}) + \log(C)\} - \sum_{j=1}^{q-1} \Theta_j\tau_j \ .$$

This shows that the stable value of the mean, with $E(R_{iq}) = E(R_{i-1q})$, occurs when this mean equals

$$E(R_q) = \{\Theta_0\log(C) - \sum_{j=1}^{q-1} \Theta_j\tau_j\}/\{1 - \Theta_0\} \ . \tag{6}$$

providing that Θ_0 is not equal to 1.

As far as variation is concerned, equation (3) shows that the change in the variation of R_q from one generation to the next is given by

$$\text{Var}(R_{iq}) = \Theta_0^2\text{Var}(R_{i-1q}) + \sum_{j=1}^{q-1} A_j \ , \tag{7}$$

where A_j is given by equation (4). Here $\text{Var}(R_{i1}) = \text{Var}(R_{i-1q})$ since R_{i1} and R_{i-1q} only differ by the constant $\log(C)$. It follows that if the variance is stable so that $\text{Var}(R_{iq}) = \text{Var}(R_{i-1q}) = \text{Var}(R_q)$ then

$$\text{Var}(R_q)\{1 - \Theta_0{}^2\} = \sum_{j=1}^{q-1} A_j , \tag{8}$$

which is only possible if $\Theta_0{}^2 < 1$ since both sides of the equation must by positive. If it can occur, the stable variance is

$$\text{Var}(R_q) = \sum_{j=1}^{q-1} A_j/(1 - \Theta_0{}^2) . \tag{9}$$

If $\Theta_0{}^2 \geq 1$ then the population will be undergoing an 'explosion' and the whole concept of key factor analysis becomes meaningless. If $\Theta_0{}^2 < 1$ then equation (8) shows that the relative importance of random variation in stage u (in comparison to the other stages) in determining the long term variance of R_q depends on the value of A_u as a percentage of ΣA_j. On the other hand, if the survival in stage u has a strong influence on the variation in the final stage because it reduces the variation produced in earlier stages then this should show up in ΣA_j changing substantially when the survival rate in the stage is made constant by changing δ_u and $\text{Var}(\epsilon_u)$ to zero. The change to ΣA_j when this is done is given by equation (5) with $A_0 = 0$. It is convenient to express this change as a percentage of ΣA_j.

In order to use the equations to detect key factors it is necessary to estimate the δ and $\text{Var}(\epsilon)$ values. This can be done for each stage using standard linear regression methods with the minor modification that the estimate of $\text{Var}(\epsilon_j)$ should be the residual sum of squares divided by $G-1$ instead of $G-2$ as would usually be done (Manly, 1979).

Estimates from the above equations for the winter moth data are shown in Table 4. The sum of the A_j values, 0.084, is the estimated contribution to the variance of R_q from the survival in stages 1 to 6. The major contribution is A_1 at 0.053 (63.2% of the total), which suggests that k_1 is the key factor. The other A_j values suggest that k_2, k_3, and k_4 are not at all important while k_5 and k_6 make a moderate contribution. The B_1 value confirms the importance of k_1. If the survival in this stage became constant then ΣA_j would apparently become about 0.031 (36.8% of the present value). Making k_2 to k_4 constant would have very little effect while making k_6 constant would make a minor change. However k_5 seems to be having a strong stabilizing effect since making this constant would increase ΣA_j by about 51.8%.

Table 4. Estimated population parameters for the winter moth population

j	τ_j	δ_j	Θ_j	A_j	A_j %	B_j	B_j %
0	—	—	0.55	—	—	—	—
1	0.53	0.16	0.66	0.053	63.2	0.031	36.8
2	−0.01	0.02	0.67	0.000	0.5	0.086	102.2
3	0.06	−0.02	0.66	0.000	0.2	0.082	97.5
4	0.02	0.01	0.66	0.000	0.1	0.085	100.7
5	0.10	0.31	0.97	0.017	19.8	0.128	151.8
6	0.12	0.03	1.00	0.014	16.2	0.073	89.5

Total 0.084

Since the absolute value of Θ_0 is less than 1, the population seems stable and equations (6) and (9) can be used to predict the long term mean and variance of R_7. From equation (6) it is calculated that $E(R_7) = 0.797$, which corresponds to the mean number entering the stage being approximately $10^{0.797} = 6.3$. This seems realistic considering the observed numbers shown in Table 1. From equation (9) it is calculated that the stable variance of R_7 is 0.1214. This also seems a realistic value since the observed variance of R_7 for the data is 0.1405.

The Simulation of Population Data

Simulation provides an obvious means for determining the properties of different methods for analyzing data. The model described in the previous section, with k values being linearly related to population density, provides a simple approach for doing this. More complicated models that take into account delayed density—dependent mortality, the action of predators and parasites, etc., are certainly possible. However, these models are necessarily very specific to a population in one place in one period and, as such, do not provide a general method for generating key factor data.

Simulations depend upon whether generations are related or not. With related generations the number entering stage 1 in generation i is equal to the number entering stage q in generation i—1 multiplied by a constant C. A set of population frequencies can therefore be determined by: choosing an initial number N_1 entering stage 1 in generation 1; generating k_1, k_2, ..., k_{q-1} using equation (1) with random ϵ_j values, and hence determining R_2, R_3, ..., R_q; finding R_1 in generation 2 as $R_q + \log(C)$; generating k values for generation 2 using equation (1), and hence determining R values; etc. In this way simulated data can be produced for as many generations as required. In the simulations described below, the ϵ_j values were made normally distributed. Since equation (1) allows negative k values these were replaced by zero whenever they occurred.

When the data from successive generations are independent, the same scheme can be used to determine population frequencies except that the $R_1 = \log(N_1)$ value for each generation is a random value from a normal distribution with an appropriate mean and variance. However, in certain cases it may be more appropriate to use the observed R_1 values as the starting points for generations and hence make the simulated data conditional on these.

Most real data sets consist of population frequencies with superimposed sampling errors. These errors may be attached to the population frequencies, to estimated survival rates, or to a combination of both. If the nature of these errors is understood, and estimates of variances are available, then it will be a simple matter to include the errors in the simulation.

In analyzing real data there are two types of simulation that will produce useful results. A null model can be used for which δ_j values are zero, τ_j values are equal to observed mean k_j values, and the variances of ϵ_j values are set equal to the observed variances of k_j values. The distribution of estimates produced by simulations with this model will indicate whether the estimated δ_j values for the real data are such as could easily have arisen by chance. In other words, a test for statistically significant density—dependence becomes possible. It is also possible to check whether the B_j values for the real data provide any evidence of 'hidden' key factors. The key factors indicated by A_j values may or may not seem the same for the real data and the null model.

The alternative to using the null model will usually be the model for which the τ_j and δ_j values and the variance of ϵ_j are set equal to the estimates from the real data. The distribution of estimates produced from simulations with this model should give a good guide to the accuracy of the real data estimates in terms of biases and standard errors.

Both the null model and the estimated model have been simulated for the winter moth population. In both cases 500 independent realizations of 19 generations were produced starting with the observed number in stage 1, generation 1 of 4365 individuals.

The most important results from the null model simulations are shown in part (a) of Table 5. These are the percentages of the 500 generated populations for which an estimated parameter was larger than the estimate found with the real data. We see, for example, that 92.6% of the estimated Θ_0 values were larger that the real data estimate of 0.55. Since small values indicate population stability, this indicates that the stability level estimated for the real data is not very likely to occur with data from a model that is not stable. Moreover, the simulated values of B_5 as a percentage of ΣA_j are all less than the real data value, and only 1% of the simulated τ_5 values are as large as the real value. The data value of A_1 as a percentage of ΣA_j is exceeded by 62.4% of the simulated values, but this only reflects the fact that k_1 is the key factor for the null model.

Table 5. Simulation results for the winter moth population.

(a) Percentages of simulated sets of data producing estimates greater than the estimates for the observed data (500 simulations with the null model).

j	τ_j	δ_j	Θ_j	A_j	A_j %	B_j	B_j %
0	—	—	92.6	—	—	—	—
1	59.0	36.6	98.8	91.0	62.4	93.2	37.6
2	94.2	2.6	98.6	75.6	23.8	96.0	3.0
3	3.6	99.2	98.8	96.6	72.6	96.8	99.0
4	81.6	14.8	98.8	92.0	50.8	96.2	15.2
5	94.8	1.0	74.6	96.6	72.0	41.4	0.0
6	71.0	25.4	—	20.6	2.4	97.0	78.2

(b) Results from 500 simulated sets of data when the simulation model has parameter values equal to the estimates obtained from the real data: biases (simulation means − values used in generating data).

j	τ_j	δ_j	Θ_j	A_j	A_j %	B_j	B_j %
0	—	—	−0.05	—	—	—	—
1	−0.20	0.07	−0.01	0.000	−0.1	−0.003	−0.1
2	0.01	−0.01	−0.01	0.000	−0.1	−0.004	−0.4
3	0.00	0.00	0.01	0.000	0.0	−0.003	0.2
4	0.00	0.00	−0.01	0.000	0.0	−0.003	0.0
5	−0.02	0.01	−0.01	−0.001	1.1	−0.003	4.4
6	0.00	0.00	—	−0.002	−0.9	−0.001	3.9

(c) Results from 500 simulated sets of data when the simulation model has parameter values equal to the estimates obtained from the real data: standard deviations.

j	τ_j	δ_j	Θ_j	A_j	A_j %	B_j	B_j %
0	—	—	0.15	—	—	—	—
1	0.57	0.21	0.09	0.023	10.9	0.007	10.9
2	0.02	0.01	0.09	0.000	0.2	0.025	1.7
3	0.01	0.01	0.09	0.000	0.1	0.024	1.0
4	0.01	0.01	0.09	0.000	0.1	0.025	1.1
5	0.12	0.07	0.08	0.006	7.6	0.042	27.5
6	0.09	0.08	—	0.004	6.6	0.024	15.7

The simulations with τ_j, δ_j and $\text{Var}(\epsilon_j)$ values equal to the estimates from the real data produced the estimates of biases and standard errors shown in parts (b) and (c) of Table 5. These indicate that the A_j and B_j are estimated with moderate accuracy, although they may have some minor biases. However, the estimates of τ_1, and δ_1 are strongly biased. Also, Θ_0 is biased in the direction of indicating more stability than really exists.

A further idea of the reliability of the key factor analysis on the real data is given by knowing that for 97.2% of the simulated sets of data A_1 was the largest A_j value, A_5 was largest for 1.8% of sets, and A_6 was largest for the remaining 1.0% of sets. Also, for 97.6% of the simulated sets of data, B_5 was the largest B_j value, for 1.4% B_6 was largest, and for 0.6% B_4 was largest. It seems fair enough to regard this as an indication of the extent to which chance may affect the choice of key factors.

The performance of Podoler & Rogers's (1975) and Smith's (1973) regression method for choosing key factors is indicated by the results in Table 6. It may be recalled that this method it involves regressing k_j values on K values and choosing the key factor as the stage with the largest regression coefficient. The corresponding k_j value is then removed from K and the process repeated to find the second most important factor, etc. From Table 6 it can be seen that k_1 was correctly chosen as the key factor for 98% of simulated sets of data, the remaining 2.0% suggesting k_5. The correct choice for the order of the remaining factors (in the sense of agreeing with what is given by the simulation parameters) has also been obtained most of the time.

Table 6. Percentages of times that different stages are chosen as the first, second, ..., fifth key factor using the Podoler–Rogers–Smith regression method on simulated winter moth data.

Key factor	1	2	3	4	5	6
1	98.0*	0.0	0.0	0.0	2.0	0.0
2	2.0	0.0	0.0	0.0	97.6*	0.4
3	0.0	0.0	0.0	0.0	0.4	99.6*
4	0.0	6.0	94.0*	0.0	0.0	0.0
5	0.0	91.6*	6.0	2.4	0.0	0.0

*
Key factor for the model used in the simulation

To sum up, it seems from the estimates and simulations that k_1 is the key factor, k_5 is density–dependent, and the apparent stability of the population is fairly unlikely to have occurred by chance. Nevertheless, the estimation of Θ_0 seems to be biased in the direction of indicating more stability than really exists, and the estimation of the density–dependent effect of k_1 seems to be biased in the direction of indicating more density–dependence than really exists.

These conclusions are essentially in agreement with those of Varley et al. (1973). However Den Boer (1986) brought to light some aspects of the winter moth data that call into question whether the conclusions just reached can really be justified. In particular, he argued that pupal predation (k_5) increases population variation rather than reducing it. His work has led to some controversy (Den Boer, 1988; Latto & Hassell, 1988; Poethke & Kirchberg, 1988).

The most compelling piece of evidence that Den Boer (1986) presented was the values shown in Table 7. Here $LR = \log_{10}$(highest density/lowest density) is the logarithmic range over the 19 generations of data and the variances are of $R_2 = \log$(larvae numbers) and $R_7 = \log$(adult numbers), the LR values and variances simply being alternative measures of variation. The values for k_5 fixed are obtained by recalculating population numbers with all k_5 fixed at the observed mean with other k values as for the original data, starting with the observed number in stage 1, generation 1. Since fixing k_5 has reduced all the measures of variation it is difficult to argue that this factor has a regulating influence!

Table 7. The effects of fixing k_5 at the mean value for the winter moth data.

	LR(larvae)	Var(R_2)	LR(adults)	Var(R_7)
Original data	1.56	0.260	1.34	0.140
k_5 fixed	1.39	0.176	1.03	0.128
Change	−0.17	−0.084	−0.31	−0.012

There are several points that need to be raised when attempting to reconcile Den Boer's results with the conclusions that have been reached from the analysis given earlier. To begin with, it is important to remember that the observed data are just one realization of a process that must be considerably affected by random events. It is entirely plausible that fixing k_5 will increase variation on average, even although it does not appear to do this for the available data. Clearly, to determine average behaviour more realizations are needed, but these are not available naturally. One apparent solution to this problem involves generating data by computer simulations. Unfortunately, if this is done then the results obtained will depend entirely on the assumptions made.

For example, if the winter moth population is simulated with k_5 set equal to its observed mean and other k values density dependent then according to equation (1) the long term values of Var(R_2) and Var(R_7) must be higher than they would be with k_5 density–dependent. Thus using estimated population parameters, without fixing k_5, equation (9) shows that the stable value for $\text{Var}(R_7)$ is 0.1214, while the stable variance of R_2 is

$$\begin{aligned}
\text{Var}(R_2 &= \text{Var}\{(1-\delta_1)R_1+\epsilon_1\} \\
&= (1-\delta_1)^2\text{Var}(R_1) + \text{Var}(\epsilon_1) \\
&= (1-\delta_1)^2\text{Var}(R_7) + \text{Var}(\epsilon_1) \\
&= (1-0.161)^2 0.1214 + 0.1297 \\
&= 0.2151.
\end{aligned}$$

The observed values $\text{Var}(R_7) = 0.1405$ and $\text{Var}(R_2) = 0.2519$ are in reasonable agreement with these calculated values. Setting $\delta_5 = \text{Var}(\epsilon_5) = 0$ in equation (9) shows that making k_5 constant would result in the long term variances of R_7 and R_2 being increased to 0.3625 and 0.3847, respectively. However, the population would still be stable since the new value of Θ_0 would be 0.805, which is less than 1. Hence it seems that the density–dependence of k_5 is not essential for regulating the population. Obviously if we believe this model then there is no need to carry out simulations to decide what happens if k_5 is fixed.

On the other hand, a population without any form of density–dependent mortality, that is not controlled by outside factors such as weather, must build up potential variation from generation to generation and cannot be stable. Again, it is not necessary to carry out simulations to determine this. Looked at from this point of view, the simulations of Latto & Hassell (1988), Poethke & Kirchberg (1988) and Den Boer (1988) do not tell us anything that is not obvious.

However, this does not mean that simulations are completely worthless in assessing Den Boer's claim that pupal predation does not have a stabilizing effect. What can be done is to simulate populations using estimated parameters, and see what is the effect of fixing k_5 values at their mean in exactly the same way as has been done by Den Boer for the real data. This will not give the same results as simulating data with k_5 fixed for two reasons. First, with the proposed simulations the mean of k_5 will vary with each set of data rather than being fixed at the same value for all sets. Second, the other k values are not allowed to adjust in a density–dependent way to the changes in density brought about by fixing k_5. Thus the precise effects of Den Boer's modification to data is difficult to predict, although it seems that it should increase variation on average.

There is also another reason why simulation is valuable. The model makes it possible to predict the long term variances for R_2 and R_7 but these are not the same as the variances that can be expected for the 19 generations actually observed. At the start of the first generation the number in stage 1 was unusually large so that the observed data can be expected to show more variation than 19 generations in general. Hence in assessing the observed variation against a model it is important to see what data the model generates when it starts from the same density in stage 1.

Table 8 provides a summary of the results obtained with 500 simulated sets of data. The first row shows the mean increases in LR values and variances found by fixing k_5. The second and third rows show the minimum and maximum changes seen, and the fourth row shows the percentages of sets of data that had smaller changes than those observed for the real data (taking into account the sign).

There was a good deal of variation in the results for different sets of data. In many cases fixing k_5 reduced variation. Rather surprisingly, fixing k_5 resulted in a mean decrease of 0.08 in the LR value for larvae, which is certainly not what is expected from the model. However, for adults the LR value and the variance are both increased substantially, which does agree with expectations. For variances and LR (adults) the changes for the real data are exceeded by most of the simulated values and are significantly low at about the 5% level on a one sided test.

Table 8. The effects of fixing k_5 values at mean values for 500 simulated sets of data.

	LR(larvae)	Var(R_2)	LR(adults)	Var(R_7)
Mean increase	−0.08	0.010	0.39	0.117
Minimum change	−0.95	−0.192	0.46	−0.063
Maximum change	1.17	0.350	1.55	0.644
Exceeding real data	41.2%	94.0%	92.0%	97.0%

Clearly there are some aspects of the real winter moth data that do not agree very well with what is expected from the density–dependence of k_5. Den Boer has good grounds to doubt the assertion that pupal predation is the main regulator of the population, although the mechanism that inhibits the regulation that should take place is not clear. One relevant point noted by Den Boer (1988) is the apparently non–random sequence of differences between observed and expected k_5 values from the regression of these values on R_5. This is related to the spiralness of the plot of k_5 values against the R_5 values (Fig. 3), which may reflect delayed density–dependent mortality. The question of whether or not the pattern of spiralness is significant is considered in the next section.

Any conclusions drawn from the winter moth data must be tentative. Pupal predation may have some stabilizing effect (since it is difficult to see how a density–dependent factor can do anything else), but it is less than what is expected on the basis of the observed regression between k_5 and density (presumably because of inhibition by other non–random effects). Also the decrease in variation found by Den Boer by fixing k_5 may be exaggerated to some extent by stochastic effects.

One limitation with the simulation results mentioned above is the absence of any allowance for sampling errors. The main problem with doing this is the lack of information about the magnitude of these errors. However, what can be done is to see how sensitive the results are to errors of a moderate size. To this end all of the simulations were repeated with independent random errors being added to population frequencies. These errors were normally distributed with a mean of zero and standard deviations equal to 5% of generated frequencies, so that the 'observed' value of a population count N was usually within about 10% (two standard errors) of the true value. Adding errors in this way makes some small differences to the simulation results, but does not change conclusions.

Testing for Delayed Density–Dependent Mortality

The final matter to be considered concerns the measurement of the 'spiralness' of plots of k values against population density, and how an observed amount of spiralness can be tested to determine whether it is significant in the statistical sense. As mentioned earlier, one explanation for an anti–clockwise spiral is delayed density–dependent mortality due to a Nicholsonian parasite. Hence significant spiralness may be evidence of this phenomenon. By its nature, this type of effect can only occur when the successive generations of a population are related.

To some extent a spiralness index must be arbitrary. However, one method to calculate an index involves noting that a spiral or circular pattern implies that the angle Θ formed by plotting three consecutive k values (as indicated on Fig. 4) will be nearly constant. Hence a measure of the variation in Θ values for successively plotted k values is an index of spiralness.

Let Θ_1 be the angle formed by plotting the k values for generations i, i+1 and i+2 for i = 1, 2, ..., G–2, where G is the number of generations of data available. Then the standard way for measuring the mean direction and the variation in Θ values (Mardia, 1982) involves calculating

$$C = \sum_{i=1}^{G-2} \cos(\Theta_1)/(G-2) \quad \text{and} \quad S = \sum_{i=1}^{G-2} \sin(\Theta_1)/(G-2)$$

and taking the mean direction as Θ_0 and the variation as τ, where

$$C = \tau \cos(\Theta_0) \quad \text{and} \quad S = \tau \sin(\Theta_0) .$$

This approach overcomes the problem of $0°$ and $360°$ meaning the same direction so that the simple average of $180°$ is not appropriate. For measuring spiralness it is the value

$$\tau = \sqrt{(C^2 + S^2)}$$

that is important. This lies between 0 and 1, with $\tau = 1$ indicating no variation in Θ_1 values (perfect spiralness) and small values indicating a good deal of variation (no spiralness). The value of Θ_0 indicates whether the spiralness is anti–clockwise ($0° < \Theta_0 < 180°$) or clockwise ($-180° < 0°$ or, equivalently, $180° < \Theta_0 < 360°$).

To assess the significance of observed values of τ, the distribution obtained from simulated data can be considered. This will indicate the mean and standard deviation that is liable to occur by chance alone without delayed density–dependent mortality, and the probability of obtaining a value as large as that observed for this null hypothesis case. The latter probability is the 'significance' of the observed value in the statistical sense. It is appropriate to carry out the simulations using the values of τ_j, δ_j and $\text{Var}(\epsilon_j)$ estimated from the data to be tested in order to make the generated data as close as possible to the real data (except for any possible delayed density–dependent effects).

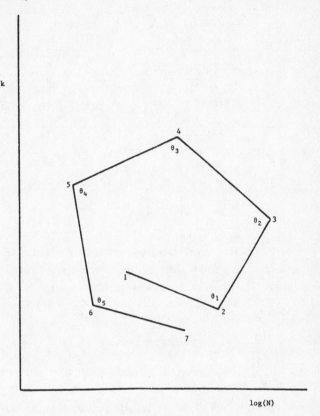

Fig. 4. Angles formed between plots of successive k values that can be used to test for spiralness.

The values of Θ_0 and τ for the winter moth data are shown in Table 9. The apparent anti–clockwise spiral plots for k_1 and k_5 are reflected in high Θ_0 values of 93.9° and 80.1°, and high τ values of 0.45 and 0.47. The table also shows simulation means, standard deviations, and significance levels for the τ values, based on 500 simulated sets of data. The observed τ values are all within about one standard deviation of the simulation means and it turns out that the observed τ values for stages 1 and 5 are not really exceptionally high, being exceeded by 51.8% and 10.6%, respectively, of the simulated values. The conclusion must therefore be that there is no real evidence of delayed density–dependent mortality with these data.

Table 9. Summary of information on spiralness indices for the winter moth data.

	Stage					
	1	2	3	4	5	6
Observed Θ_0	93.9°	−4.5°	51.5°	2.6°	80.1°	−79.3°
Observed τ	0.45	0.12	0.12	0.07	0.47	0.15
Simulation* mean	0.45	0.22	0.21	0.21	0.29	0.30
Standard deviation	0.12	0.14	0.14	0.14	0.14	0.14
Significance level	51.8%	72.6%	68.8%	85.6%	10.6%	86.8%

*Simulation means and standard deviations, and significance levels, are estimated from 500 simulated sets of data.

Acknowledgements

The research reported in this paper was supported partially by grants from the United States – New Zealand Cooperative Science Program and the University of Otago Research Committee. Thanks are due to Dr. P. J. Den Boer for his comments on the paper.

References Cited

Den Boer, P. J. 1986. Density dependence and the stabilization of animal numbers. 1. The winter moth. *Oecologia* 69: 507–12.

Den Boer, P. J. 1988. Density dependence and the stabilization of animal number. 3. The winter moth reconsidered. *Oecologia* (in press).

Latto, J. & Hassell, M. P. 1988. Do pupal predators regulate the winter moth? *Oecologia* 74: 153–5.

Manly, B. F. J. 1977. The determination of key factors from life table data. *Oecologia* 31: 111–7.

Manly, B. F. J. 1979. A note on key factor analysis. *Res. Pop. Ecol.* 21 30–39.

Mardia, K. V. 1982. Directional distributions. *Encycl. Statist. Sci.* 2: 381–6

Morris, R. F. 1957. The interpretation of mortality data in studies of population dynamics. *The Canad. Entomol.* 89: 49–69

Morris, R. F. 1959. Single factor analysis in population dynamics. *Ecology* 40: 580–8

Nicholson, A. J. 1933. The balance of animal populations. *J. Anim. Ecol.* 2: 132–78.

Nicholson, A. J. & Bailey, V. A. 1935. The balance of animal populations. *Proc. Zool. Soc. London* 1935: 551–598

Podoler, H. & Rogers, D. 1975. A new method for the identification of key factors from life table data. *J. Anim. Ecol.* 44: 85–115.

Poethke, H. J. & Kirchberg, M. 1988. On the stabilizing effect of density–dependent mortality factors. *Oecologia* 74: 156–8.

Smith, R. H. 1973. The analysis of intra–generation change in animal populations. *J. Anim. Ecol.* 42: 611–22.

Southern, H. N. 1970. The natural control of a population of tawny owls (*Strix aluco*). *J. Zool. London* 162: 197–285.

Varley, G. C. & Gradwell, G. R. 1960. Key factors in population studies. *J. Anim. Ecol.* 29: 399–401

Varley, G. C. & Gradwell, G. R. 1970. Recent advances in insect population dynamics. *Ann. Rev. Entomol.* 15: 1–24.

Varley, G. C., Gradwell, G. R. & Hassel, M. P. 1973. *Ins. Pop. Ecol.* Blackwell, Oxford.

Are Natural Enemy Populations Chaotic?

J. C. Allen[1]

ABSTRACT It is demonstrated that three simple types of natural enemy population models: host–parasitoid, predator–prey, and host–pathogen can easily have quasiperiodic and chaotic dynamical behavior. The most important ingredients are nonlinearity, time lag and periodic forcing all of which seem realistic and, in fact, quite likely. The time–asymptotic final state of the predator–prey system was observed to depend not only on parameters but on initial conditions as well. The system could be periodic or chaotic for the same parameters simply by shifting the initial predator and prey densities. When spatial heterogeneity and natural selection are considered, it seems likely that chaotic dynamics could occur and persist at the "local" population level. Thus, complex dynamical behavior in these populations could, in fact, arise from simple deterministic processes.

This paper will show how the parameter space of typical natural enemy population models can affect qualitative dynamics or motion of the system through time. Three basic types of systems will be considered: insect host–parasitoid models in discrete time, continuous time predator–prey models and continuous time host–pathogen models. These categories are fairly representative of the more common natural enemy models and serve as good examples of nonlinear dynamical systems. Several rather fundamental problems exist with nonlinear dynamical systems which can affect our philosophy and our approach to their study and understanding. It is now a well known result that many apparently simple systems can have alarmingly complicated motion and sensitivity to initial conditions (May 1974a, 1976; May & Oster 1976; Holden 1987). It is useful to have a descriptive "taxonomy" of the possibilities (Schaffer & Kot 1985, 1987), so some definitions of dynamic behaviors from these references will be repeated here.

Fixed Point

A point in the variable space upon which the system is motionless. In a difference equation system, these are usually found by equating the variables to constants in successive time periods. In a differential equation system, they are found by setting derivatives to zero.

Point Attractor

A fixed point in the variable space to which the system is attracted from some region around the point called a basin of attraction. The approach to the point may be smooth (asymptotic) or by damped oscillation like a playground swing being slowed by friction. Such points are also referred to as stable points.

---

[1]Department of Entomology and Nematology, University of Florida, Gainesville, FL 32611.

Limit Cycle

The system does not come to rest but is attracted into cyclic motion with a constant amplitude and period. The system is then said to have an attracting or stable cycle.

Toroidal Flow

In this case the motion is on the surface of a torus (i.e., a doughnut–like surface). This sort of motion results when a cyclic system is subjected to an additional cyclic force of a different frequency. Two possibilities exist. The motion can be periodic in which case the path winds around the hole in the center of the torus an integer number of times while winding around the axis another integer number of times. That is, the orbit closes on itself and repeats exactly. If the orbit does not close on itself, it will never repeat the same path, and then the motion is said to be quasiperiodic. Such an orbit will eventually cover the entire torus. Cyclic natural enemy systems in cyclic environments seem to be good candidates for this sort of behavior.

Strange Attractor

This is often defined as "none of the above". The system is drawn into a restricted region of the variable space, but in that region the motion is neither periodic nor quasiperiodic. Initial conditions which are arbitrarily close together are observed to diverge from one another at an exponential rate. A tiny cloud of initially very close points soon becomes completely dispersed over the entire attractor. The paths of the initial points separate rapidly and become uncorrelated making long–term prediction impossible. Such systems are said to be chaotic, and the rate of exponential divergence on the attractor is measured by a positive Lyapunov exponent (Jensen 1987; Wolf 1987; Grassberger 1987).

Chaotic behavior in simple models suggests that complex behavior of high dimensional systems might be understood in terms of simple low–dimensional representations. This has produced a plethora of papers on chaotic behavior in such diverse fields as physics (Chirikov 1979; Jensen 1987; Berry 1983), chemistry (Turner et al. 1981; Simoyi et al. 1982), biochemistry (Rapp 1987), electronics (Parker & Chua 1987; Wu 1987; Matsumoto 1987), physiology (Glass & Mackey 1979), cardiac dynamics (Glass et al. 1983, 1987), economics (Jensen & Urban 1984; Day 1983), and even the international arms race (Saperstein 1984). In what follows, some examples will be used to show that natural enemy population systems are prime candidates for chaotic and quasiperiodic motion.

Discrete Time Host–Parasitoid Systems

Cyclic behavior has been the hallmark of discrete time host–parasitoid models where an insect parasitoid kills its host as part of its reproductive cycle. The celebrated Nicholson–Bailey model (1935), which was oscillatory and unstable, was rescued from its unstable demise by the addition of parasite competition (Hassell & Varley 1969), but introducing density–dependence in the host population produces potentially chaotic dynamics (Beddington et al. 1975). At this stage of development, the model can be written as

$$h_{t+1} = h_t \, exp(r(1-h_t/k)-Qp_t^{1-m})$$

$$p_{t+1} = Nh_t(1-exp(-Qp_t^{1-m}))$$

where h_t and p_t are host and parasitoid populations at time t, r is host reproductive rate, k is environmental carrying capacity of the host, Q is a measure of the parasitoid's searching ability, m ($0<m<1$) is a measure of competitive interference between parasitoids and N is the number of parasitoids produced from each attacked host. As Beddington et al. (1978) have pointed out, there is a problem with units in the term Qp_t^{1-m}. In order to obtain a dimensionless argument for the exponential, a term of the form $(ap_t)^{1-m}$ is required where a (area parasite $^{-1}$) cancels with p_t (parasites area $^{-1}$) to produce a dimensionless number. Making this correction and substituting $x_t = h_t/k$ and $y_t = p_t/Nk$, we arrive at a dimensionless form

$$x_{t+1} = x_t \, exp(r(1-x_t)-(\gamma y_t)^{1-m}) \tag{1a}$$

$$y_{t+1} = x_t(1-exp(-(\gamma y_t)^{1-m})) \tag{1b}$$

having only 3 parameters: r, m, and γ ($\gamma=aNk$). A fixed point of the sort $x_{t+1} = x_t = x^*$ and $y_{t+1} = y_t = y^*$ can be found by solving

$$\gamma x^*(1-exp(-r(1-x^*))) = (r(1-x^*))^{1/(1-m)} \tag{2a}$$

iteratively for x^* and then calculating y^* from

$$y^* = x^*(1-exp(-r(1-x^*))) \ . \tag{2b}$$

One can then follow the usual stability analysis trail which leads (after considerable algebra) to the characteristic equation

$$\lambda^2 - [1-r(1-m(1-x^*))+\phi]\lambda + [(1-rx^*)\phi+(1-m)r^2x^*(1-x^*)] = 0 \tag{3}$$

where $\phi = \gamma(1-m)x^*[r(1-x^*)]^{m/(1-m)} = (1-m)r(1-x^*)/[1-exp(-r(1-x^*))]$. Fortunately, this agrees with the characteristic equation given by Beddington et al. (1975) for the $m=0$ case. The fixed point (x^*, y^*) will be stable in nearby state space if the largest root (λ) of eq. (3) is less than 1 in absolute value. One can apply the Schur–Cohn criteria to eq. (3) (May 1974b, p. 219–220) as did Beddington et al. (1975, 1978). This would be handy if the coefficients in eq. (3) were relatively simple functions of 2 or 3 parameters, but they are complicated functions of 3 parameters and a fixed point which must be determined by iterative solution of eq. (2a). For these reasons, the computer was enlisted to find x^* from eq. (2a) for a given (r, m, γ) point, and

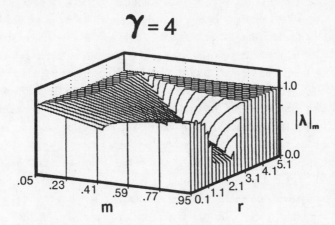

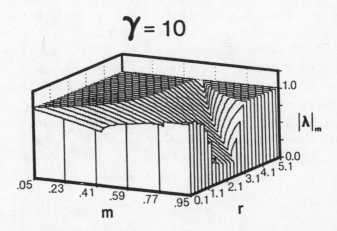

Fig. 1. Largest root, $|\lambda|_m$, of the characteristic equation (eq. (3)) for the host–parasitoid model, eq. (1). The model is stable for $|\lambda|_m < 1$ (the "valley" in the Fig.) and unstable when $|\lambda|_m > 1$ (the "plateau" in the Fig.). The "stable valley" is a function of r, m, and γ.

this result was then used to find the root of largest absolute value, $|\lambda|_m$, of eq. (3) directly from the quadratic formula. In Fig. 1 $|\lambda|_m$ is plotted over the (r, m) plane for $\gamma = 4$ and $\gamma = 10$, setting $|\lambda|_m = 1$ if $|\lambda|_m > 1$. This produces a flat "plateau" where the fixed point is unstable and a "stable valley" where $|\lambda|_m < 1$ which moves over the (r, m) plane as γ changes. For (r, m) combinations on the unstable plateau, periodic, quasiperiodic or chaotic dynamics are possible.

Using this graphical technique the dynamics of eq. (1) are examined when $\gamma = 4$ and $\gamma = 10$ for $r = 3.5$. Fig. 1 shows that this should produce a transition from unstable to stable to unstable again as m is increased from 0 to 1. When $\gamma = 4$ the system is chaotic for small $m-$ values (Fig. 2a,b) which give way to what appears to be quasiperiodic toroidal flow as m increases (Fig. 2c,d,e) and finally an attracting point (Fig. 2f) at about $m = 0.18$ as one enters the "stable valley" of parameter values. Stable, attracting point behavior continues as m increases until $m \simeq 0.706$ where the attracting point "bifurcates" into a 2 point attracting cycle which persists as m approaches 1 (Fig. 2g,h). There appears to be no "chaos" on this side of the stable valley.

For higher attack rates ($\gamma = 10$) similar behavior occurs (Fig. 3). The main difference is that chaotic dynamics persists to much higher m values, and the strange attractor is very close to zero suggesting extinction of one or both species. Simply said, a more effective parasitoid (higher γ) will be more likely to cause extinction of the system in the absence of spatial heterogeneity. In a spatially distributed system, high γ values could be tolerated since "local" extinction could be tolerated, and chaotic dynamics on a local scale could persist.

Increasing r or γ tends to destabilize the system with corresponding proliferation of chaotic and quasiperiodic behavior for low $m-$values, periodic behavior for high $m-$values and stable behavior (the "stable valley" of Fig. 1) in between. It might be argued that the host reproductive rate ($r = 3.5$) in the examples is unrealistically high, but for an efficient parasitoid ($\gamma = 10$), Fig. 1 and simulation show that even for a modest r of 2, chaotic dynamics persists until $m \simeq 0.15$. The existence of high $m-$values under field situations has been argued by some (Diamond 1974; $1/3 < m < 1$), but most authors have argued for low $m-$values or a reinterpretation of m (Beddington et al. 1978; Hassell & May 1973; Free et al. 1977). The existence of widespread "avoidance" behavior of previously attacked hosts by parasitoids is now the subject of a large and active literature (Salt 1961; Lenteren Van 1981; Bakker et al. 1985), and this behavior reduces m and clustering of attacks on hosts, making the observed distribution of parasitoid attacks appear more random. Thus, it seems likely that $m-$values will tend to be on the low side while natural selection would favor high r and γ values. This puts us squarely in the chaotic dynamic region, and suggests that chaos is a real possibility in systems of this kind.

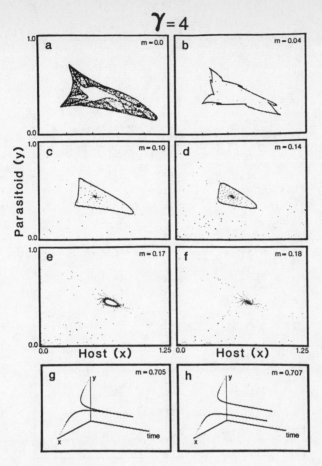

Fig. 2. Host–parasitoid dynamics (eq. (1)) when r = 3.5, γ = 4 for increasing m–values. These behaviors follow a transect across the "stable valley" of Fig. 1.

Predator and Prey: Cyclic Systems in Cyclic Environments

In this section the dynamics of continuous time predator–prey interactions are examined with only the most rudimentary concessions to reality: a cyclic prey reproductive rate and a time lag in the predator's reproductive response to prey density. Chaos has been demonstrated in a similar predator–prey model with cyclic prey reproduction but without the time lag used here (Inoue & Kamifukumoto 1984). To keep things as simple as possible, the time lag is represented by a one–stage "distributed" delay with mean τ_o and the prey reproductive rate by $r(1+\delta\cos(2\pi t_o/T_o))$, so we write after Lotka–Volterra tradition:

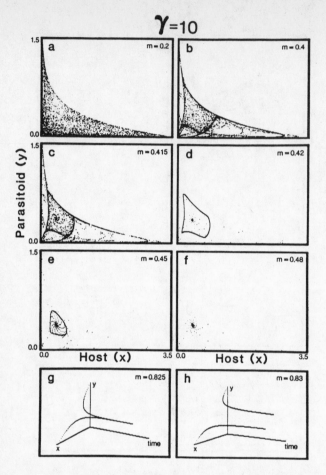

Fig. 3. Same as Fig. 2 except $\gamma = 10$ (a more effective parasitoid). The strange attractor (a,b,c) persists to much higher m–values than when $\gamma = 4$, and it appears that this parasitoid could cause local host extinction.

$$\dot{x}_0 = r[1+\delta cos(2\pi t_0/T_0)]x_0(1-x_0/k) - \gamma_0 x_0 z_0 \qquad \text{(prey)}$$

$$\dot{y}_0 = (x_0-y_0)/\tau_0 \qquad \text{(delay)}$$

$$\dot{z}_0 = \epsilon\gamma_0 y_0 z_0 - \nu_0 z_0 \qquad \text{(predator)}$$

These can be simplified to dimensionless form by letting $x_0 = xk$, $y_0 = yk$, $z_0 = z\epsilon k$, and $t_0 = t/r$ (for an example of this see Nisbet & Gurney 1982, p. 105)

$$\dot{x} = [1+\delta cos(2\pi t/T)]x(1-x) - \gamma xz \qquad (4a)$$
$$\dot{y} = (x-y)/\tau \qquad (4b)$$
$$\dot{z} = \gamma yz - \nu z \qquad (4c)$$

where $\gamma = \epsilon k\gamma_o/r$, $\nu = \nu_o/r$, $T = rT_o$, and $t = rt_o$.

In the absence of periodic forcing ($\delta = 0$), eqs. (4) have a fixed point $x^* = y^* = \nu/\gamma$, $z^* = (\gamma-\nu)/\gamma^2$, and the usual linear analysis produces the characteristic equation

$$\lambda^3 + (1/\tau+\nu/\gamma)\lambda^2 + (\nu/\tau\gamma)\lambda + (\nu/\gamma)(1-\nu/\gamma) = 0 \quad . \qquad (5)$$

Applying May's (1974b) criteria to the coefficients indicates a stable system iff

$$\tau < \gamma/(\gamma^2-\nu(1+\gamma)) \quad . \qquad (6)$$

For a positive fixed point ($z^* > 0$), we require that $\gamma > \nu$, and for a finite, non–negative delay (τ) one must have $\gamma^2 > \nu(1+\gamma)$. The surface (τ) from eq. (6) is illustrated for positive τ in Fig. 4. The system is stable (attached to x^*, y^*, z^*) near the point for delay times below the surface and unstable for delay times above the surface. This is demonstrated in Fig. 5 for $\gamma = 0.5$, $\nu = 0.1$ for increasing τ. For these values, eq. (6) requires $\tau < 5$ for a stable system. Fig. 5 illustrates an attracting point for $\tau = 1,3$ but an attracting quasiperiodic cycle for $\tau = 5.1$.

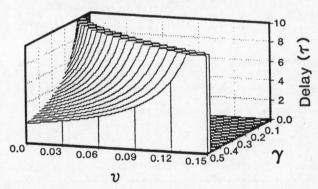

Maximum Delay for Stability

Fig. 4. Stability criterion (eq. (6)) for the predator–prey model (eq. (4)) w/constant prey reproductive rate ($\delta = 0$). Delays in the predator reproductive response below the surface are stable, above the surface, unstable.

If periodic forcing enters the picture ($\delta = 0.5$), one finds bizzare and peculiar behavior from this simple system. Periodic, quasiperiodic or chaotic dynamics can occur depending upon parameters <u>and</u> initial conditions Fig. (6). Many nonlinear dynamical systems have more than one time–asymptotic final state (attractor), and different initial conditions can lead to different final states (Grebogi et al. 1987; Mees and Sparrow 1987). This system is no exception. Illustrated in Fig. 6 is quasiperiodic behavior unaffected by an initial condition (i.c.) shift (Fig. 6a,b,c), chaotic behavior transformed to periodic behavior by an i.c. shift (Fig. 6d,e,f), chaotic behavior which remained chaotic under an i.c. shift (Fig. 6g,h,i) and a smaller quasiperiodic attractor transformed to a much larger periodic attractor by an i.c. shift (Fig. 6j,k,l). It would be fascinating to map the final behavior space as a function of initial conditions as Grebogi et al. (1987) did for the forced pendulum (see cover of Science 232: 585). Such studies go beyond our present purpose but have important implications for man–made perturbations of ecological systems including biological control introductions. The main point is that real world predator–prey systems could apparently have enormously complicated behavior which would be difficult to understand without some knowledge of nonlinear dynamics and its analytical techniques for attacking data sets (see e.g., Schaffer and Kot 1985, 1987; Roux et al. 1983; Schaffer 1985; Buchler and Kovacs 1987).

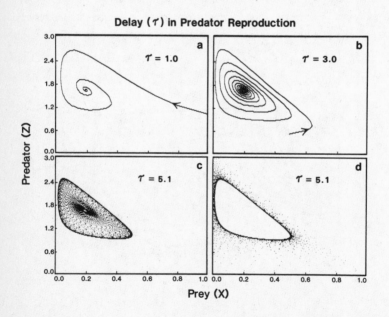

Fig. 5. Dynamics of the predator–prey model (eq. (4)) with constant prey reproductive rate ($\delta = 0$) for $\gamma = 0.5$, $\nu = 0.1$. From eq. (6) the system is stable for $\tau < 5$ (a,b). A quasiperiodic cycle (c,d) for $\tau = 5.1$.

A Host–Pathogen System: A Cyclic System With Cyclic Transmission

Consider a simple disease model

$$\dot{S} = rS(1-S/k) - \gamma_0 SI$$

$$\dot{E} = \gamma_0 SI - E/\tau_0$$

$$\dot{I} = E/\tau_0 - \nu_0 I$$

where S is the population of susceptible hosts, E represents exposed hosts in a latent (time lag) phase and I is the infectious population of pathogen. γ_0, τ_0, and ν_0 are the transmission rate, latent period and mortality rate of the pathogen. By letting $x = S/k$, $y = E/k$, $z = I/k$ and defining a new time unit $t = rt_0$, one arrives at a simpler dimensionless system

$$\dot{x} = x(1-x) - \gamma xz \tag{7a}$$
$$\dot{y} = \gamma xz - y/\tau \tag{7b}$$
$$\dot{z} = y/\tau - \nu z \tag{7c}$$

where $\gamma = \gamma_0 k/r$, $\tau = r\tau_0$ and $\nu = \nu_0/r$. This system has a fixed point $x^* = \nu/\gamma$, $y^* = \nu\tau(1-\nu/\gamma)/\gamma$, z^* $(1-\nu/\gamma)/\gamma$ where $\nu < \gamma$ for a feasible point (y^*, $z^* > 0$). Applying linear analysis to eqs. (7) the characteristic equation is

$$\lambda^3 + (\nu+1/\tau+\nu/\gamma)\lambda^2 + (\nu/\gamma)(\nu+1/\tau)\lambda + \nu(1-\nu/\gamma)/\tau = 0.$$

whose coefficients indicate a stable system (using May's criteria again) iff $\nu < \gamma$ (already required for a feasible fixed point), and

$$\tau < 2/\{[(3\nu-\gamma+\nu/\gamma)^2 - 4\nu^2(1+1/\gamma)]^{1/2} - (3\nu-\gamma+\nu/\gamma)\} \tag{8}$$

which is plotted in Fig. 7 and bears a strong resemblance to the predator–prey system criterion (Fig. 4). A check on eq. (8) for $\gamma = 0.5$, $\nu = 0.05$ is illustrated in Fig. 8. For these values eq. (8) gives (approximately) $\tau < 4.6$ for a stable system. Fig. 8 shows agreement in that the system has an attracting point when $\tau = 4$ and an attracting quasiperiodic cycle when $\tau = 5$.

In many real situations it seems justified to suppose that disease transmission (γ) may be approximately cyclic due to weather effects (wetness, temperature, etc.). In the spirit of this, let $\gamma = \gamma(t) = \gamma(1+\delta\cos(2\pi t/T))$ in eq. (6) with forcing amplitude δ and period T. Periodic transmission rates in even simpler "human" epidemic models (w/o host reproduction) have been shown to produce chaotic dynamics (Schwartz & Smith 1983; Aron & Schwartz 1984), and chaos

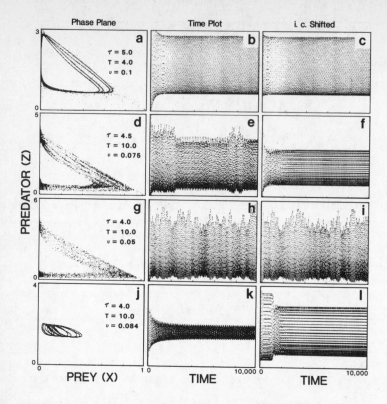

Fig. 6. Dynamics of predator–prey model (eq. (4)) with periodic prey reproductive rate ($\delta =$ 0.5) for $\gamma = 0.5$. Notice that an initial conditon (i.c.) shift *alone* can change *qualitative* behavior (chaotic vs. periodic) (e,f) and quasiperiodic vs. periodic (k,l).

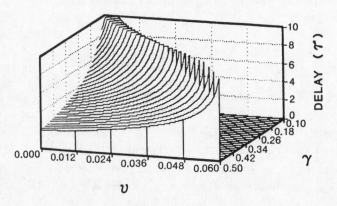

Fig. 7. Stability condition (eq. (8)) for the host–pathogen model (eq. (7)). Delays (τ) in disease development ("latent period") below the surface are stable, above the surface, unstable.

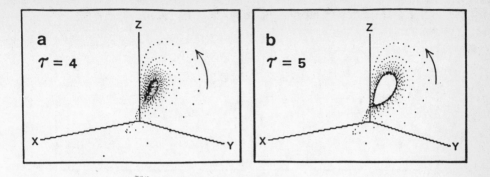

Fig. 8. Dynamics of the host–pathogen model (eq. (7)) with constant transmission rate ($\delta = 0$) for $\gamma = 0.5$, $\nu = 0.05$. The system is stable for $\tau < 4.6$ from eq. (8). A stable *point* occurs at $\tau = 4$ (a) and a quasiperiodic *cycle* at $\tau = 5$ (b).

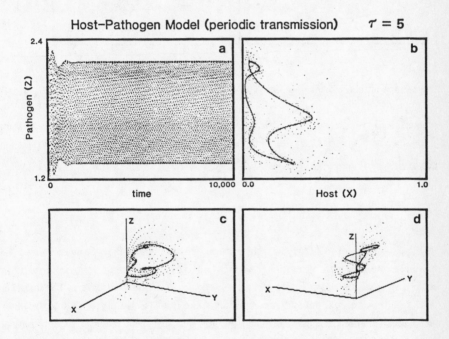

Fig. 9. Dynamics of the host–pathogen model (eq. (7)) with periodic transmission rate ($\delta = 0.5$) for $\gamma = 0.5$, $\nu = 0.05$ and T $= 9+\sqrt{2}$ for $\tau = 5$. Quasiperiodic cycle.

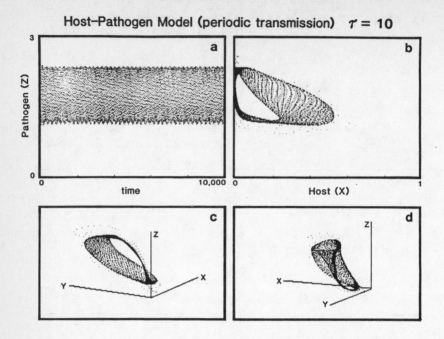

Fig. 10. Same as Fig. 9 except $\tau = 10$. A chaotic strange attractor.

has been detected in <u>data</u> on human epidemics (Schaffer 1985). It is therefore no surprise that in addition to quasiperiodic invariant cycles (Fig. 9) this model has a beautiful chaotic strange attractor as well (Fig. 10).

Summary and Conclusions: Could Real Systems Be Like This?

The dynamical behavior of fairly simple natural enemy models has been shown to be surprisingly complex. Even small concessions to reality such as time lag and periodic forcing can lead to dynamical gymnastics which would completely mystify the casual observer. The question remains: to what extent are real natural enemy systems like this? Certainly the kinds of things which produce chaos (nonlinearity, time lag, and periodic forcing) seem realistic enough and, in fact, quite likely. But could a species have chaotic dynamics without danger of extinction? The answer appears to be yes if the strange attractor is "strongly" attracting and does not go dangerously close to zero (see Figs. 2a, 10).

When the spatial dimension is added to such systems, even local extinction could be easily tolerated if movement rates were high. It is quite conceivable in spatially heterogeneous situations that natural selection might even favor parameters which are chaotic in the "local" dynamic window (i.e., high reproductive and attach rates). And thus complex, mystifying behavior in the dynamics of natural enemy populations could originate from rather simple deterministic processes—a refreshing turn of events and one which deserves further study and attention.

Acknowledgements

I am grateful to Drs. J. W. Jones and G. H. Smerage of the Department of Agricultural Engineering, University of Florida and Drs. S. H. Kerr and H. L. Cromroy of the Department of Entomology and Nematology, University of Florida for reviews of the manuscript. Helpful discussion with Drs. L. S. Block of the Department of Mathematics, University of Florida and J. A. Logan of the Natural Resource Ecology Laboratory, Colorado State University, was appreciated. Any errors which remain are my own.

References Cited

Aron, J. L, & B. I. Schwartz. 1984. Seasonality and period doubling bifurcations in an epidemic model. *J. Theor. Biol.* 110: 665–679.

Bakker, K., J. J. M. van Alphen, F. H. D. van Batenburg, N. van den Hoeven, H. W. Nell, W. T. F. H. van Strien–van Liempt and T. C. J. Turlings. 1985. The function of host discrimination and superparasitization in parasitoids. *Oecologia* 67: 572–576.

Beddington, J. R. 1975. Mutual interference between parasites or predators and its effect on searching efficiency. *J. Anim. Ecol.* 44: 331–340.

Beddington, J. R., C. A. Free & J. H. Lawton. 1975. Dynamic complexity in predator–prey models framed in difference equations. *Nature* 255: 58–60.

Beddington, J. R., C. A. Free & J. H. Lawton. 1978. Characteristics of successful natural enemies in models of biological control of insect pests. *Nature* 273: 513–519.

Berry, M. V. 1983. Semi–classical mechanics of regular and irregular motion. In G. Looss, R. H. G. Hellman, and R. Strom [eds]. Chaotic behavior of deterministic systems. North Holland Press.

Buchler, J. R. & G. Kovacs. 1987. Period doubling bifurcations and chaos in W virginis models. *Astrophys. J.* 320: L57–L62.

Chirikov, B. V. 1979. A universal instability of many–dimensional oscillator systems. *Phys. Rep.* 52: 263–379.

Day, R. H. 1983. The emergence of chaos from classical economic growth. *Quart. J. Econ.* 98: 202.

Diamond, P. 1974. Area of discovery of an insect parasite. *J. Theor. Biol.* 45: 467–471.

Free, C. A., J. R. Beddington & J. H. Lawton. 1977. On the inadequacy of simple models of mutual interference for parasitism and predation. *J. Anim. Ecol.* 46: 543–554.

Glass, L., A. Shrier & J. Belair. 1987. Chaotic cardiac rhythms. In A. V. Holden [ed.]. Chaos. Princeton Univ. Press, Princeton, NJ.

Glass, L., R. Guevara, A. Shrier & R. Perez. 1983. Bifurcation and chaos in a periodically stimulated cardiac oscillator. *Physica* 7D: 89–101.

Glass, M. & M. C. Mackey. 1979. Pathological conditions resulting from instabilities in physiological control systems. *N. Y. Acad. Sci.* 316: 214–235.

Grassberger, P. 1987. Estimating the fractal dimensions and entropies of strange attractors. In A. V. Holden [ed.]. Chaos. Princeton Univ. Press, Princeton, NJ.

Grebogi, C., E. Ott & J. A. Yorke. 1987. Chaos, strange attractors and fractal basin boundaries in nonlinear dynamics. *Science* 238: 632–638.

Hassell, M. P. & G. C. Varley. 1969. New inductive population model for insect parasites and its bearing on biological control. *Nature* 223: 1133–1137.

Hassell, M. P. & R. M. May. 1973. Stability in insect host–parasite models. *J. Anim. Ecol.* 42: 693–726.

Holden, A. V. 1987. Chaos. Princeton Univ. Press, Princeton, NJ.

Inoue, M. & H. Kamifukumoto. 1984. Scenarios leading to chaos in a forced Lotka–Volterra model. *Prog. Theor. Phys.* 71: 930–937.

Jensen, R. V. 1987. Classical chaos. *Am. Sci.* 75: 168–181.

Jensen, R. V.& R. Urban. 1984. Chaotic price behavior in a nonlinear cobweb model. *Econ. Lett.* 15: 235–240.

Lenteren, J. C. van. 1981. Host discrimination by parasitoids. In D. A. Nordlund, R. L. Jones, and W. J. Lewis [eds.]. Seriochemicals: their role in pest control. Wiley and Sons, NY.

Matsumoto, T. 1987. Chaos in electronic circuits. *Proc. IEEE* 75: 1033–1057.

May, R. M. 1974a. Biological populations with nonoverlapping generations: Stable points, stable cycles, and chaos. *Science* 186: 645–647.

May, R. M. 1974b. Stability and complexity in model ecosystems. Princeton Univ. Press, Princeton, NJ.

May, R. M. 1976. Simple mathematical models with very complicated dynamics. *Nature* London 261: 459–467.

May, R. M. & G. Oster. 1976. Bifurcations and dynamic complexity in simple ecological models. *Am. Nat.* 110: 573–599.

Mees, A. & C. Sparrow. 1987. Some tools for analyzing chaos. *Proc. IEEE* 75: 1058–1070.

Nicholson, A. J. & V. A. Bailey. 1935. The balance of animal populations. Part I. *Proc. Zool. Soc.* London 3: 551–598.

Nisbet, R. M. & W. S. C. Gurney. 1982. Modelling fluctuating populations. Wiley–Interscience, NY.

Parker, T. S. & L. O. Chua. 1987. Chaos: a tutorial for engineers. *Proc. IEEE* 75: 982–1008.

Rapp, P. E. 1987. Oscillations and chaos in cellular metabolism and physiological systems. In A. V. Holden [ed.]. Chaos. Princeton Univ. Press, Princeton, NJ.

Roux, J. C., R. H. Simoyi & H. L. Swinney. 1983. Observation of a strange attractor. *Physica* 8D: 257–266.

Salt, G. 1961. Competition among insect parasitoids. *Symp. Soc. Exp. Biol.* 15: 96–119.

Saperstein, A. M. 1984. Chaos–a model for the outbread of war. *Nature* London 309: 303–305.

Schaffer, W. M. 1985. Can nonlinear dynamics elucidate mechanisms in ecology and epidemiology? *IMA J. Math. Appl. Med. Biol.* 2: 221–252.

Schaffer, W. M. & M. Kot. 1985. Do strange attractors govern ecological systems? *Bioscience* 35: 342–350.

Schaffer, W. M. & M. Kot. 1987. Differential systems in ecology and epidemiology. In A. V. Holden [ed.]. Chaos. Princeton Univ. Press, Princeton, NJ.

Schwartz, I. B. & H. L. Smith. 1983. Infinite subharmonic bifurcation in an SEIR epidemic model. *J. Math. Biol.* 18: 233–253.

Simoyi, R. H., A. Wolf & H. L. Swinney. 1982. One–dimensional dynamics in a multicomponent chemical reaction. *Phys. Lett.* 49: 245–248.

Turner, J. S., J. C. Roux, W. D. McCormick & H. L. Swinney. 1981. Alternating periodic and chaotic regimes in a chemical reaction – experiment and theory. *Phys. Lett.* 85A: 9–12.

Wolf, A. 1987. Quantifying chaos with Lyapunov exponents. In A. V. Holden [ed.]. Chaos. Princeton Univ. Press, Princeton, NJ.

Wu. S. 1987. Chua's circuit family. *Proc. IEEE* 75: 1022–1032.

Demographic Framework for Analysis
of Insect Life Histories

James R. Carey[1]

ABSTRACT: The purpose of this paper is to generalize both cohort and population phenomena. All cohort properties are viewed as 'events' and divided into two types: i) non–renewable—events that cannot be repeated (eg. death); and ii) renewable—events that can be repeated (eg. reproduction). A framework for examining averages and heterogeneity of each type is presented. The foundation for examining population properties is the stable population model. An organizational framework for the conventional population parameters is presented as well as two example extensions: i) a two–sex model; and ii) harvesting models. A brief discussion at the end emphasizes the importance of understanding the more generic properties of insect life histories.

A great deal of attention has been given to developing and applying demographic methods for evaluating evolutionary trade–offs in insect life history traits (eg. Birch 1947; Cole 1954; Lewontin 1965; Stearns 1977). However, little emphasis has been placed on developing either a general framework for collecting and analyzing the traits as biological components of stand–alone interest or on application of population models other than the Lotka equation. That is, most analyses are concerned only with summary data derived from the life table or the stable population model.

The objectives of this paper are two fold. First, to present a demographic framework for analysis of cohort life histories that generalizes traits as events. Second, to offer an organizational scheme for parameters derived from the Lotka equation and introduce demographic population models that extend stable theory. I include a two–sex model and harvesting models for insect mass rearing as examples of extensions of stable theory.

Framework and Analysis

Cohort Analysis

The first step in developing a general framework for cohort analysis is to formalize the insect life course. An example of this formalization is given diagramatically in Fig. 1. There are two kinds of traits depicted in this diagram both of which are generic. The first type are those associated with growth or development. Here an insect develops through various stages and the exit of one stage and entry into another is termed an event. These types of events are called

---

[1]Department of Entomology, University of California, Davis, CA 95616

non—renewable in that they can occur only once (Pressat 1985). For example, an egg can hatch only once, a larvae can pupate only once and an individual of any age can die only once.

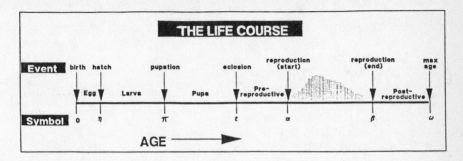

Fig. 1. Diagramatic sketch of the life course of a holometabolous insect.

The second trait depicted in Fig. 1 is reproduction which represents, more generally, any event that can be repeated. These events are termed *renewable* and, besides reproduction, include such things as mating, migration and most aspects of behavior.

By specifying event order, all renewable events can be re—defined as a non—renewable event. For example, a female can lay her 10th egg only once. The importance of this abstract perspective is that all life history phenomena at the cohort level can be viewed as dichotomous and analyzed in a life table (quantal) context.

Non—renewable Events

The life table provides the basic foundation for analyzing all non—renewable events. Mortality can serve as a metaphor for a wide range of phenomena by viewing it as an exit from one dichotomous state and entry into the other. When an individual dies it exits the live state and enters the dead state. Similarly, all larvae that pupate (thus continue to live) are leaving the larval state and entering the pupal state and all individuals that mate for the first time enter the non—virgin state for the first time and thus exit the virgin state (see Carey 1986).

The five basic life table functions are: i) fraction remaining in the designated state at age x, denoted l_x; ii) fraction of the cohort at age x remaining in the current state from x to x+1, denoted p_x where $p_x = l_{x+1}/l_x$; iii) the fraction of the cohort at age x exiting the current state from x to x+1, denoted q_x where $q_x = 1 - p_x$; iv) fraction of the total cohort exiting the current state in the interval x to x+1, denoted d_x where $d_x = l_{x+1} - l_x$; and v) expected number of days in the current state that remain to the average female at age x, denoted e_x where $e_x = \Sigma \, xd_x$. Example applications of life table methods are given in Fig. 2a for medfly survival and in Fig. 2b for medfly pupation. The same life table functions apply to both cases.

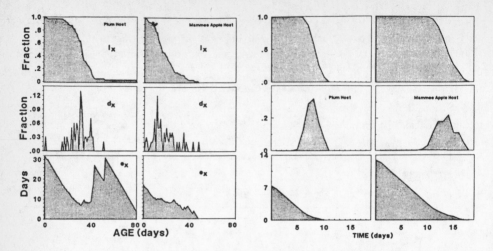

Fig. 2 a,b. Life table functions applied to medfly adult survival and pupation for medflies reared on two different larval host species (Carey, 1984; 1986; Krainacker et al., 1987).

If all individuals die (i.e. exit current state) at exactly the same age, the l_x schedule is rectangular whereas if all individuals have exactly the same probability of dying at each age (i.e. all p_x's are identical) the l_x schedule is a geometric decrease. These represent the two idealized and extreme cases of heterogeneity. A measure of heterogeneity that has received much recent attention in the classical demographic literature (e.g., Keyfitz 1985) is termed entropy, H. The formula for this measure using life table notation is

$$H = \{\Sigma\, e_x d_x\}/e_0$$

According to Goldman & Lord (1986) the sum of products $e_x d_x$ in the numerator can be viewed as: i) the weighted average of life expectancies at age x; ii) the average days of future life that are lost by the observed deaths; or iii) the average number of days an individual could expect to live, given a second chance on life. The denominator is the expectation of life at birth, e_0, and thus converts the absolute effect to a relative effect.

Entropy, H, can be interpreted as (see Vaupel 1986): i) the proportional increase in life expectancy at birth if every individual's first death were averted; ii) percentage change in life expectancy produced by a reduction of one percent in the force of mortality at all ages; and iii) the number of days lost due to death per number of days lived. This last interpretation is the reason why entropy is important in the current context. That is, it provides a quantitative measure for survival (or exit) pattern which, in turn, is an index of variation or heterogeneity. If H = 0 then all deaths occur at exactly the same age and if H = 1 then the l_x schedule is exponentially declining. The intermediate value, H = 0.5, suggests that the l_x schedule is linear (see Fig. 3).

The medfly adult survival curves given in Fig. 2a yielded entropy values of H = .31 and expectation of life of about 32 days for adults reared from plum hosts and H = .54 and expectation of life of about 17 days for adults reared from mammee apple hosts. Note that all but about 5% of plum reared adults were dead by age 50 days and all mammee apple–reared adults were dead by this age. Therefore, the primary differences in host–specific life expectancies had to do with differences in the shapes of the survival schedules—the schedule for plum–reared adults was more rectangular than that for mammee apple–reared adults.

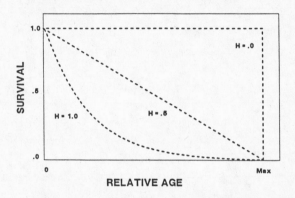

Fig. 3. Idealized shape of survival schedule (l_x) and associated values of entropy, H.

Renewable Events

Let g_x denote the average number of events per individual age x and l_x denote the fraction of the cohort living to age x. Then the various types of averages for the event in question are given in Table 1. The gross and net reproductive schedules for female medflies reared from plum hosts and banana hosts are given in Fig. 4. Using age–specific reproduction from these schedules as g_x, the average gross, net, daily rates and mean ages are summarized in Table 2. Note that each comparative measure of reproduction reveals something different about reproduction in the medfly reared from the two fruit hosts. They show, for example, that the reason medflies reared from plums lay more eggs in their lifetime than do those reared from mammee apple is partly due to their longer life span and partly due to their higher daily productivity. That is, both gross and net fecundity in plum–reared flies was 2.5–fold higher than mammee apple–reared flies yet their daily reproduction was only 1.5–fold higher.

To illustrate the notion of reproductive heterogeneity, consider reproduction in an hypothetical cohort of four individuals over a 4–day period given in Table 3. While average daily reproduction is the same over all age classes, it is clear that differences in reproduction exist among the cohort members. That is: i) individual #1 consistently produced only 1 offspring/day and individual #4 consistently produced two offspring/day; and ii) individuals #2 and #3 were inconsistent on a daily basis but were consistent when their laying is considered over the entire 4–day period. Differences in daily offspring production of individuals within a cohort can be

Table 1. Formulae for determining event intensity and event timing.

	Description	Formula
Event Intensity		
	Gross Rate	$\Sigma\, g_x$
	Net Rate	$\Sigma\, l_x g_x$
	Daily Rate	$\Sigma\, l_x g_x / \Sigma\, l_x$
Event Timing		
	Mean Age Gross Schedule	$\Sigma\, x g_x / \Sigma\, g_x$
	Mean Age Net Schedule	$\Sigma\, x l_x g_x / \Sigma\, l_x g_x$

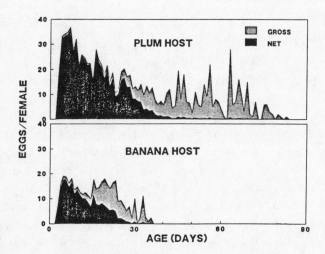

Fig. 4. Gross and net reproduction of medfly females reared from two different larval hosts (data from Krainacker et al. 1987).

expressed as *daily parity* which expresses the fraction of the cohort whose offspring production falls into one–of–several preselected reproductive classes. For example, every day one–half of the hypothetical cohort above produced a single offspring and one–half produced two offspring.

A second parity measure involves the cumulative daily reproduction of individuals and is thus termed *cumulative parity*. This measure expresses the fraction of the cohort whose previous and current cumulative offspring production falls into one–of–several preselected reproductive classes. Daily parity provides insight on the daily consistency of the cohort while cumulative parity reveals the long term consistency of individuals.

Examples of reproductive heterogeneity are given in Fig. 5 for the oriental fruit fly, *Dacus dorsalis*, and the melon fly, *D. cucurbitae* (from Carey et al., 1988). Three aspects of this Fig. may be noted. First, a large percentage (>30%) of the oriental fruit fly cohort lay over 50 eggs per day from 2 to 3 weeks of age while only about 5 to 10% of the melon fly cohort lay over 50 eggs/day. Second, no more than 40% of melon flies lay on any given day but up to 90% of the oriental fruit flies consistently laid every day once each reached reproductive maturity after around 2 weeks. Third, over 80% of the oriental fruit flies had laid a sum total of at least 600 eggs by day 30 while only about 40% of the melon flies had laid this many after a month. Thus oriental fruit flies are not only capable of laying large number of eggs on any given day as reflected in daily parity, but are also capable of doing this every day for a long period of time.

Table 2. Summary of reproductive traits for medflies reared as larvae from plum and banana hosts (data from Krainacker et al. 1987).

	Description	Plum Host	Banana Host
Reproductive Intensity			
	Gross Reproductive Rate	1019.2	407.9 eggs/female
	Net Reproductive Rate	618.6	241.8 eggs/female
	Daily Reproductive Rate	19.2	12.7 eggs/female
Reproductive Timing			
	Mean Age Gross Schedule	29.0	16.6 days
	Mean Age Net Schedule	15.8	12.4 days

Population Analysis

The Lotka model (Lotka 1922) and the parameters associated with it such as the intrinsic rate of increase or stable age distribution serves as the starting point for almost all population analyses. Departures from this standard analysis can be grouped into 1-of-2 categories: i) special properties of the stable model—these include such things as sensitivity analysis; and ii) extensions of stable theory—these typically include models that eliminate one or more of the standard assumptions of stable theory such as varying vital rates (e.g., Tuljapurkar 1986; Pollard 1973) or adding other dimensions besides age such as sex (e.g., Goodman 1954) or space (e.g., Rogers 1975).

Table 3. Hypothetical reproductive patterns of 4 insects (#1 to #4) over 4 age intervals (0 to 3).

	Individual				
AGE	#1	#2	#3	#4	AVE.
0	1	1	2	2	1.5
1	1	2	1	2	1.5
2	1	1	2	2	1.5
3	1	2	1	2	1.5
TOTALS	4	6	6	8	6.0

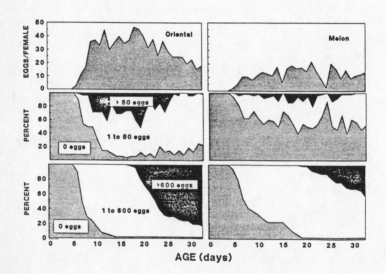

Fig. 5. Reproductive patterns in the oriental fruit fly and the melon fly. Top graphs give the mean eggs per day, middle give daily parity and bottom give cumulative parity (from Carey et al., 1988).

Conventional Stable Model

The Lotka equation serves as an analytical foundation for determining the properties of closed, one–sex populations with fixed vital rates. These properties include the following: i) *growth rates*—intrinsic rate of increase (r), intrinsic birth rate (b) and intrinsic death rate (d); ii) *growth times*—doubling time (DT) and mean generation time (T); iii) *populayion age structure*—proportion at each age in stable population; and iv) *mean ages of individuals or events*—mean age in population, mean age of reproduction, mean age of deaths and mean age of any age property or event. Formulae for all of these parameters are given in most introductory ecology texts and also in Carey (1982). Population parameters for two tephritid fruit flies are summarized in Table 4. Note that the main differences between species' intrinsic rates (and thus doubling times) are due to differences in birth rates as reflected in the intrinsic rates of birth and on the mean ages of reproduction. Also large differences exist in their stable stage distributions—32% in the egg stage for the oriental fruit fly but 11% for the melon fly and 4% in the adult stage for the oriental but 15% for the melon fly. These differences in population parameters reflect basic difference in their cohort parameters and life histories.

Table 4. Organizational framework for presenting basic life hostory parameters derived from the Lotka equation. Data from Carey (1988) on the oriental fruit fly, *Dacus dorsalis* and the melon fly, *D. cucurbitae.*

Parameter	Symbol	Oriental	Melon	Units
GROWTH RATES				
Intrinsic rate	r	.16	.12	Y_z
Finite rate	λ	1.17	1.12	per day
Intrinsic birth rate	b	.21	.15	Y_z
Intrinsic death rate	d	.05	.04	Y_z
GROWTH TIMES				
Doubling time	DT	4.3	6.9	days
Mean Generation time	T	37.4	46.9	days
AGE STRUCTURE				
Eggs	–	32	11	%
Larvae	–	49	32	%
Pupae	–	15	42	%
Adults	–	4	15	%
MEAN AGES				
Mean age in stable population	– a	6.0	7.1	days
Mean age reproduction	A	32.6	46.3	days

Two–sex Model

In a two–sex population the dynamics of the male sub–population is the same as the female's if age– and sex–specific survival are the same and if the primary sex ratio is unity. If any of these sex–specific factors differ, a two–sex model is needed to better understand the determinants of population sex ratio. A model developed by Hamilton et al. (1986) allows for sex–specific survival differences and shifts in primary sex ratio with maternal age is

$$SR = s \left[\Sigma\ e^{-rx}l_m(x)\right] / \left[\Sigma\ e^{-rx}l_f(x)\right]$$

where

$$s = \Sigma\ e^{-rx}l_f(x)m_m(x)$$

and SR denotes sex ratio, $l_f(x)$ and $l_m(x)$ denote survival of females and males, respectively and $m_f(x)$ and $m_m(x)$ denote female production of female and male offspring at age x, respectively. The results of this model over a range of r–values are given in Fig. 6. The main result of the model is this. If primary sex ratio is conditional on maternal age as is the case with many haplodiploids such as tetranychid spider mites, then overall sex ratio will be conditional on population growth rate, r, and no standard sex ratio will exist. This is because no standard and constant population growth rate exists in the field. The reason for the constantly changing skew in sex ratio is because of the effect of growth rate on the age population distribution. In growing populations young females are more abundant than old females. Thus cohort differences in primary sex ratio between young and old females will become more exaggerated.

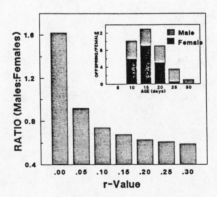

Fig. 6. Relationship between population sex ratio and population growth rate in the twospotted spider mite. Inset gives female production of male and female offspring (redrawn from Hamilton et al. 1985).

Harvesting

Two harvesting models exist that apply to arthropod mass—rearing. The first was published by Carey & Vargas (1985) for rearing tephritid fruit flies. This model applies to arthropods where each stage can be easily separated while rearing and a single stage is then harvested (e.g., dipteran pupae). If h denotes the fraction of the target stage that must be harvested to confer population replacement and d the age at which reproductives are discarded, then h is the solution to the equation

$$1 = (1 - h) \sum_{x=1}^{\delta} l_x m_x$$

$$h = 1 - [\sum_{x=1}^{\delta} l_x m_x]^{-1} .$$

Details and formula for computing per female production rates of the target stage and overall factory age structure are given in the original paper (Carey & Vargas 1985; also Carey 1988). Results of applying this type of harvesting model to two tephritid fruit fly species are given in Fig. 7 for per capita pupal production. Two main points merit comment. First, per capita pupal production for the oriental fruit fly is about twice that or the melon fly. Thus to rear equal numbers of each would require twice the amount of space for melon fly adults relative to oriental fruit fly adults, all else being equal. Second, the optimal discard age for the oriental fruit fly is at about 1 month of age but for the melon fly could range from 20 to 40 days with no clear loss in maximum per capita pupal production.

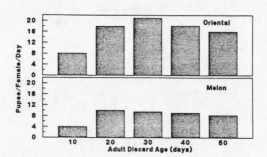

Fig. 7. Trade—offs between adult discard age and per capita pupal production in two fruit fly species (redrawn from Carey & Vargas 1985).

The second model considers the fraction of the total population that must be harvested to confer population replacement and therefore is appropriate for addressing rearing problems for arthropods that must be held in colonies of mixed ages (eg. mites; aphids). The harvesting model for this case is (Carey & Krainacker 1988):

$$1 = \sum_{x=1}^{\omega} (1-h)^x \lambda^{-x} l_x m_x$$

$$h = (\lambda - 1)/\lambda$$

$$= 1 - (\lambda)^{-1}$$

where $\lambda = e^r$. Here h is the fraction of each age class that is removed daily. Note that the harvest rate for the first model is 1 minus the inverse of the per generation rate of change (net reproductive rate) while harvest rate in the second model is 1 minus the inverse of the daily rate of increase (λ). A simple extension for the second model regards the fraction removed after t days of population growth, denoted h_t. That is

$$h_t = 1 - (\lambda)^{-t}$$

Because the same fraction of each age class is removed from the population, the age structure of the factory population will be identical to the stable age distribution for the unconstrained (unharvested) case. The per capita production rate will simply be h or h_t. An application of this model to Tribolium life history data is given in Fig. 8. Note that with unconstrained growth the population will increase by 15.64–fold over a 30–day period at which time 94% of all age classes can be harvested and 6% used for renewal. The stable age distribution is always maintained at 26% eggs, 67% larvae, 3% pupae and 4% adults.

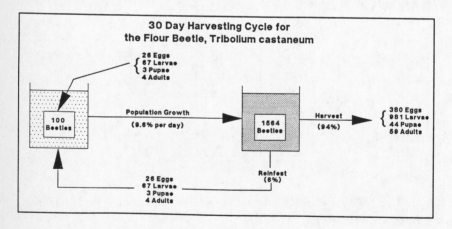

Fig. 8. *Tribolium* harvesting cycle and rates where a fixed fraction of all stages are removed every 30 days (data from Leslie & Park 1949).

Conclusions

The main point of this paper has been to introduce the notion that demographic properties of insect life histories can be addressed at two levels. The first is the cohort level where the life history attributes of individuals are generalized as events which, in turn, are categorized as either non—renewable or renewable. Renewable events can be viewed as non—renewable events by specifying order and therefore all events can be analyzed using life table methods. The second is the population level where cohort attributes (i.e. vital rates) are examined using stable population theory. In this simplest case, populations are structured only by age. The general case is multistate demographic theory where, for example, populations can be structured by sex, by birth type (eg. morph) or by region. In all cases, the analytical framework and the concept can be traced to the basic Lotka model. In general, simple and transparent frameworks for analyzing insect life histories have considerable merit. Ecologists stand to gain from a greater familiarity with the human demographic literature where many of the original concepts used in the field originated.

References Cited

Birch, L. C. 1948. The intrinsic rate of natural increase of an insect population. *J. Anim. Ecol.* 17: 15–26.

Carey, J. R. 1982. Demography and population dynamics of the Mediterranean fruit fly. *Ecol. Model.* 16: 125–150.

Carey, J. R. 1984. Host—specific demographic studies of the Mediterranean fruit fly, *Ceratitis capitata. Ecol. Entomol.* 9: 261–270.

Carey, J. R. 1986. Interrelations and applications of mathematical demography to selected problems in fruit fly management. In: Pest Control: Operations and Systems Analysis in Fruit Fly Management (eds. M. Mangel, J. Carey and R. Plant), Springer—Verlag, Berlin pp 227–262.

Carey, J. R. 1988. Demographic analysis of fruit flies. In: World Crop Pests. Fruit Flies: Their Biology, Natural Enemies and Control (eds. G. Hooper and A. Robinson), Elsevier Science Publishers, Amsterdam (in press)

Carey, J. R. & D. Krainacker. 1988. Demographic analysis of tetranychid spider mite populations: extensions of stable theory. *Exp. and Appl. Acarol.* 4: 191–210.

Carey, J. R. & R. Vargas. 1985. Demographic analysis of insect mass rearing: case study of three tephritids. *J. Econ. Entomol.* 78: 523–527.

Carey, J. R., P. Yang & D. Foote. 1988. Demographic analysis of insect reproductive levels, patterns and heterogeneity: case study of three Hawaiian tephritids. *Entomol. exper. & appl.* 46: 85–91.

Cole, L. 1954. The population consequences of life history phenomena. *Qtr. Rev. Biol.* 29: 103–137.

Goldman, N. & G. Lord. 1986. A new look at entropy and the life table. *Demography* 23: 275–282.

Goodman, L. 1953. Population growth of the sexes. *Biometrics* 9: 212–225.

Hamilton, A., L. W. Botsford & J. R. Carey. 1986. Demographic analysis of sex ratio in the twospotted spider mite, Tetranychus urticae. *Entomol. Exper. Appl.* 41: 147–151.

Keyfitz, N, 1985. Applied Mathematical Demography. Springer–Verlag, New York, 441 pp.

Krainacker, A. D., J. R. Carey & R. Vargas. 1987. Effect of larval host on the life history parameters of the Mediterranean fruit fly, Ceratitis capitata. *Oecologia* 73: 583–590.

Leslie, P. H. & R. Park. 1949. The intrinsic rate of natural increase of *Tribolium castaneum*. *Ecology* 30: 469–477.

Lewontin, R. C. 1965. Selection for colonizing ability. <u>In</u> The Genetics of Colonizing Species. A. G. Baker and G. L. Stebbins [eds.]. pp. 77–94.

Lotka, A. J. 1922. The stability of the normal age distribution. *Proc. Nat. Acad. Sci.* (Wash.) 8: 339–345.

Pollard, J. H. 1973. Mathematical Models for the Growth of Human Populations. Cambridge University Press, Cambridge. pp. 186.

Pressat, Roland. 1985. The Dictionary of Demography. Edited by Christopher Wilson. Bell and Bain Ltd, Glasgow. pp. 243.

Rogers, A. 1975. Introduction to Multiregional Mathematical Demography. John Wiley & Sons, New York. pp. 203.

Stearns, S. C. 1977. The evolution of life history traits: A critique of the theory and a review of the data. *Ann. Rev. Ecol. Syst.* 8: 145–171.

Tuljapurkar, S. D. 1986. Demography in stochastic environments. II. Growth and convergence rates. *J. Math. Biol.* 24: 569–581.

Vaupel, J. W. 1986. How change in age–specific mortality affects life expectancy. *Pop. Studies* 40: 147–157.

Stochastic Differential Equations As
Insect Population Models

Brian Dennis[1]

ABSTRACT Stochastic differential equations are a potentially important class of models for describing insect population dynamics. Their advantages include ease of use, relative tractability, ease of understanding, and the potential for approximating many types of stochastic variation affecting insect populations. This paper is an exposition for quantitative ecologists on parameter estimation for one–dimensional stochastic differential equations. Stochastic versions of the exponential growth model and the logistic model are developed in detail as examples. Topics discussed include transition distributions and moments, stationary distributions, maximum likelihood estimates, conditional least squares estimates, maximum quasi–likelihood estimates, jackknifing, multiple stable and unstable equilbria, and deterministic chaos.

Life is stochastic. Ecologists have long observed that the abundances of natural populations, and of insect populations in particular, are highly variable (Allee et al. 1949, p. 319, Andrewartha & Birch 1954, p. 358). Field estimates of insect populations typically show large temporal and spatial fluctuations over and above pure sampling errors. Even replicate laboratory populations started under similar initial conditions often display widely varying outcomes, as demonstrated by classic experiments on arthropod systems discussed in most ecology texts.

Traditionally, ecological modelers have used simple differential or difference equations for summarizing the general forces regulating population growth. This deterministic approach in ecology has a rich history dating back to Verhulst's logistic model in the nineteenth century. Occasionally some ecologists have raised questions about the wisdom of ignoring random components of population growth. Most notable were the insect population biologists who warned of the inherent vacillations in field data during the rancorous debates on density dependence vs. independence during the 1950's. The deterministic approach still predominates in population modeling, to the extent that unpredictable fluctuations in insect populations are now fashionably hypothesized to be the result of deterministic forces producing "chaotic" behavior.

Admittedly, stochastic models are sometimes proposed for describing insect population abundances. The mathematical ecology literature contains numerous explanatory discussions of various stochastic processes presented as possible candidates for population models (May 1974a, Goel & Richter–Dyn 1974, Pielou 1977, Ricciardi 1977, Nisbet & Gurney 1982). Almost never, however, are such models actually used to analyze real data sets, with the exception of non–biological time series models. Stochastic models exist mostly as concepts in ecology rather than as serious testable hypotheses about the forces affecting population growth. One reason for

---

[1]College of Forestry, Wildlife and Range Sciences, University of Idaho, Moscow, Idaho 83843.

this state of affairs is that it has not been clear to ecologists just how to apply such models. It has not been clear to statisticians, for that matter; methods for estimating parameters and testing hypotheses in stochastic processes are only recently receiving comprehensive treatment in the statistics literature (for instance, Basawa & Prakasa Rao 1980).

It is the intent of this paper to provide an exposition for quantitative ecologists on the use of stochastic differential equations (SDE's) as population models. SDE's, known also as diffusion processes, offer great potential in population analysis, since they have many desirable statistical properties and are easy to understand, to apply, and to test. I concentrate on one–species models and show explicit ways of estimating parameters in these models from data. As examples, I develop SDE versions of the exponential growth model and the logistic growth model.

The first section of this paper reviews the main statistical properties of SDE's needed for parameter estimation. I present without proof the relevant results for subsequent use in estimation. The next section discusses parameter estimation for time series data using the full time–dependent statistical properties of SDE's. Maximum likelihood (ML) estimates and conditional least squares (CLS) estimates are developed. In the third section, I present SDE–based analysis methods for populations fluctuating around a stable equilibrium. Instead of focusing on a deterministic fixed point equilibrium, the section advocates estimating parameters for a stationary probability distribution for population size. The last section discusses some related topics and points out problems for further research. The topics include: maximum quasi–likelihood estimation, jackknifing, sampling variability, systems with multiple stable and unstable equilibria, and deterministic "chaos" models.

Statistical Properties of SDE's

Deterministic models of single species populations are often in the form of an ordinary differential equation (ODE):

$$dN(t) = N(t)g(N(t))dt. \qquad (1)$$

Here $N(t)$ represents a measure of population abundance (density, biomass, numbers, etc.), and $g(N(t))$ represents the per–unit–abundance growth rate. Two examples frequently seen are: (a) the exponential growth model defined by $g(N(t)) = r$ (constant), and (b) the logistic growth model defined by $g(N(t)) = r - (r/k)N(t)$. The logistic may be regarded as an approximation to a more detailed growth model near a stable equilibrium population abundance (Dennis & Patil 1984, Dennis & Costantino 1988).

Many stochastic versions of (1) can be constructed, but a type of stochastic differential equation (SDE) has potential for describing many features of population fluctuations in a relatively simple fashion. The stochastic version of (1) discussed in this paper is the following SDE:

$$dN(t) = N(t)[g(N(t))dt + \sigma h(N(t))dW(t)]. \qquad (2)$$

Here $dW(t)$ has a normal distribution with a mean of zero and a variance of dt, $h(N(t))$ is a positive–valued function, and σ is a positive constant. The formulation describes the addition of random perturbations to the per–unit–abundance growth rate, with the function $h(N(t))$ describing any dependence on $N(t)$ of the magnitude of the fluctuations. A useful form for ecological applications is $h(N(t)) = 1$, corresponding to "multiplicative noise", that is, the scale of fluctuations in the overall growth rate $N(t)g(N(t))$ is proportional to $N(t)$. Population trajectories under this model may be simulated by generating a normal (independent) random variable $dW(t)$ during a small time increment dt, calculating the differential $dN(t)$ using (2), and then computing the new population size as $N(t + dt) = N(t) + dN(t)$, etc.

Mathematically, the differential $dN(t)$ is rigorously defined by either an Ito or a Stratonovich stochastic integral (see Soong 1973, Karlin & Taylor 1981, or Horsthemke & Lefever 1984). The above simulation method corresponds to the Ito interpretation, which will be assumed in this paper. The differences between Ito and Stratonovich calculi have generated a lot of colorful copy in the mathematical ecology literature (for instance, Turelli 1977, Feldman & Roughgarden 1975). The differences from the standpoint of modeling are to some extent semantic (Braumann 1983a, Dennis & Patil 1984) and are not of concern to this paper.

The two examples of SDE's considered here arise from the exponential growth and the logistic growth deterministic models. The exponential growth SDE is defined by

$$dN(t) = N(t)rdt + \sigma N(t)dW(t), \qquad (3)$$

and the SDE version of logistic growth becomes

$$dN(t) = N(t)[r - (r/k)N(t)]dt + \sigma N(t)dW(t). \qquad (4)$$

Both models use $h(N(t)) = 1$.

A stochastic process $N(t)$ defined by an SDE in the form (2) is known as a diffusion process. Such diffusion processes have the Markov property and are continuous functions of time (see Karlin & Taylor 1981). Two functions are particularly important for obtaining statistical properties of diffusion processes. They are the infinitesimal mean, denoted $m(n)$, and the infinitesimal variance, $v(n)$, given by

$$m(n) = \lim_{\Delta t \to 0}(1/\Delta t)E[N(t+\Delta t) - N(t)|N(t) = n] = ng(n); \qquad (5)$$

$$v(n) = \lim_{\Delta t \to 0}(1/\Delta t)E[\{N(t+\Delta t) - N(t)\}^2|N(t) = n] = \sigma^2 n^2[h(n)]^2. \qquad (6)$$

The infinitesimal mean and variance for the exponential SDE (3) are $m(n) = rn$ and $v(n) = \sigma^2 n^2$, while those for the logistic SDE (4) become $m(n) = n[r - (r/k)n]$ and $v(n) = \sigma^2 n^2$.

Several properties of SDE's or diffusion processes make them valuable for modeling applications. First, the Markov property allows the formulation of explicit likelihood functions for fitting the models to data. Second, many statistical properties such as transition distributions and moments, or approximations thereof, are straightforwardly derived; these properties are useful for fitting the models to data or for studying the dynamic behavior of the models. Third, many other types of stochastic processes, such as birth–death processes, stochastic difference equations, or branching processes, can be approximated by SDE's through scaling techniques (see Karlin & Taylor 1981, p. 168). Finally, if $N(t)$ is a diffusion process, then a transformation $X(t) = f(N(t))$ is also a diffusion process, provided $f(N(t))$ is a continuous, strictly increasing (or decreasing) function. The infintesimal mean and variance of $X(t)$ are given by

$$m_X(x) = v_N(n)f''(n)/2 + m_N(n)f'(n), \tag{7}$$

$$v_X(x) = v_N(n)[f'(n)]^2, \tag{8}$$

where $n = f^{-1}(x)$ (Karlin & Taylor 1981, p. 173). This property often permits the transformation of a novel diffusion process into a known process with well–studied statistical properties.

All the essential properties of a diffusion process $N(t)$ governed by an SDE (2) are embodied in the transition probability density function (pdf) of the process. The transition pdf, denoted $p(n,t|n_0)$, is a pdf with time t and initial population abundance $N(0) = n_0$ appearing as parameters. The area under the transition pdf between a and b gives the probability that the population is in the interval $(a,b]$ at time t, given that $N(0) = n_0$:

$$\Pr[a < N(t) \leq b] = \int_a^b p(n,t|n_0)dn. \tag{9}$$

The transition pdf is a solution to a partial differential equation known as the Fokker–Planck or forward equation,

$$\partial p/\partial t = (1/2)\partial^2[vp]/\partial n^2 - \partial[mp]/\partial n, \tag{10}$$

where $p = p(n,t|n_0)$, $v = v(n)$, and $m = m(n)$. The solution must obey the initial condition $p(n,0|n_0) = \delta(n - n_0)$ (i.e. $\Pr[N(0) = n_0] = 1$); $\delta(x)$ is the Dirac delta function which is zero everywhere except for a "spike" of infinite height at $x = 0$ such that the area under $\delta(x)$ is 1. When appropriate, the solution $p(n,t|n_0)$ must also obey boundary conditions relating to integrability and to reflection or absorption of the process at the edge of its range. The Fokker–Planck equation has been solved for many specific SDE models; solution details and examples are provided by Goel & Richter–Dyn (1974), Karlin & Taylor (1981), Gardiner (1985), and Risken (1984).

The transition pdf or suitable approximation is needed to fit an SDE to time series data using ML estimation.

The transition pdf for the exponential growth SDE (3) is easily obtained using the transformation $X(t) = \log N(t)$. The transformation formulas (7) and (8) yield constant infinitesimal mean and variance for the process $X(t)$:

$$m_X(x) = r - \sigma^2/2, \tag{11}$$

$$v_X(x) = \sigma^2. \tag{12}$$

These are the infinitesimal moments of a Wiener process (Bownian motion) with drift. A well–known result gives a normal transition distribution for $X(t)$ (e.g. Ricciardi 1977, p. 58): $X(t)$ ~ normal($x_0 + (r - \sigma^2/2)t$, $\sigma^2 t$), where $x_0 = \log n_0$. Thus, the distribution of $N(t)$ is lognormal with transition pdf given by

$$p(n,t|n_0) = [n(\sigma^2 t 2\pi)^{1/2}]^{-1} \exp\{-[\log n - \log n_0 - (r - \sigma^2/2)]^2/(2\sigma^2 t)\},$$

$$0 < n < \infty. \tag{13}$$

This highly skewed distribution starts as a spike at n_0 and spreads rapidly as t increases.

The process $N(t)$ governed by the logistic SDE (4) can be transformed into a process with a linear infinitesimal mean through the Bernoulli transformation $X(t) = 1/N(t)$. The infinitesimal moments (7) and (8) for $X(t)$ are

$$m_X(x) = (r/k) - (r - \sigma^2)x, \tag{14}$$

$$v_X(x) = \sigma^2 x^2. \tag{15}$$

As Prajneshu (1980) pointed out, these infinitesimal moments correspond to a process introduced by Wong (1964). Wong (1964) provided an expression for the transition pdf of $X(t)$, and Prajneshu (1980) transformed the pdf to obtain the transition pdf of $N(t)$.

Unfortunately, the resulting transition pdf for the logistic SDE is extremely complicated, involving an integral of functions of complex variables. Computing it is not out of the question, but is hardly within the scope of routing insect population analyses. Fortunately, though, one can obtain suitable approximations for the transition pdf amenable to computing using perturbation methods (details are beyond the scope of this paper; perturbation methods are discussed by Gardiner 1985). One such approximaton is displayed later in this paper (equation (27)). The transition pdf so approximated starts as a spike at n_0 and resembles an S–shaped ridge converging ultimately to a stationary distribution.

If the deterministic population trajectory goverend by (1) approaches a stable point equilibrium, a corresponding SDE (2) may possess a limiting stationary distribution. As t becomes large, the transition pdf $p(n,t|n_0)$ may approach a pdf, denoted $p(n)$, that does not depend on t or n_0. The form of the pdf is given by

$$p(n) = Cexp\{(2/\sigma^2)\int((1/n)g(n)/[h(n)]^2)dn - 2\log n - 2\log h(n)\} \qquad (16)$$

(see Dennis & Patil 1984). The constant C is found by setting the area under the curve $p(n)$ equal to one (if the area is infinite, then a stationary distribution for the process does not exist). The exponential growth SDE (3) does not have a stationary distribution, but the logistic SDE (4) does have a stationary distribution. It is a straightforward exercise to use (16) to obtain a stationary gamma distribution, with pdf given by

$$p(n) = [\beta^{\alpha}/\Gamma(\alpha)]n^{\alpha-1}e^{-\beta n}, \ 0 < n < \infty, \qquad (17)$$

for the logistic SDE. Here $\alpha = (2r/\sigma^2) - 1$, $\beta = 2r/(k\sigma^2)$. Just as the deterministic logistic approximates more detailed deterministic models, the gamma distribution (17) can be regarded as an approximate stationary distribution for more detailed SDE models (Dennis & Patil 1984).

Moments of $N(t)$ and other distributional properties of $p(n,t|n_0)$ are useful for summarizing the statistical behavior of the process through time. Moments or other expected values are also needed for estimating parameters with the CLS method. The expected value of a function, $f(N(t))$, of a diffusion process given that $N(0) = n_0$ is itself a function of n_0 and t:

$$E[f(N(t))|N(0) = n_0] = \int_{-\infty}^{+\infty} f(n)p(n,t|n_0)dn = u(n_0,t). \qquad (18)$$

Setting $f(n) = n$ gives the time–dependent mean of $N(t)$, $f(n) = n^2$ gives the second moment, etc. Such expectations can be computed directly from (18) using the transition pdf. Alternatively, $u(n_0,t)$ satisfies a partial differential equation known as the backward equation:

$$\partial u/\partial t = (v(n_0)/2)\partial^2 u/\partial n_0{}^2 + m(n_0)\partial u/\partial n_0. \qquad (19)$$

The function $u(n_0,t)$ is obtained by solving (19) subject to the condition $u(n_0,0) = f(n_0)$. A derivation of the backward equation from the definition of $u(n_0,t)$ (18) is provided by Karlin & Taylor (1981, p. 214).

For the exponential growth SDE, the νth moment of $N(t)$ defined by $E[(N(t))^{\nu}|N(0) = n_0] = u_{\nu}(n_0,t)$ is obtained straightforwardly from (18) using the lognormal transition pdf (13) and $f(n) = n^{\nu}$:

$$u_\nu(n_0,t) = n_0^\nu \exp\{[r - (\sigma^2/2)]\nu t + (\sigma^2/2)\nu^2 t\}. \tag{20}$$

The mean $E[N(t)|N(0) = n_0]$ of the process given by

$$u_1(n_0,t) = n_0 e^{rt} \tag{21}$$

corresponds to the solution of the deterministic model. However, the mean is not necessarily a good characterization of the behavior of the process. In fact, if $0 < r < \sigma^2/2$, it is easy to demonstrate using the transition pdf (13) that $\Pr[0 < N(t) \leq \epsilon] \to 1$ as $t \to \infty$ for arbitrarily small $\epsilon > 0$ (see, for instance, Dennis & Patil 1988). In biological terms, the SDE predicts virtually certain extinction of the population if σ^2 is large compared with r.

For the transformed process $X(t) = 1/N(t)$, where $N(t)$ is defined by the logistic SDE (4), a recursion expression relates the νth moment of $X(t)$ to the $(\nu-1)$th moment. Derivation of this relationship is beyond the scope of this paper but is based on a moment result for SDE's (Goel & Richter–Dyn 1974, p. 46). Letting $E[(X(t))^\nu|X(0) = x_0] = E[(N(t))^{-\nu}|N(0) = n_0] = u_{-\nu}(n_0,t)$, the relationship is

$$u_{-\nu}(n_0,t) = \exp\{(\nu\sigma^2/2)[\nu - 1 - 2(r-\sigma^2)/\sigma^2]t\}\{n_0^{-\nu}$$

$$+ \nu(r/k)\int_0^t \exp\{-(\nu\sigma^2/2)[\nu - 1 - 2(r-\sigma^2)/\sigma^2]w\}u_{-\nu+1}(n_0,w)dw\}. \tag{22}$$

In particular, the mean of $X(t)$ is found by setting $\nu = 1$ and noting that $u_0(n_0,t) = 1$:

$$u_{-1}(n_0,t) = (r/k)/(r-\sigma^2) + [1/n_0 - (r/k)/(r-\sigma^2)]\exp[-(r-\sigma^2)t]. \tag{23}$$

An immediate consequence of (23) is that the harmonic mean of $N(t)$ grows according to a logistic equation. The harmonic mean of $N(t)$ is defined as $1/u_{-1}(n_0,t)$:

$$1/E[1/N(t)|N(0) = n_0] = k(1 - (\sigma^2/r))/\{1 +$$

$$(1/n_0)[k(1 - (\sigma^2/r)) - n_0]\exp[-(r-\sigma^2)t]\}. \tag{24}$$

In other words, the harmonic mean is a solution of a logistic ODE except with a loss term $\sigma^2 n$ subtracted: $dn/dt = rn - (r/k)n^2 - \sigma^2 n$. It is interesting that this nonlinear SDE (4) preserves a logistic–type trajectory for one of its measures of central tendency.

The mean of $N(t)$ defined by $u_1(n_0,t) = E[N(t)|N(0) = n_0]$ does not obey a logistic equation. The mean of $N(t)$ has been derived by Hamada (1981) and is a complicated expression involving numerous intractable integrals. An approximation for $u_1(n_0,t)$ can be obtained from the backward equation (19) using singular perturbation methods:

$$u_1(n_0,t) \approx k/[1 + ((k{-}n_0)/n_0)e^{-rt}] + (\sigma^2/2)(k/r)[1 + ((k{-}n_0)/n_0)e^{-rt}]^{-3} *$$

$$\{[1 - (2k/n_0)]e^{-2rt} + (2k/n_0)e^{-rt} - 1 - 2r[(k{-}n_0)/n_0]te^{-rt}\}. \tag{25}$$

Wiesak (1988) has given a rigorous justification of this approximation as well as for the following one (26). The mean of $N(t)$ has the familiar sigmoid shape, but is less than the solution of the deterministic logistic. Additionally, an approximation for the second moment is

$$u_2(n_0,t) \approx \{k/[1 + ((k{-}n_0)/n_0)e^{-rt}]\}^2$$

$$+ (\sigma^2/2)(k/r)^2 2r[1 + ((k{-}n_0)/n_0)e^{-rt}]^{-4}\{[-(4k/n_0) + (5/2)]e^{-2rt}$$

$$+ [(4k/n_0){-}2]e^{-rt}{-}(1/2)+r[((k{-}n_0)/n_0)e^{-rt}{-}1]((k{-}n_0)/n_0)te^{-rt}\}. \tag{26}$$

With these two moments, one can approximate the transition pdf for the stochastic logistic with a time–dependent gamma distribution having matching moments. Let $\alpha(n_0,t) = u_1^2/[u_2 - u_1^2]$, $\beta(n_0,t) = u_1/[u_2 - u_1^2]$, where u_1 and u_2 are given by (25) and (26). Then the transition pdf given by

$$p(n,t|n_0) = [\beta^{\alpha}/\Gamma(\alpha)]n^{\alpha-1}e^{-\beta n}, \quad 0 < n < \infty, \tag{27}$$

where $\alpha = \alpha(n_0,t)$ and $\beta = \beta(n_0,t)$, satisfies the initial condition $p(n,0|n_0) = \delta(n{-}n_0)$, has first two moments identical to (25) and (26), and approaches the exact stationary gamma pdf (17) as t becomes large.

Time–Dependent Analysis

Suppose an insect population is observed at times $0, t_1, t_2, ..., t_q$. the recorded observations of population size will be denoted $n(0) = n_0$, $n(t_1) = n_1$, ..., $n(t_q) = n_q$, and the time intervals (not necessarily equal) between observations denoted $t_1 - 0 = \tau_1$, $t_2 - t_1 = \tau_2$, ..., $t_q - t_{q-1} = \tau_q$. A recommended way of fitting an SDE to such observations is ML estimation.

ML estimation typically requires an approximate or exact transition pdf for the process $N(t)$ governed by the SDE (2). The SDE will generally contain one or more unknown parameters; the vector of unknown parameters will be denoted by θ. The initial population size n_0 can be regarded as fixed in many population studies; the ML estimates developed here are consequently conditioned on n_0. The likelihood function $\ell(\theta)$ is defined as the joint pdf for $N(t_1)$, $N(t_2)$, ...,

$N(t_q)$, given $N(0) = n_0$, evaluated at observations n_1, n_2, ..., n_q. Since $N(t)$ is a Markov process with stationary transition probabilities (pdf of n_i given n_{i-1} depends only on τ_i, not t_{i-1}), the likelihood function is a product of transition pdfs:

$$\ell(\theta) = \prod_{i=1}^{q} p(n_i, \tau_i | n_{i-1}; \theta). \tag{28}$$

Here $p(n_i, \tau_i | n_{i-1}; \theta) \equiv p(n_i, \tau_i | n_{i-1})$ is the transition pdf defined by (9) evaluated at n_i, τ_i, and n_{i-1} (likelihood of system moving to n_i from n_{i-1} in a time interval of τ_i); the above notation emphasizes the dependence on the unknown parameters in θ. The ML estimates, $\hat{\theta}$, of the parameters in θ are the parameter values jointly maximizing $\ell(\theta)$ or $\log \ell(\theta)$.

ML estimation for the exponential growth SDE (3) was studied by Braumann (1983b) for the case of equal time intervals between observations: $\tau_1 = \tau_2 = ... = \tau_q$. It is straightforward to generalize his results for unequal intervals. The SDE has two unknown parameters: r and σ^2. It is somewhat more convenient to reparameterize by letting $\mu = r - \sigma^2/2$ and finding estimates of μ and σ^2 instead. Using the transition pdf (13), the log–likelihood function becomes

$$\log \ell(\mu, \sigma^2) = \sum_{i=1}^{q} \log p(n_i, \tau_i | n_{i-1}; \mu, \sigma^2) = -\sum_{i=1}^{q} \log[n_i(\tau_i 2\pi)^{1/2}] - (q/2)\log \sigma^2$$

$$- [1/(2\sigma^2)] \sum_{i=1}^{q} (1/\tau_i)[\log(n_i/n_{i-1}) - \mu\tau_i]^2. \tag{29}$$

It is an easy exercise to set partial derivatives of $\log \ell(\mu, \sigma^2)$ with respect to μ and σ^2 equal to zero and solve for the ML estimates:

$$\hat{\mu} = \left\{ \sum_{i=1}^{q} \log(n_i/n_{i-1}) \right\} / \sum_{i=1}^{q} \tau_i = [\log(n_q/n_0)]/t_q; \tag{30}$$

$$\hat{\sigma}^2 = (1/q) \sum_{i=1}^{q} (1/\tau_i)[\log(n_i/n_{i-1}) - \hat{\mu}\tau_i]^2. \tag{31}$$

The ML estimate for r becomes $\hat{r} = \hat{\mu} + (\hat{\sigma}^2/2)$.

One can obtain information on the distributions of $\hat{\mu}$ and $\hat{\sigma}^2$ by recalling that $X(t) = \log N(t)$ is Brownian motion with drift. Then, let $Y_i = \log[N(t_i)/N(t_{i-1})] = X(t_i) - X(t_{i-1})$. The variables Y_1, Y_2, ..., Y_q are increments of Brownian motion with drift, and are therefore normal, independent, and stationary (e.g. Ricciardi 1977). In fact, if $\underline{Y} = [Y_1, Y_2, ..., Y_q]'$ and $\underline{\tau} = [\tau_1, \tau_2, ..., \tau_q]'$ are defined as column vectors, the distribution of $\underline{Y}$ becomes a multivariate normal:

$$\underline{Y} \sim \text{normal}(\mu\underline{\tau}, \sigma^2 T). \tag{32}$$

Here $T = \text{diag}(\underline{\tau})$ is a matrix with the elements of $\underline{\tau}$ on the main diagonal and zeros elsewhere. Let $G = \text{diag}(\sqrt{\tau_1}, ..., \sqrt{\tau_q})$, that is, $T = G'G$. A transformation of $\underline{Y}$ produces an ordinary normal linear model (e.g. Graybill 1976, p. 207):

$$\underline{Y}^* = G^{-1}\underline{Y} \sim \text{normal}(\mu[\sqrt{\tau_1}, ..., \sqrt{\tau_q}]', \sigma^2 I). \tag{33}$$

This is seen to be a model for a simple linear regression without intercept. In practice, one transforms the data by $y_i = \log(n_i/n_{i-1})$, $i = 1, ..., q$. The regression approach uses $y_1/\sqrt{\tau_1}$, $y_2/\sqrt{\tau_2}, ..., y_q/\sqrt{\tau_q}$ as values of the "dependent variable", $\sqrt{\tau_1}, ..., \sqrt{\tau_q}$ as values of the "independent variable", and a linear regression without intercept is performed. The formula (30) for μ is recognized as the slope parameter estimate, and $\hat{\sigma}^2$ (31) is the (biased) ML estimate of the error variance parameter. The unbiased estimate is $q\hat{\sigma}^2/(q-1)$. The usual linear model theory yields the distributions of $\hat{\mu}$ and $\hat{\sigma}^2$: $\hat{\mu} \sim \text{normal}(\mu, \sigma^2/t_q)$, and $q\hat{\sigma}^2/\sigma^2 \sim \text{chisquare}(q-1)$.

Though the approximate transition pdf (27) for the logistic SDE is tractable, closed formulas for the ML estimates of r, k, and σ^2 cannot be obtained. Instead, the likelihood function (28) must be maximized numerically using one of various iterative algorithms (see Press et al. 1986) and a computer. Matrix programming languages such as GAUSS, SAS/IML, or APL make the calculation of ML estimates a fairly straightforward task.

When ML estimation is impractical, an alternative estimation method is conditional least squares. CLS estimates of parameters for an SDE model do not have all the statistical qualities of ML estimates. CLS estimates, like ML estimates, are consistent (i.e. $\hat{\theta}$ tends to be "closer" to θ as the sample size becomes large), asymptotically unbiased, and have asymptotic normal distributions (Klimko & Nelson 1978). However, CLS estimates tend to be less efficient (i.e. they have larger variances) than ML estimates. In addition, practical experience suggests that a bias is often present in CLS estimates for smaller samples. On the other hand, there are some Gauss–Markov style optimality results for certain CLS estimates arising from the theory of estimating equations (Godambe 1985).

The main practical advantage of CLS estimates is ease of calculation. They can often be computed for SDE models using linear or nonlinear regression packages. They make convenient "starter" values for iterative ML calculations.

CLS estimates are based on time–dependent moments or other expected values. Suppose one can write the time–dependent expected value of a function, $X(t) = f(N(t))$, of a diffusion process $N(t)$:

$$E[X(t)|X(0) = x_0] = E[f(N(t))|N(0) = n_0]$$

$$= u(n_0, t) \equiv u(n_0, t; \theta). \tag{34}$$

One could obtain the form of u by solving the backward equation (19). Due to the Markov property, $u(n_{i-1}, \tau_i; \theta)$ is then the expected value of $f(N(t_i))$ given $N(t_{i-1}) = n_{i-1}$. CLS estimates arise from a sum of squared differences between observed values, $f(n_i)$, and their conditional expected values, $u(n_{i-1}, \tau_i; \theta)$:

$$s(\theta) = \sum_{i=1}^{q} [f(n_i) - u(n_{i-1}, \tau_i; \theta)]^2. \tag{35}$$

CLS estimates are the values of parameters in θ that jointly minimize $s(\theta)$. Note that many different CLS estimates may be available for a given SDE, based on different forms for the function $f(N(t))$.

CLS estimates for two of the parameters in the logistic SDE (4) can be constructed from the mean of $X(t) = 1/N(t)$ given by (23). Let $\beta_1 = (r/k)/(r-\sigma^2)$, $\beta_2 = r - \sigma^2$, and $f(n) = 1/n$, so that

$$u(n_{i-1}, \tau_i; \beta_1, \beta_2) = E[1/N(t_i) | N(t_{i-1}) = n_{i-1}]$$

$$= \beta_1(1 - e^{-\beta_2 \tau_i}) + (1/n_{i-1})e^{-\beta_2 \tau_i}. \tag{36}$$

One could perform a nonlinear regression to find the CLS estimates, i.e. the values of β_1 and β_2 minimizing

$$s(\beta_1, \beta_2) = \sum_{i=1}^{q} [(1/n_i) - u(n_{i-1}, \tau_i; \beta_1, \beta_2)]^2. \tag{37}$$

The values $1/n_i$, $i = 1, ..., q$, would be entered as the "dependent variable" in a computer package, with the model to be fit given by (36).

It is interesting to note that minimizing (37) reduces to a simple linear regression of $1/n_i$ on $1/n_{i-1}$ when the time intervals between observations are equal. When $\tau_1 = \tau_2 = ... = \tau_q = \tau$, then (36) can be written as $\theta_1 + \theta_2(1/n_{i-1})$, where $\theta_1 = \beta_1(1 - e^{-\beta_2 \tau})$, $\theta_2 = e^{-\beta_2 \tau}$.

As an alternative, one could estimate all three parameters r, k, and σ^2 with the CLS method through use of the approximate first moment of $N(t)$ given by (25) using the untransformed data, n_i, $i = 1, ..., q$ as the dependent variable, and $u_1(n_{i-1}, \tau_i)$ (from (25)) as the model to be fit.

Equilibrium Analysis

The data considered in this section consist of observed sizes of an insect population, or ensemble of populations, fluctuating around an equilibrium. The main idea is to estimate parameters and test the fit of a stationary distribution for population size, instead of concentrating on estimating a fixed point equilibrium. The data are a time series (or group of time series) of the form $n_1 = n(t_1)$, $n_2 = n(t_2)$, ... , $n_q = n(t_q)$. The methods described here are better when the intervals between observations are large, but the intervals can be small if there are many observations over a long period of time.

The statistical methods are based on the fact that the transition pdf, $p(n,t|n_0;\theta)$ approaches a stationary pdf, $p(n;\theta)$ as t becomes large, for some SDE models (16). Thus, as the intervals $\{\tau_i\}$ between observations increase, the time–dependent likelihood function (28) would approach a product of stationary pdfs:

$$\ell(\theta) = \prod_{i=1}^{q} p(n_i;\theta). \tag{38}$$

One computes ML estimates of the parameters in θ by maximizing $\ell(\theta)$ or log $\ell(\theta)$.

If a chisquare goodness of fit test is desired, or if the time intervals between observations are small, use of a multinomial likelihood is recommended instead of (38). The investigator partitions the positive real line into m abundance classes: $(0, s_1]$, $(s_1, s_2]$, ..., (s_{m-1}, ∞), where $0 < s_1 < ... < s_{m-1} < \infty$. Grouped data denoted by y_1, y_2, ..., y_m are formed; y_1 is the number of observations that are less than or equal to s_1, y_2 is the number of observations that are greater than s_1 but less than or equal to s_2, etc. Define $\pi_j(\theta)$ as the area under the stationary pdf between s_{j-1} and s_j, j = 1, ..., m (where $s_0 = 0$ and $s_m = +\infty$):

$$\pi_j(\theta) = \int_{s_{j-1}}^{s_j} p(n;\theta)dn. \tag{39}$$

Since a diffusion process with a stationary pdf is ergodic, $\pi_j(\theta)$ represents the long–run proportion of time the process spends in the interval $(s_{j-1}, s_j]$. The multinomial likelihood function is

$$\ell(\theta) = C \prod_{j=1}^{m} [\pi_j(\theta)]^{y_j}, \tag{40}$$

where $C = q!/[y_1! \, y_2! \, ... \, y_m!]$. The ML estimates are obtained by computing the values of the paramters in θ which maximize $\ell(\theta)$ or log $\ell(\theta)$. Goodness of fit testing can be accomplished with the Pearson statistic, X^2 or the likelihood ratio statistic, G^2:

$$X^2 = \sum_{j=1}^{m} [y_j - q\pi_j(\hat{\theta})]^2 / [q\pi_j(\hat{\theta})], \tag{41}$$

$$G^2 = \sum_{j=1}^{m} y_j \log \{y_j / [q\pi_j(\hat{\theta})]\}. \tag{42}$$

A term in G^2 is understood to be zero if the corresponding count y_j is zero. The statistics X^2 and G^2 have identical large–sample chisquare distributions with degrees of freedom given by m − (# parameters estimated in θ) − 1. The chisquare approximation is adequate when $q\pi_j(\theta) \geq 5$ for at least 80% of the abundance classes (and this should be kept in mind when constructing the classes). If the time intervals between observations are small, then properties of the $\check{G}^2$ statistic are unknown at this time. However, under such circumstances X^2 is known to reject the null hypothesis (that the model fits) somewhat too often (Gleser & Moore 1985).

The multinomial likelihood (40) is easily maximized using nonlinear regression packages. The procedure is to use the y_j values as observations on the "dependent variable". Corresponding to each y_j value, $q\pi_j(\theta)$ is computed with programming statements as the model to be fit. Also, weights of $1/[q\pi_j(\theta)]$ are computed (and recomputed every iteration) for each y_j value. This setup "tricks" the nonlinear least squares (Gauss–Newton) algorithm into maximizing the multinomial likelihood (40) (see Jennrich & Moore 1975).

Further details and many numerical examples of stationary distribution analysis have been presented recently by Dennis & Costantino (1988).

Discussion

This section discusses various topics related to SDE analysis and points out problems for further research. The topics include other approaches to statistical inference for SDE's, incorporating sampling variability, models with multiple stable/unstable equilibria, and distinguishing stochasticity from deterministic chaos.

Additional Approaches to Inference.

One alternate approach to parameter estimation for SDE's is through the concept of quasi–likelihood. Quasi–likelihood is finding many uses in statistical theory, particularly in the literature of generalized linear models (McCullagh & Nelder 1983). Suppose x is a vector of observations arising from some stochastic model with a mean vector given by $E[X] = \mu$ and a variance–covariance matrix of $\eta V(\mu)$, where η is a positive constant. The quasi–likelihood function $\ell^*(\mu)$ is defined by a set of partial derivatives:

$$\partial\ell^*(\mu)/\partial\mu = V^{-1}(\mu)(x - \mu). \tag{43}$$

If $\mu = \mu(\theta)$, i.e. μ depends further on a vector θ of underlying parameters, then the maximum quasi–likelihood (MQL) estimate of θ is the solution to $\partial\ell^*(\mu(\theta))/\partial\theta = 0$. By letting $D(\theta) = \partial\mu(\theta)/\partial\theta$ be a matrix of partial derivatives, applying the vector derivative chain rule, and using (43), we have

$$D^T(\theta)V^{-1}(\mu(\theta))[x - \mu(\theta)] = 0 \qquad (44)$$

as the system of equations for the MQL estimates of the parameters in θ.

For an SDE model, one can think of a quasi–likelihood as approximating a product of transition pdfs (28). The elements of the vector x in (44) are taken as the observations $f(n_1)$, $f(n_2)$, ..., $f(n_q)$ of a diffusion process $X(t) = f(N(t))$, the corresponding elements in $\mu(\theta)$ are found from $E[f(N(t_i))|N(t_{i-1}) = n_{i-1}] = u_1(n_{i-1},\tau_i;\theta)$ (34), and V becomes a diagonal matrix of conditional variances: $E[\{f(N(t_i))\}^2|N(t_{i-1}) = n_{i-1}] - [u_1(n_{i-1},\tau_i;\theta)]^2 = u_2(n_{i-1},\tau_i;\theta) - [u_1(n_{i-1},\tau_i;\theta)]^2$, say. Also, D becomes a matrix (q rows) of partial derivatives of $u_1(n_{i-1},\tau_i;\theta)$, i = 1, ... q, with respect to each parameter in θ. Thus, the estimating equations (44) for θ resemble those resulting from finding CLS estimates by minimizing (35), except that the terms in (44) are weighted by (the reciprocals of) the conditional variances.

The calculations to solve (44) can be accomplished with iterative reweighted least squares. From a current set of parameter values, θ_1, the algorithm computes improved values, θ_2, according to

$$\theta_2 = \theta_1 + (D^TV^{-1}D)^{-1}D^TV^{-1}(x - \mu), \qquad (45)$$

where D, V, and μ are evaluated at θ_1. In nonlinear regression packages, this amounts to using $f(n_1)$, ..., $f(n_q)$ as observations on the dependent variable, $u_1(n_0,\tau_1;\theta)$, ..., $u_1(n_{q-1},\tau_q;\theta)$ as the model to be fit, with weights of $1/w_1$, ..., $1/w_q$ computed at each iteration, where $w_i = u_2(n_{i-1},\tau_i;\theta) - [u_1(n_{i-1},\tau_i;\theta)]^2$.

MQL appears to be a promising inference approach for stochastic population models. Much research remains to be done concerning the statistical properties of MQL estimates applied to SDE models; I would recommend that Monte Carlo studies be undertaken to examine MQL estimates in comparison to ML and CLS estimates, for specific models such as the stochastic logistic.

Another approach to statistical inference for SDEs involves jackknifing. Recent research by Lele (1988) indicates that jackknifing may be a highly useful way of handling a long–standing problem in stochastic processes: how to estimate the variance of parameter estimates. Lele has shown that jackknifing the linear estimating equations leads to a consistent estimate of the variance–covariance matrix for the parameter estimates. His results applied to SDE models are sketched here briefly. ML, CLS, and MQL estimates of parameters in SDE models are found by solving equations of the form

$$H(\theta) = \sum_{i=1}^{q} h(n_i,n_{i-1},\tau_i;\theta) = 0. \qquad (46)$$

For example, in ML estimation h is a vector of partial derivatives of the log–transition pdf, log $p(n_i,\tau_i|n_{i-1};\theta)$, with respect to parameters in θ. Let $\hat{\theta}$ denote the estimate resulting from (46). One finds as many as q additional estimates, denoted $\hat{\theta}_j$, j = 1, 2, ..., q, by solving

$$H_j(\theta) = \sum_{i \neq j}^{q} h(n_i, n_{i-1}, \tau_i; \theta) = 0 \qquad (47)$$

where $H_j(\theta)$ represents $H(\theta)$ except with $h(n_j, n_{j-1}, \tau_j; \theta)$ deleted from the sum. The jackknife estimate of θ is

$$(JK)\hat{\theta} = \hat{\theta} - [(q-1)/q]\sum_j (\hat{\theta}_j - \hat{\theta}). \qquad (48)$$

We may expect an improvement in any finite sample bias of $\hat{\theta}$ by using $(JK)\hat{\theta}$. The jackknife estimate of $\Sigma(\theta)$, the variance–covariance matrix of $\hat{\theta}$ (or $(JK)\hat{\theta}$), is

$$(JK)\hat{\Sigma}(\theta) = [(q-1)/q]\sum_j (T_j - \overline{T})^2, \qquad (49)$$

where $T_j = \hat{\theta}_j - \hat{\theta}$ and $\overline{T} = (\Sigma_j T_j)/q$. Large sample theory provides that $\hat{\theta}$ (or $(JK)\hat{\theta}$) has an asymptotic multivariate normal distribution with mean vector θ and variance–covariance matrix $\Sigma(\theta)$.

In practice, (48) and (49) will be computer–intensive, since they require solving not just one system of nonlinear equations, but up to q + 1 of them! However, the benefits would appear to be well worth the trouble. Lele's results apply to linear estimating equations in general, opening up many valuable applications in spatial analysis, stochastic processes, and statistical distribution modeling.

Sampling Variability.

A problem that has been glossed over in the previous discussions is the question of sampling variability. The abundances of field populations must typically be estimated with samples. The variability from sampling produces variability in the parameter estimates beyond that inherent in the SDE.

One model for incorporating sampling variability is a compound Poisson model. Suppose $Y(t)$ represents the number of insects appearing in a sample at time t. The compound Poisson model would assume a Poisson distribution for $Y(t)$ with a mean of $\lambda N(t)$, where $N(t)$ is itself a stochastic process defined, for example, by an SDE (2), and the proportionality constant λ reflects sampling effort. The sampled abundances $y(t_1)$, $y(t_2)$, ..., $y(t_q)$ would constitute a realization of $Y(t)$ and not $N(t)$.

In principle, one can easily write down a probabilistic model (i.e. a joint pdf, and hence a likelihood function) for the sampled abundances. In practice, the expression involves numerous repeated integrals and is not likely to be very useful. Instead, there are ways of dealing with sampling variability in applications. The first is to ignore it. One fits the SDE model directly to the observations using the methods described earlier. This is a reasonable approach if large samples (e.g. many hundreds) of insects appear in each sample, since the variability from sampling would then be small. For instance, under Poisson sampling, if 400 or more insects appear in a sample, the estimated coefficient of variation from sampling is under 5%. The second is to broaden the conceptual interpretation of the SDE to include sampling. One regards the sample observations (say, numbers of insects caught in pheromone traps at times t_1, t_2, ...) as being generated by a stochastic difference equation having variance components due to population fluctuations and sampling; one then uses an SDE merely as an approximation to that process. The procedure involves fitting the SDE directly to the sample observations; the resulting larger value of the parameter σ^2 conceptually reflects variability due to sampling as well as stochastic population fluctuations. Garcia (1983) incorporated sampling variability into an SDE model of forest growth by transforming the model to a Gaussian (Ornstein–Uhlenbeck) process and incorporating normally distributed sampling error. However, one of the variance parameters was nearly non–identifiable (i.e. data provided little information for its estimation) in his applications.

This problem of how to account for sampling variability is not peculiar to SDE models; it must be confronted with virtually all dynamic models of population abundances. SDE's are proposed here mainly for situations in which actual population fluctuations are the prime source of variability in the observed data. It is my contention that such situations are more numerous in ecological studies than has been previously acknowledged.

Multimodal Models.

Several forest insect systems, including gypsy moth and spruce budworm, have been hypothesized to have two or more stable equilibria (Takahashi 1964, Campbell & Sloan 1978, Ludwig et al. 1978, Berryman 1978). The insects are thought to be held in check at a low–abundance endemic equilibrium by a complex of many predator species. If for some reason the insects increase in abundance beyond a threshold value, however, reproduction gains outpace predation losses. The insects then continue increasing until reaching an upper, epidemic equilibrium where population size is regulated by sheer lack of resources (due to defoliation). Deterministic models in the form of (1) have been proposed to describe such systems (see review by May 1977).

Stochastic forces are likely to play an important role in such systems. Stochastic population fluctuations could provide the initial population increases necessary to move away from a lower stable equilibrium into a basin of attraction to an upper stable equilibrium. Such population outbreaks would occur seemingly at random.

SDE models in the form of (2) can be constructed from deterministic models with multiple stable and unstable equilibria (see Dennis & Patil 1984). One of the more interesting predictions from these SDE models involves the stationary pdf (16) for population abundance. For moderate noise levels, the stationary pdf from such a model may display multiple modes and antimodes corresponding to (though not equal to) the underlying stable and unstable equilibria. For higher noise levels, the underlying equilibria structure becomes obscured, and a stationary pdf, if it exists at all, may have fewer modes than the number of stable equilibria.

If data on population abundance existed for systems suspected of multiple equilibria, multimodal stationary pdfs such as those listed by Dennis & Patil (1984) could be fitted using the methods discussed in this paper.

Chaos.

Ever since the papers by May (1974b, 1976) and May & Oster (1976) appeared, it has been well—known among mathematical ecologists that simple difference equation models of population growth can display complicated behavior seemingly indistinguishable from a random process. (The same is true for nonlinear differential equation systems of three or more species, as discussed for example, by Schaffer & Kot (1986); the discussion here is restricted to one—species systems.) Indeed, mathematical ecologists had long been steeped in a deterministic traditon, yet had been increasingly confronted with fluctuating population data; the "chaos" hypothesis of population regulation is now regarded as an important contending explanation of unpredictable data.

Chaos can in fact be classified as a type of stochastic behavior. Current thinking by Diaconis and others on the meaning of "randomness" (see Research News, Science Vol. 231, 7 March 1986, p. 1068) views as random a system with output behavior extremely sensitive to initial conditions. That perennial random system, a coin toss, is in principle a deterministic system. However, a tiny change in, say, the initial angular and/or upward velocity of the coin can cause a drastic change in the system output (heads or tails); thus the system may be regarded as random. Another such system is a string of pseudo—random numbers generated on a computer. Change the seed number slightly and a wholly different sequence emerges. Similar things happen in a chaotic deterministic model. Model trajectories differing only slightly in initial conditions diverge from each other exponentially (see review by Grebogi et al. 1987). It is reasonable, then, to contemplate the use of stochastic—based analysis methods on possibly chaotic time series data to see what statistical properties are present.

The statistical properties of SDE models such as the stochastic logistic (4) differ substantially from those of deterministic chaos models (such as the models catalogued by May & Oster 1976). For instance, the concept of a stationary distribution can be applied to the chaotic behavior of a difference equation model. The typical difference equations used as population models possess so—called invariant measures; that is, the long—run abundance frequencies of a chaotic population approach a limiting stationary distribution (see Lasota & Mackey 1985). For instance, the simple difference equation given by $n_{t+1} = 4n_t(1 - n_t)$ has a "stationary distribution" of $p(n) = \pi^{-1}[n(1 - n)]^{-1/2}$, $0 < n < 1$. Stationary distributions for other chaos models can seldom be obtained analytically, but it is straightforward to iterate any given model until limiting relative frequencies are obtained. Such exercises carried out to date in my knowledge typically produce U—shaped, multimodal, or irregular stationary distributions for population abundance. By constrast, the logistic SDE produces a unimodal mound—shaped or J—shaped distribution.

Also, a common feature of chaotic behavior is the presence of quasiperiodicity. A chaotic system may have time intervals of seemingly periodic behavior followed by irregularity, or periodicity in which the amplitudes and frequencies undergo gradual precession. "Windows" of actual periodic behavior seem to be abundant in parameter sets corresponding to chaotic regimes (see Grebogi et al. 1987). One—species SDE models, by contrast, do not produce periodic behavior unless periodic forcing terms are included in the models. Time series methods such as spectral analysis can help determine if periodic components are present in the data.

Acknowledgements

This research was supported by the Forest, Wildlife and Range Experiment Station of the University of Idaho (publication no. 362), by a grant from the University of Idaho Research Council (#681—Y508), and by a grant from the USDA—ARS Rangeland Insect Laboratory (#00580401700139).

References Cited

Allee, W. C., A. E. Emerson, O. Park, T. Park & K. P. Schmidt. 1949. Principles of Animal Ecology. W. B. Saunders, Philadelphia.

Andrewartha, H. G. & L. C. Birch. 1954. The Distribution and Abundance of Animals. University of Chicago Press, Chicago.

Basawa, I. V. & B. L. S. Prakasa Rao. 1980. Statistical Inference for Stochastic Processes. Academic Press, New York.

Berryman, A. A. 1978. Towards a Theory of Insect Epidemiology. *Res. Pop. Ecol.* 19: 181–196.

Braumann, C. A. 1983a. Population Growth in Random Environments. *Bull. Math. Biol.* 45: 635–641.

Braumann, C. A. 1983b. Population Extinction Probabilities and Methods of Estimation for Population Stochastic Differential Equation Models. Pp. 553–559. In R.S. Bucy and J.M.F. Moura [eds.], Nonlinear Stochastic Problems. D. Reidel, Dordrecht, Holland.

Campbell, R. W. & R. J. Sloan. 1978. Numerical Bimodality among North American Gypsy Moth Populations. *Environ. Entomol.* 7: 641–646.

Dennis, B. & R. F. Costantino. 1988. Analysis of Steady—state Populations with the Gamma Abundance Model: Application to Tribolium. *Ecology* 69 (in press).

Dennis, B. & G. P. Patil. 1984. The Gamma Distribution and Weighted Multimodal Gamma Distributions as Models of Population Abundance. *Math. Biosci.* 68: 187–212.

Dennis, B. & G. P. Patil. 1988. Applications in Ecology. Chapter 12 pp. 303–330. In E.L. Crow and K. Shimizu [eds.], Lognormal Distributions: Theory and Applications. Marcel Dekker, New York.

Feldman, M. W. & J. Roughgarden. 1975. A Population's Stationary Distribution and Chance of Extinction in a Stochastic Environment with Remarks on the Theory of Species Packing. *Theor. Popul. Biol.* 7: 197–207.

Garcia, O. 1983. A Stochastic Differential Equation Model for the Height Growth of Forest Stands. *Biometrics* 39: 1059–1072.

Gardiner, C. W. 1985. Handbook of Stochastic Methods for Physics, Chemistry, and the Natural Sciences. Second edition. Springer–Verlag, Berlin.

Gleser, L. J. & D. S. Moore. 1985. The Effect of Positive Dependence on Chi–squared Tests for Categorical Data. *J. R. Stat. Soc.* B47: 459–465.

Godambe, V. P. 1985. The Foundations of Finite Sample Estimation in Stochastic Processes. *Biometrika* 72: 419–428.

Goel, N. S. & N. Richter–Dyn. 1974. Stochastic Models in Biology. Academic Press, New York.

Graybill, F. A. 1976. Theory and Application of the Linear Model. Wadsworth, Belmont, California.

Grebogi, C., E. Ott & J. A. Yorke. 1987. Chaos, Strange Attractors, and Fractal Basin Boundaries in Nonlinear Dynamics. *Science* 238: 632–638.

Hamada, Y. 1981. Dynamics of the Noise–induced Phase Transition of the Verhulst Model. *Progr. Theor. Phys.* 65: 850–860.

Horsthemke, W. & R. Lefever. 1984. Noise–induced Transitions. Springer–Verlag, Berlin.

Jennrich, R. I. & R. H. Moore. 1975. Maximum Likelihood Estimation by Means of Nonlinear Least Squares. *Proc. Stat. Comp. Am. Stat. Assoc.* 52–65.

Karlin, S. & H. M. Taylor. 1981. A Second Course in Stochastic Processes. Academic Press, New York.

Klimko, L. A. & P. I. Nelson. 1978. On Conditional Least Squares Estimation for Stochastic Processes. *Ann. Stat.* 6: 629–642.

Lasota, A. & M. C. Mackey. 1985. Probabilistic Properties of Deterministic Systems. Cambridge University Press, Cambridge.

Lele, S. 1988. Jackknifing Linear Estimating Equations: Asymptotic Theory in Stochastic processes. *J. Royal Statist. Soc.* B (in press).

Ludwig, D., D. D. Jones & C. S. Holling. 1978. Qualitative Analysis of Insect Outbreak Systems: the Spruce Budworm and Forest. *J. Anim. Ecol.* 47: 315–332.

May, R. M. 1974a. Stability and Complexity in Model Ecosystems. Princeton University Press, Princeton, New Jersey.

May, R. M. 1974b. Biological Populations with Nonoverlapping Generations: Stable Points, Stable Cycles and Chaos. *Science* 186: 645–647.

May, R. M. 1976. Simple Mathematical Models with very Complicated Dynamics. *Nature* 261: 459–467.

May, R. M. 1977. Thresholds and Breakpoints in Ecosystems with a Multiplicity of Stable States. *Nature* 269: 471–477.

May, R. M. & G. F. Oster. 1976. Bifurcations and Dynamic Complexity in Simple Ecological Models. *Am. Natur.* 110: 573–599.

McCullagh, P. & J. A. Nelder. 1983. Generalized Linear Models. Chapman and Hall, London.

Nisbet, R. M. & W. S. C. Gurney. 1982. Modelling Fluctuating Populations. John Wiley & Sons, New York.

Pielou, E. C. 1977. Mathematical Ecology. John Wiley & Sons, New York.

Prajneshu. 1980. Time–dependent Solution of the Logistic Model for Population Growth in Random Environment. *J. Appl. Prob.* 17: 1083–1086.

Press, W. H., B. P. Flannery, S. A. Teukolsky & W. T. Vetterling. 1986. Numerical Recipes. Cambridge University Press, Cambridge.

Ricciardi, L. M. 1977. Diffusion Processes and Related Topics in Biology. Springer–Verlag, Berlin.

Risken, H. 1984. The Fokker–Planck Equation. Springer–Verlag, Berlin.

Schaffer, W. M. & M. Kot. 1986. Differential Systems in Ecology and Epidemiology. Chapter 8 pp. 158–178. In A. V. Holden [ed.], Chaos. Princeton University Press, Princeton, New Jersey.

Soong, T. T. 1973. Random Differential Equations in Science and Engineering. Academic Press, New York.

Takahashi, F. 1964. Reproduction Curve with Two Equilibrium Points: a Consideration of the Fluctuation of Insect Population. *Res. Pop. Ecol.* 6: 28–36.

Turelli, M. 1977. Random Environments and Stochastic Calculus. *Theor. Pop. Biol.* 12: 140–178.

Wiesak, K. 1988. Asymptotic Solution of a Stochastic Logistic Equation with a Small Diffusion Coefficient. Ph.D. Thesis, University of Idaho, Moscow, Idaho.

Wong, E. 1964. The Construction of a Class of Stationary Markoff Processes. Pp. 264–276. In R. Bellman [ed.], Stochastic processes in Mathematical Physics and Rngineering. American Mathematical Society, Providence, Rhode Island.

Intensive Study and Comparison of Single Species Population Simulation Models

Li Dianmo and Liu Chang[1]

ABSTRACT This paper reviews and compares several general single species models that can be used as part of a larger ecosystem model for further research.

Detailed studies of single species population dynamics are important to ecology to understand population behavior and to analyze large, complex ecosystems. In pest management, single species population dynamics can be a basic component of the system (Haynes et al., 1980; Koenig & Tummala, 1980; Lee et al., 1976). There are many simulation models of population dynamics (Gage & Sawyer, 1979; Stinner et al., 1974; Wenke, 1974) but most are so particular that they are not useful for other insects or animals. Here we discuss some models that can be incorporated into a larger model.

Description of Models

The ssp Process

Here ssp stands for the simple simulation process. This is the easiest and simplest way to simulate population dynamics.

Suppose a population varies according to the equation

$$\frac{dX(t)}{dt} = (a(t)-p(t)) \cdot X(t) , \tag{1}$$

where $a(t)$ is the pure reproduction rate (natural birth rate minus natural death rate) of individuals without predation, and $p(t)$ is a single predation rate by predators. When the age structure of the population is stable so that $a(t)$ and $p(t)$ are constant, then the solution is found as

$$X(n+1) = e^{(a-p)} \cdot X(n) , \tag{2}$$

---

[1]Institute of Zoology, Academy of Science of China.

using discrete time, with a small time interval.

Now let us suppose that population densities between two generations can be calculated by using linear interpolation as

$$X(n+\Delta t) = \Delta t \cdot X(n+1) + (1-\Delta t) \cdot X(n) , \quad (0 \leq \Delta t \leq 1) . \tag{3}$$

Suppose Δt in equation (3) is not calendar time, but a proportion of completed development for one generation. Thus let Drate = DD/TDD, which is a proportion of completed development of one generation in one day, where DD = $T-T_0$, when $T \geq T_0$, and DD = 0, when $T < T_0$. Here T is the daily temperature and T_0 is the base temperature for development, so that DD is the degree days on base temperature. Also, TDD is the total degree days for this species to complete one generation. Equation (3) can then be modified to

$$X(n+1) = \text{Drate} \cdot XN(n) + (1-\text{Drate}) \cdot X(n) \tag{4}$$

where $XN(n)$ is the population density of the next generation for the present population $X(n)$, and $n, n+1$ in brackets indicate simulation time units.

The Account Process

Coulman et al. (1972) established this model which is based on a life table approach and contains all the parameters of natural population growth. The basic functional representations are proposed for the ith class. The symbols are as follows: Y = population, B = birth, D = natural death, M = maturation, P = predation, S = supply, n = day, p_i = fraction lost to predation, d_i = fraction lost to natural death, and b_i = brood size ratio of the number of young to the number of adults. The unit for all capital letter symbols is the number of individuals. The following equations then apply:

$$Y_i(n+1) = Y_i(n) + S_{i-1}(n) - P_i(n) - M_i(n) \tag{5}$$

$$P_i(n) = p_i(n) \, Y_i(n) \tag{6}$$

$$M_i(n) = \prod_{j=0}^{q(T)-1} (1-p_i(n-j)) \, S_{i-1}(n-q(T)) \tag{7}$$

$$D_i(n) = d_i(n) \, M_i(n) \tag{8}$$

$$S_i(n) = M_i(n) - D_i(n) \tag{9}$$

$$B_i(n) = b_i(n) \, S_i(n) \tag{10}$$

The implication of maturation M_i and class life period $q(T)$ deserves special attention. The value of $q(T)$ is the duration in days of the ith class and is a function of environmental temperature (T). The number that enter a class i at time $n-q(T)$ is $S_{i-1}(n-q(T))$. The number maturing at time n, namely $M_i(n)$, are those individuals from this group which remain, having survived the effects of predation over the past $q(T)$ days.

The DTVMPL Process

DTVMPL stands for Distributed Time Variable Maturation Processes with Predation Losses. Distributed delays have found widespread application in the modeling of aggregative bahavior in large systems. Delays are inherent in almost all biological processes (gestation delays, maturation delays, etc.). The distributed delay model which has been used in this application is always a time—invariant one (Manetsch, 1978).

A kth order time—invariant distributed delay is defined by the following first—order differential equations:

$$\frac{dr_1}{dt} = \frac{K}{DEL}(X(t)-r_1(t)) ,$$

$$\frac{dr_2}{dt} = \frac{K}{DEL}(r_1(t)-r_2(t)) , \tag{11}$$

$$\frac{dr_k}{dt} = \frac{K}{DEL}(r_{k-1}(t)-r_k(t)) ,$$

where the input is $X(t)$ and the output is $r_k(t) = y(t)$. The variables $r_1(t)$, $r_2(t)$, ..., $r_k(t)$ are the intermediate rates of the distributed delay. The storage $Q_i(t)$ in the ith stage of the delay process is

$$Q_i(t) = \frac{DEL}{K}r_i(t) . \tag{12}$$

Accordingly, the converse flow entity is

$$\frac{dQ_i(t)}{dt} = r_{i-1}(t)-r_i(t) . \tag{13}$$

Hence, considering the kth order time—variant distributed delay, the duration of delay is the function of time $DEL(t)$ and equation (12) is modified to

$$Q_i(t) = \frac{DEL(t)}{K}r_i(t) . \tag{14}$$

Hence

$$\frac{dQ_i(t)}{dt} = \frac{DEL(t)}{K}\frac{dr_i(t)}{dt} + \frac{dDEL(t)}{K}r_i(t) \ . \tag{15}$$

Then from equation (13) and (15)

$$\frac{dr_i(t)}{dt} = \frac{K}{DEL(t)}(r_{i-1}(t)-r_i(t)(1+\frac{dDEL(t)}{K})) \ , \tag{16}$$

which is the basic equation of time—variant distributed delay. If predation losses are included this becomes

$$\frac{dQ_i(t)}{dt} = r_{i-1}(t)-r_i(t)-p_i(t)\ Q_i(t) \ , \tag{17}$$

where $p_i(t)$ is the instantaneous predation rate by predators and

$$\frac{dr_i(t)}{dt} = \frac{K}{DEL(t)}(r_{i-1}(t)-r_i(t)\ (1+p_i(t)\cdot\frac{DEL(t)}{K}+\frac{dDEL(t)}{K})) \ . \tag{18}$$

In most insect populations the time required for maturation from one growth stage (instar) to another is directly related to the ambient temperature. Therefore daily ambient temperature is used to calculate an instantaneous value of the maturation delay from

$$DEL(t) = \frac{TDD}{Max((T(t)-T_0),0)} = \frac{TDD}{f(t)} \tag{19}$$

It is noted that DEL (t) = ∞ when $(T(t)-T_0) \leq 0$ and $DEL(t) = TDD$ when $(T(t)-T_0) = 1$, as required.

Age Specified Model

This is a general model that simulates population growth in insects grouped both by stages and age classes (Tummala et al., 1984). The age structure within each stage is incorporated by dividing each stage into ten age classes. If we assume that $N_{ij}(t)$ represents the number of individuals in stage i and age class j, then the age class dynamic changes within this stage can be represented by

$$N_i(t+1) = M'(a)\ N_i(t)+L(a)\ N_{i-1}(t) \tag{20}$$

where

$$N_i(t) = (N_{i1}(t), N_{i2}(t), ..., N_{i10}(t))^T,$$

and

$$M'(a) = (m_{ij}(a)), \text{ and } L(a) = (l_{ij}(a))$$

are 10 x 10 matrices that are functions of temperature a.

The population numbers in stage i, age class K, are denoted by $N_{i,k}(t+1)$. These are obtained by noting that

$$1) \quad N_{i,j}(t) \xrightarrow{P_i} N^A_{i,k}(t+1)$$

$$2) \quad N_{i-1}(t) \xrightarrow{P_{i-1}} N_{i,0}(t+1) \xrightarrow{P_i} N^B_{i,k}(t+1) \qquad (21)$$

$$3) \quad N_{i,k}(t+1) = N^A_{i,k}(t+1) + N^B_{i,k}(t+1)$$

when p_i is the fraction of total development of stage i at a given temperature in a given interval Δt, $p_i = \Delta t/D$ and $\Delta \leq p_i < 1$.

Here D represents the total development time of the insect in a given stage expressed in simulation units. The details of model development are omitted here. Interested readers may refer to Tummala et al. (1984).

Poisson Process

In general, most population dynamics models either do not consider the random behavior of individuals development carefully or make the random process of individuals too complex. Under the reasonable hypothesis that individual development behavior follows a Poisson process, a new age class model can be developed (Liu, 1987).

Let $D(t)$ be the level of development. The development increment within interval Dt is then

$$\int_t^{t+\Delta t} R(\tau)d\tau = D(t+\Delta t)-D(t) = j \qquad (22)$$

where, $R(t)$ is the development rate. The probability of developing an increment between each time interval, t_1-t_0, t_2-t_1, ... is then an independent random event. If the time interval is short enough, the probability of increment will follow the Poisson distribution

$$P\{D(t+1) - D(t) = j\} = e^{-r\Delta t} \frac{(r\Delta t)^j}{j!}. \tag{23}$$

The probability of insect from i age class developing into j age class in a time interval Δt is then

$$P_{ij} = P\{D(t+\Delta t) - D(t) = j-i\} = e^{-r\Delta t} \cdot \frac{(r\Delta t)^{j-1}}{(j-1)!}. \tag{24}$$

Suppose X and Y are two adjacent stages of development. Let the X stage be divided into two age classes X_1 and X_2. Then according to the Leslie Matrix

$$X_1(t+1) = P_{11}X_1(t)$$

$$X_2(t+1) = P_{12}X_1(t) + P_{22}X_2(t) \tag{25}$$

$$Y(t+1) = P_{13}X_1(t) + P_{23}X_2(t)$$

and

$$\sum_{j=i}^{n+1} P_{ij} = 1, \quad i = 1, 2, ..., n, \ j \geq i.$$

We analyze the probability of individuals finishing the X stage development and moving into the Y stage after each simulation time Δt as follows

Simulation time	State		Probability
	x_1	x_2	Y
0	1	0	0
1	$e^{-\tau'}$	$e^{-\tau'}r'$	$1-e^{-\tau'}-e^{-\tau'}r'$
2	$(e^{-\tau'})^2$	$2(e^{-\tau'})^2r'$	$e^{-\tau'}((r'+1)-e^{-\tau'}-2e^{-\tau'}r')$
K	$(e^{-\tau'})^k$	$K(e^{-\tau'})^k r'$	$(e^{-\tau'})^{k-1}((k-1)r'+1-e^{-\tau'}-Ke^{-\tau'}r')$

The probability of individuals entering Y stage in time i is then

$$P_i = (e^{-\tau'})^{i-1}((i-1)r'+1-e^{-\tau'}-e^{-\tau'}r') \tag{26}$$

where $r' = r \cdot \Delta t$.

Now, if the mean value (μ) and variance (δ^2) of the development duration are known then the following equations can be used to estimate the parameter r and Δt (τ):

$$\sum_{i=1}^{\infty} (\tau i) P_i = \mu$$

$$\sum_{i=1}^{\infty} (\tau i)^2 P_i = \delta^2 - \mu^2 .$$

(27)

Further, we can also consider the loss rate d in development:

$$X_1(t+1) = d_1 P_{11} \cdot X_1(t),$$

(28)

and

$$X_2(t+1) = P_{12} \cdot X_1(t) + d_2 \cdot P_{22} \cdot X_2(t) .$$

Assuming $d_1 = d_2$, parameter estimation is easy.

Comparison of the Models

Among these five models, ssp is the simplest to use, and is the most flexible. It can be easily modified to incorporate other theoretical population formula.

The Account Model, based on the life cycle of an insect, is a straightforward method, so it is more intutitive in the biological sense.

The age specific model, may be regarded as a generalization of the models proposed by Lewis (1942), Leslie (1943), and Lefkovitch (1965). Lewis and Leslie treated a species population as being divisible into age groups. Lefkovitch felt it would be impossible to look at individuals and estimate their age, and used stage groupings instead of age groupings.

The age specific model groups insects by stages and, therefore, may give more accurate simulation results. For example, in many references, predation rates are treated either as a constant within one stage or as a transition rates at the ecdyses (Wenke, 1974). These are both extreme viewpoints. Predation rate as an age specified function within a stage might be more realistic. In this case the age specified model will be a powerful simulation tool.

The DTVMPL process considers distributed delay. In this process, parameters DEL and D are significant to the real world process being modeled. Here DEL is either the expected value of the transit time of an individual entity through the given process or the mean of the probability density function describing the transit times of the population of entities passing through the process. The parameter K specifies a member of the Erlang family of density functions to describe the transit times of individual entities in a particular application of the distributed delay model.

The discrete Poisson model deals with the behavior of individual development of a group of free entities with random forward movement. The model provides us with detailed explanation of why the population development is not even, and how this unevenness is formed since the continuous developing flow can be treated as discrete jumps when this flow is observed withlimited times. The Poisson model is similar to the boxcar train model (Wit & Goudriaan, 1974) but the former avoids considering the transition happening only between two neighboring age classes, and the parameter estimation is easier than with age specific models. The Poisson model may be a powerful tool for us to research and model population dynamics, but this one can only be used simply because we have to add other important aspects of the development process to make the simulation realistic.

References Cited

Coulman, G. A., S. R. Reice & R. L. Tummala. 1972. Population modeling: a systems approach. *Science* 175: 518–521.

Gage, S. H. & A. J. Sawyer. 1979. A simulation model Eastern spruce budworm population in a balsam fir stand. *Model. and Simul.* 10: 1103–1114.

Haynes, D. L. , R. L. Tummala & T. L. Ellis. 1980. Ecosystem management for pest control. *Bioscience* 30: 690–696.

Koenig, H.E. & R. L. Tummala. 1972. Principles of ecosystem design and management. *IEEE Trans. Syst. Man. Cybern.* 2: 449–459.

Lee, K. Y., R. O. Barr, S. H. Gage, & A. H. Kharkar. 1976. Formulation of a mathematical model for insect pest control – the cereal leaf beetle problem. *J. Theor. Biol.* 59: 33–76.

Lefkovitch, L. P. 1985. The study of population growth in organisms grouped by stages. *Biometrics*: 1–18.

Leslie, P. H. 1945. On the use of matrices in certain population mathematics. *Biometrika* 33: 183–212.

Lewis, E. G. 1942. On the generation and growth of a population. *Sankhya* 6: 93–96.

Liu, C. 1987. A Poisson process model of insect populations dynamics development. M.D. Thesis, Institute of Zoology, Academy of Sciences of China.

Manetsch, T. J. 1978. Time–varying distributed delays and their use in aggregative models of large systems. *IEEE Trans. Syst. Man. Cybern.* 6: 547–553.

Stinner, R. E., R. L. Raff & J. R. Bradley. 1974. Population dynamics of Heliothis zea (Baddie) and H.virescens in North Carolina: a simulation model. *Environ. Entom.* 3: 163–171.

Tummala, R. L., D. Li & D. L. Haynes. 1984. General models for simulating population growth in insects grouped both by stages and age classes. *IEEE Trans. Syst. Man. Cybern.* 14: 339–345.

Wenke, W. W. 1974. Identification of viable biological strategies for pest management by simulation studies. *IEEE Trans. Syst. Man. Cybern.* 4: 379–386.

Wit, G. T. & J. Goudriaan. 1974. Simulation of ecological process. Wageningen, Centre for Agricultural publication and Documentation.

Potential Use of an Engineering–Based Computer Simulation Language (SLAM) for Modeling Insect Systems

W. Scott Fargo and W. David Woodson[1]

ABSTRACT   The engineering–based simulation language, SLAM (Simulation Language for Alternative Modeling) is proposed as a method for developing models of insect systems.   The advantages of the language are its ease of use, availability, convenience, and versatility.   The use of SLAM would allow the scientists involved in the modeling effort to concentrate on biological processes rather than computer programming.   An example is given in which the insect population on a single plant is modeled over a 60 day period.   The model is very simple yet performs quite well. Results of the simulation run compare favorably with field collected data.   The significance of the example is in illustrating the potential of SLAM in modeling insect systems.   Much of the information required to develop models for many insects already exists.   The use of a simulation language such as SLAM could greatly simplify and speed the development and implementation of these models.

Our objective in this presentation is to explore the use of the Simulation Language for Alternative Modeling (SLAM) in simulating insect systems.   SLAM is an engineering–based computer simulation language originally developed for modeling industrial manufacturing processes.   After a short time, however, applications were realized in risk analysis, evaluation of security systems, insurance work flow analysis, automated warehouse simulation, system analysis of chemical manufacturing, and a myriad of industrial engineering and management science uses (Pritsker 1986).

The broad appeal of SLAM is attributable to its ability to overcome problems associated with the use of conventional computer languages in simulation modeling.   Many simulation models written in the agricultural and biological sciences are coded in FORTRAN, BASIC, APL or similar languages.   The use of any of these languages requires that either the scientist(s) involved have a thorough understanding of the language with which they are working, or that they hire professional programmers.   Of these two alternatives, the first is time–consuming and the second, expensive.   The development of integrated pest management (IPM) systems for various agricultural commodities is increasingly becoming a simulation–based exercise.   The use of languages such as those mentioned requires an extreme duplication of effort.   This overlap involves such tasks as writing random number generators, programming distribution functions required for processes in the system, event scheduling and other time related functions. Many of these functions are reprogrammed in each simulation effort.   Additionally, since the functions and variables modeled may require either discrete or continuous models, many languages require access to mainframe libraries of mathematical techniques such as the International Mathematics

---

[1]Department of Entomology, Oklahoma State University, Stillwater, Oklahoma 74078.

and Statistics Library (IMSL). When mixed models are required, the procedural requirements and costs incurred in data processing become staggering.

SLAM has the capability to overcome all of these limitations. Most of the functions and distributions are readily available and are executed rapidly. If an additional distribution or functional process is required, it may be easily added to the program through user written functions and subroutines. SLAM has the added advantage of taking over the "bookkeeping" and time related attributes of the simulation problem. Either continuous, discrete or mixed models may be written using SLAM. A useful feature of the SLAM software is the ability to trace the flow of program execution through the simulation process. This debugging tool is extremely valuable and reduces program development time significantly.

With the SLAM software essentially relieving the scientist of the drudgery, time, and programming associated with the modeling effort, more emphasis and thought can be directed towards the underlying biological processes of the systems under study.

The use of simulation languages in modeling insect systems would substantially reduce the costs of program development both in terms of time and money. We are encouraged in this proposal by the earlier work of Barrett et al. (1978) who used the computer language GASP IV (the forerunner to SLAM) in simulating Hessian fly, *Myetiola destructor* (Say), populations.

An additional advantage of SLAM is its availability for personal computers (PC) allowing much of the initial model development to be done inexpensively. The PC version of SLAM is upwardly compatible to the mainframe version. Thus, when multiple simulations are desired or the model becomes very large, it can be uploaded to the mainframe computer for execution.

Example of SLAM's Utility

A simple example of the potential of SLAM in modeling biological processes is included in this presentation to illustrate its potential in modeling insect systems. Over the past six years much information has been collected regarding the squash bug, *Anasa tristis* (DeGeer), and squash plant, *Cucurbita pepo* L., system. Immature development of the squash bug, as well as adult fecundity, longevity, and egg viability have been examined under constant temperatures in the laboratory (Fargo 1986, Fargo & Bonjour 1988, Al–Obaidi 1987). Additionally, the field dynamics of both the plant and insect have been monitored for two seasons (Fargo et al. 1988). All of this data could be used in the development, validation, and verification of a simulation model of the system.

For this example we have modeled the development of the insect population per plant on a daily basis. Fig. 1 is a SLAM network model of the insect dynamics. The model is initiated by allowing one overwintered, mated, adult female squash bug to arrive on the plant. Her longevity is randomly drawn from a normal distribution ($\mu = 42.0$, $\sigma = 10.0$).

She is then given a value for daily fecundity selected from a triangular distribution (Low = 1.0, Mode = 2.0, High = 3.0). At this point, the age of the female is checked, and if she has

249

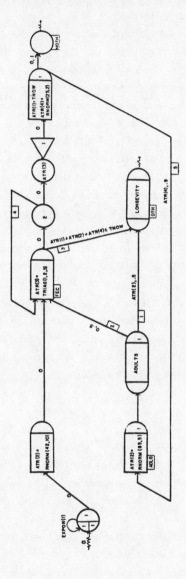

Fig. 1. SLAM network model for the squash bug system.

reached her assigned longevity she is routed to a death or termination node along Activity 3. If she is not terminated, she is allowed to lay her complement of eggs and is then routed back to the fecundity assignment node with a one–day delay, on Activity 4. She continues in this cycle until her death.

The eggs that she has oviposited are each assigned the current simulation time as their "birthday". Then each egg is assigned an individual immature (egg to adult) development time from a normal distribution ($\mu = 25.5$, $\sigma = 1.7$). Next, 10% of the immature population is routed to a termination node corresponding to egg and nymphal mortality. The remainder are sent via Activity 5 to another assignment node, the time taken to arrive being their development time.

When the insects reach the assignment node, they are given an adult longevity time taken from a normal distribution ($\mu = 84.0$, $\sigma = 5.0$). The adults are directed to a collection node where the average time between the creation of adults is computed along with the total number of adults arriving. One–half of the adults created are assumed to be females and are routed along Activity 2 to the fecundity node. The other half, representing the males, are routed along Activity 1 to the termination node. The time that it takes to arrive being the longevity of each adult male.

The population of live insects may be continuously monitored as follows: the number of females is given by the number currently in Activity 4, the number of immature insects is the total in Activity 5, and the males are in Activity 1. The total population is obtained by summing these three activity counts.

The driving program for this model is given in Fig. 2. For the many activities and distributions included in the model, the code is surprisingly brief. Many of the input lines relate to collecting run time statistics and describing the plotted output desired by the programmer. SLAM also includes a comprehensive output phase which relieves the programmer of formatting output reports and generating statistics associated with the simulation.

Results

An example of this output for the simple model presented here is given in Fig. 3. This figure shows that the simulation was allowed to run for 60 days. A total of 72 adults were created during that time (31 males and 41 females which developed from eggs). The overwintered female died after 38 days of simulation.

Fig. 4 is also a SLAM report and shows the dynamics of the system over the last 30 days of simulation. SLAM will automatically scale the plots unless a scale for each variable is assigned, as in the present case. The number of adult males has a minimum of value of three on day 30 and increases to 31 by day 60. The four females on day 30 increase to 41 over the course of the simulation. The number of immature insects begins at 43 and increases to 578 at 60 days. Summing all the life stages (eggs to adults) the total population increases from 50 to 650 over the last 30 days of simulation.

Table 1 compares these simulated results with field collected data (Fargo et al. 1988). After 30 days, the field population consisted of two adult males, two adult females, and 59 immatures

```
GEN, W. S. FARGO, A. TRISTIS DYNAMICS,9/14/87,1;
LIMITS,1,5,1250;
NETWORK;
; *** SIMULATE IMMIGRATING OVERWINTERING INSECTS ***
      CREATE,EXPON(1.),0,1,1,1;
      ASSIGN,ATR(2)=RNORM(42,10); OVWNTRNG ADULT LONGEVITY
      ACT,,,FEC;
ADLG  ASSIGN,ATR(2)=RNORM(84,5); SET ADULT LONGEVITY
      ACT,,,COL;
COL   COLCT(1),BET,ADULTS CREATED,,1;
      ACT/2,0,.5,FEC; SEND FEMALES TO OVIP CYCLE
      ACT/1,ATR(2),.5,DTH; MALES DIE;
FEC   ASSIGN,ATR(3)=TRIAG(1,2,3),1;SET DAILY FECUNDITY
; ***CHECK IF FEMALES HAVE REACHED ADULT LONGEVITY ***
      ACT/3,,ATR(1)+ATR(4)+ATR(2).LE.TNOW,DTH;FEMALE DIE
      ACT,,,GOON;
GOON  GOON,2;
      ACT/4,1,,FEC; FEMS TO FECUNDITY
      ACT,0,,UNBAT; SEND ENTITY TO UNBATCH
UNBAT UNBATCH,3,1; SEPARATE EGGS FOR PROCESSING
      ASSIGN,XX(1)=NNACT(1)+NNACT(4)+NNACT(5);
      ASSIGN,ATR(1)=TNOW; SET OVIPOSITION TIME TO CURRENT TIME
      ASSIGN,ATR(4)=RNORM(25.5,1.7),1;EGG TO ADULT TIME
      ACT/5,ATR(4),0.9,ADLG; SEND NYMPHS TO ADULTHOOD
      ACT,0,0.1,NDTH; 10% OVERALL NYMPHAL MORTALITY
;*** DEATH AND TERMINATE NODE ***
NDTH  TERM;
DTH   COLCT,INT(1),TIME TO DEATH;
      TERM;
      ENDNETWORK;
RECORD,TNOW,DAYS,0,P,1,30,60;
   VAR,NNACT(1),M,MALES,0,50;
   VAR,NNACT(4),F,FEMALES,0,50;
   VAR,NNACT(5),N,NYMPHS,0,700;
   VAR,XX(1),T,TOTAL,0,700;
INIT,0,60;
FIN;
```

Fig. 2. SLAM input model for SB population dynamics.

1

S L A M   I I   S U M M A R Y   R E P O R T

SIMULATION PROJECT A. TRISTIS DYNAMICS   BY W. S. FARGO

DATE   9/14/1987   RUN NUMBER   1 OF   1

CURRENT TIME   .6000E+02
STATISTICAL ARRAYS CLEARED AT TIME   .0000E+00

STATISTICS FOR VARIABLES BASED ON OBSERVATION

	MEAN VALUE	STANDARD DEVIATION	COEFF. OF VARIATION	MINIMUM VALUE	MAXIMUM VALUE	NO.OF OBS
ADULTS CREATED	.475E+00	.608E+00	.128E+01	.237E-02	.315E+01	71
TIME TO DEATH	.380E+02	.000E+00	.000E+00	.380E+02	.380E+02	1

FILE STATISTICS

FILE NUMBER	ASSOC NODE LABEL/TYPE	AVERAGE LENGTH	STANDARD DEVIATION	MAXIMUM LENGTH	CURRENT LENGTH	AVERAGE WAIT TIME
1		.000	.000	0	0	.000
2	CALENDAR	133.515	162.243	650	650	2.600

REGULAR ACTIVITY STATISTICS

ACTIVITY INDEX/LABEL	AVERAGE UTILIZATION	STANDARD DEVIATION	MAXIMUM UTIL	CURRENT UTIL	ENTITY COUNT
1 MALES DIE	6.6200	8.3592	31	31	0
2 SEND FEMALES	.0000	.0000	1	0	41
3 FEMALE DIE	.0000	.0000	1	0	1
4 FEMS TO FECU	7.8508	9.3494	41	41	451
5 SEND NYMPHS	119.0443	144.7232	578	578	72

Fig. 3. Simulation summary table generated by SLAM.

PLOT NUMBER 1

```
F=FEMALES     .000E+00              .250E+02              .500E+02
N=NYMPHS      .000E+00              .350E+03              .700E+03
T=TOTAL       .000E+00              .350E+03              .700E+03
              0    10   20   30   40   50   60   70   80   90  100 DUPS
DAYS

    .3000E+02  +   MF                      +                    + MN FT
    .3100E+02  +   MF                      +                    + MN FT
    .3200E+02  +   MF                      +                    + FN FT
    .3300E+02  +   MT                      +                    + MF MN
    .3400E+02  +   MT                      +                    + MF MN
    .3500E+02  +   MNF                     +                    + FT
    .3600E+02  +    MF                     +                    + FN FT
    .3700E+02  +    MF                     +                    + MN FT
    .3800E+02  +     MF                    +                    + MN FT
    .3900E+02  +      MT                   +                    + MF MN
    .4000E+02  +      MT                   +                    + MF MN
    .4100E+02  +      MNF                  +                    + FT
    .4200E+02  +     M NF                  +                    + FT
    .4300E+02  +       MF                  +                    + MN FT
    .4400E+02  +       MNFT                +                    +
    .4500E+02  +        MF T               +                    + FN
    .4600E+02  +        MFN T              +                    +
    .4700E+02  +        MF N T             +                    +
    .4800E+02  +         M  N T            +                    + MF
    .4900E+02  +          M  N T           +                    + MF
    .5000E+02  +           M  N T +                             + MF
    .5100E+02  +          FM    N T +                           +
    .5200E+02  +           M    N  T                            + MF
    .5300E+02  +          F M     N+T                           +
    .5400E+02  +          F M     N  T                          +
    .5500E+02  +              M   F  N  T                       +
    .5600E+02  +              M +  F N  T                       +
    .5700E+02  +               M     FN   T                     +
    .5800E+02  +             + M       FN   T                   +
    .5900E+02  +               +   M     F N   T                +
    .6000E+02  +               +       M      F    T     + FN
              0    10   20   30   40   50   60   70   80   90  100 DUPS
DAYS
```

Fig. 4. Population plot generated by SLAM.

for a total population of 63 insects. These numbers increased until at 60 days the total number of insects per plant averaged 751. This population consisted of 49 males, 54 females, and 648 immatures.

Table 1. Comparison of simulated to actual squash bug populations.

LIFE STAGE	SIMULATION DURATION			
	30 DAYS		60 DAYS	
	MODEL	DATA	MODEL	DATA
Adult males	3	2	31	49
Adult females	4	2	41	54
Immatures	43	59	578	648
Total	50	63	650	751

Discussion

While this model shows the advantage of using SLAM over more usual computer languages, many ecological aspects of the system are not included in this model. Most importantly, the plant is not even present in the model, and thus insect–plant interactions are not incorporated. At present, the development, mortality, and fecundity of the insects are not related to temperature, and we know they must be. Also, individual development and mortality of each immature life stage are not presently included.

The significance of this example is that the simulation language exists, is readily available, and appears to be very promising in the modeling of insect systems. Additionally, most of the data required to develop the SLAM model for many insect systems has already been collected. The use of simulation languages such as SLAM could greatly simplify and speed the development and implementation of insect system models.

Acknowledgement

We appreciate the critical review of an earlier draft of this manuscript by J. H. Young, Department of Entomology, and L. J. Young, Department of Statistics, Oklahoma State University.

References Cited

Al–Obaidi, A. A. 1987. Reproductive bionomics of the squash bug, *Anasa tristis* (*Heteroptera*: *Coreidae*) as affected by temperature. M. S. Thesis, Oklahoma State University, Stillwater, Oklahoma.

Barrett, J. R., J. E. Foster, G. A. Wong, & E. C. Stanley. 1978. Insect population simulation. Proc. Winter Simul. Conf. Inst. Elec. and Electronic Engineers.

Fargo, W. S. 1986. Estimation of life stage development time based on cohort data. *Southw. Entomol.* 11: 89–94.

Fargo, W. S., E. L. Bonjour, & T. L. Wagner. 1986. An estimation equation for squash (*Cucurbita pepo*) leaf area using leaf measurements. *Can. J. Plant Sci.* 66: 677–682.

Fargo, W. S. & E. L. Bonjour. 1987. Developmental rate of the squash bug, *Anasa tristis* (*Heteroptera*: *Coreidae*), at constant temperatures. *Environ. Entomol.* (Accepted).

Fargo, W. S., P. E. Rensner, E. L. Bonjour, & T. L. Wagner. 1988. Population dynamics in the squash bug, *Anasa tristis* (Hemiptera: Coreidae)/squash plant, *Cucurbita pepo* (curcurbitales: Cucurbitaceae) system. *J. Econ. Entomol.* (In press).

Pritsker, A. A. B. 1986. Introduction to Simulation and SLAM II 3rd Edition. John Wiley & Sons. New York.

Modeling Southern Pine Beetle (*Coleoptera: Scolytidae*) Population
Dynamics: Methods, Results and Impending Challenges

M. P. Lih and F. M. Stephen[1]

ABSTRACT SPBMODEL is a computer simulation model that predicts southern
pine beetle, *Dendroctonus frontalis* Zimmermann (*Coleoptera: Scolytidae*), infestation
growth in currently infested pine stands over a three–month period. SPBMODEL
estimates number of currently infested trees, cumulative total number of dead trees,
and associated timber volume and dollar losses in loblolly (*Pinus taeda* L.) or
shortleaf (*P. echinata* Mill.) pine stands.
  SPBMODEL simulates reproduction, development, and mortality of
stage–specific cohorts of southern pine beetle (SPB) within and between infested
trees. Three methods were used to estimate model parameters: field data (particular
mortality rates and densities), laboratory studies (development rates), and estimation
using the model (between–tree parameters).
  SPBMODEL has been modified extensively to accommodate prospective users,
i.e. forest pest managers. The model has been made available on U.S. Forest
Service's Data General mainframe computer, and a personal computer version of the
model has been developed and distributed. The model has been tested using data
from 70 infested spots in five southern states. Those tests showed mean absolute
error of 16.7% for predicted number of dead trees over a 92–day prediction period.
The model is being used by the Forest Service as an aid in making control
recommendations in SPB infestations. SPBMODEL also provides an efficient means
of testing research hypotheses, such as the importance of insect natural enemies in
controlling SPB spot growth.
  Our current research is focused on defining mechanisms of host tree resistance
and suitability and evaluating their roles on SPB reproduction and mortality. The
data base for testing the model's performance will be expanded to include a wider
geographic area. Information on overwintering SPB populations is lacking, and
incorporation of such data would enhance the model's usefulness to resource
managers.

The southern pine beetle, *Dendroctonus frontalis* Zimmermann (*Coleoptera: Scolytidae*), is a
major pest of southern pine forests. In 1986 more than 100 million dollars in damage resulted
from southern pine beetle infestations in southern forests (Connor et al. 1987). Although
population and damage levels fluctuate greatly from year to year, the threat of southern pine
beetle epidemics is a constant concern of forest pest managers.

A southern pine beetle infestation usually originates in spring or summer. Beetle–produced
pheromones, together with host tree odors, trigger a mass attack on the tree by beetles. The resin
system is a tree's primary defense, as beetles are flushed from the tree or entrapped by
crystallizing resin. If enough beetles aggregate in a sufficiently short period of time, the tree may
be unable to resist attack. The beetles then bore through the bark to the phloem/cambium
interface, mate, and construct winding galleries with egg niches on both sides. Parent beetles may

---

[1]Department of Entomology, University of Arkansas, Fayetteville, Arkansas 72701.

then reemerge from the tree and attack a neighboring tree, drawn by attractive chemicals in the advancing edge of the infestation, often called the "active front" of the "spot".

Young larvae emerge from the eggs and begin feeding in the phloem/cambium tissues. Late–stage larvae enter the outer bark to pupate and complete development. Emerging brood adults join other adult beetles in attacking attractive host trees nearby, resulting in infestations that continue to increase in size as the season progresses. From three to nine overlapping generations of beetles may develop each year, depending on geographic locality (Payne 1980). A number of factors influence final infestation size, including regional beetle population size, arthropod natural enemy densities, environmental conditions, host tree characteristics, and stand characteristics.

Forest pest managers must decide whether to control spots using salvage or cut–and–leave operations, or to let infestations run their natural course. This decision is affected by spot size, location, accessibility, ownership, and use, as well as market factors. SPBMODEL is designed to aid this decision–making process. SPBMODEL is a computer simulation model that predicts spot growth in currently infested loblolly (*Pinus taeda* L.) or shortleaf (*P. echinata* Mill.) pine stands over a three–month period. In addition, the user may request a prediction of timber volume and monetary loss. This paper examines model structure and development, model testing procedures, and application situations. In addition, areas in need of additional research are identified.

Methods

Model Structure

The model simulates reproduction, development, and mortality of stage–specific cohorts of southern pine beetle within and between trees. A graphical representation of SPBMODEL is presented in Fig. 1. Attacking adult beetles become parent adults, lay eggs, and reemerge from the tree, where they join the pool of emerging brood adult beetles. Within the tree, eggs hatch, larvae mature and pupate, brood adults develop and emerge from the tree. Ambient temperature influences attack, reemergence, rearrival, and egg production rates, as well as development rates of immature beetles. Beetles undergo mortality as they advance from one stage to another.

Sensitivity analyses were conducted on the model to determine relative impact on model output resulting from small percentage changes in values of various parameters (Taha et al. 1980). The numbers next to the parameters in Fig. 1 show the sensitivity ranking of the parameters, with "1." indicating the most sensitive parameter (BAAMR), i.e. the parameter which has the greatest impact on model output.

Estimation of Model Parameters

An extensive data base has been compiled on within–tree southern pine beetle population variables and associated host tree and stand characteristics. The data base represents nine years

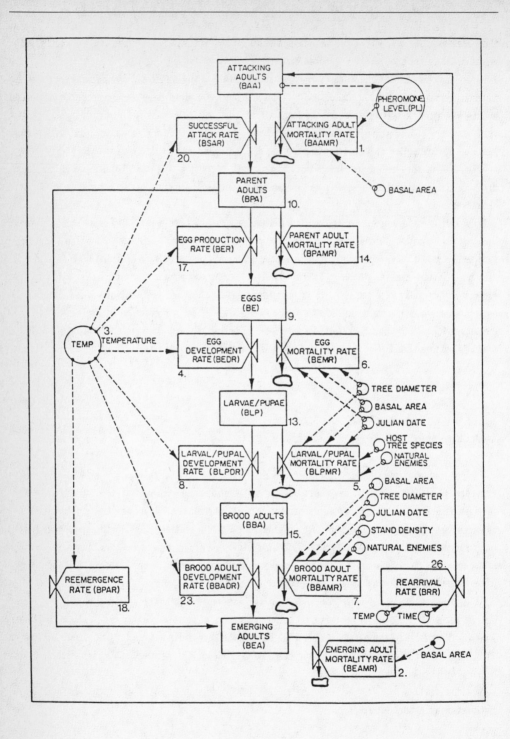

Fig. 1. Graphical representation of the southern pine beetle simulation model.

of intensive population sampling in four southern states during epidemic and endemic population conditions. These data were used to develop regression equations to predict the within–tree mortality rates of developing brood (BEMR, BLPMR, BBAMR) from tree and stand characteristics and time of year. Attack and egg densities also were predicted using stand characteristics (Table 1).

Table 1. Results of stepwise regression analyses[a] used to predict southern pine beetle population variables from tree and stand data[b].

Dependent Variable	Final Regression Equation[c]	R–SQ	P>F	df
Attack density/ 100 sq cm	$y=6.557-0.00743(PPH)+1.0536(LYRTIO)$	.62	.0001	28
Egg density/ 100 sq cm	$y=164.466+49.308(LYRTIO)-0.130(PPH)$	.37	.0023	28
Natural log of egg mortality rate	$y=-3.370-0.0369(PBA)-0.0426(DBH)+0.00864(JDAY)$	.63	.0010	19
Larval/pupal mortality rate	$y=0.417-0.0101(TBA)+0.00362(DBH)+0.00554(JDAY)$ $-0.0000146(JDAYSQ)+0.128(LYRTIO)$	.47	.0247	24
Brood adult mortality rate	$y=-0.384-0.0177(PBA)+0.0105(DBH)+0.00122(PPH)$ $+0.00509(JDAY)-0.0000111(JDAYSQ)$	.45	.0094	29

[a] PROC STEPWISE, SAS Institute Inc. 1982.

[b] Each observation is the mean of an infested spot–sampling period.

[c] PPH = pine per hectare; LYRTIO = proportion of pines that are loblolly;
PBA = pine basal area (sq m/ha); TBA = total basal area (sq m/ha);
DBH = mean tree diameter at breast height (cm); JDAY = Julian date;
JDAYSQ = JDAY*JDAY.

Parent adult mortality rate (BPAMR) is a constant. This rate is based on mean reemergence observed in a combined field/laboratory study (Cooper & Stephen 1978).

Two within–tree beetle development rates (BEDR, BBADR) are based on laboratory rearings of eggs and brood adults under different constant temperature regimes (Gagne 1980, Hines et al. 1980a, b). The data on brood adult development are modified to approximate observed optimal development under field conditions. Since data on egg development are most complete, this relationship is used as the basis for determining estimated development rates (BLPDR, BSAR, BPAR). The estimation procedure assumes that the sigmoid curvilinear relationship is appropriate for each rate being estimated. A schedule of within–tree life processes is produced by using a multiple of BEDR to estimate each unknown rate, and each rate is adjusted until the schedule conforms to field development times. This use of BEDR explains its high sensitivity when compared to other development rates. Constant temperature development data have often been used to estimate insect development rate curves and predict development times as functions of temperature. Wagner et al. (1984a, b) have recently reviewed the literature on this subject.

SPBMODEL uses an iterative methodology based on the rate summation approach (Wagner et al. 1985) to catalogue the life processes of all beetle stages except emerging adults. Using this approach, the beetle population is represented as a collection of cohorts, or groups of insects with similar developmental status. Each cohort has two attributes: size (i.e., number of insects) and cumulative developmental status. Cumulative development of each cohort is updated hourly based on development rate (or attack or production rate) associated with current ambient temperature. When cumulative development of a cohort reaches or exceeds one, cohort size is reduced by an appropriate mortality rate, and the cohort moves to the next life stage.

The population of emerging (and reemerging) adults is maintained as a pool that is depleted exponentially during the day to reflect flight. When beetles in flight land on pine surfaces they become attacking adults, and tree colonization begins. Flight occurs at temperatures above 57^{0}F (13.9^{0}C) and between 800 and 2000 hours.

Data are not available to reliably determine the values of three of the rates used in the model, i.e. attacking adult mortality rate (BAAMR), reemerged parent adult and emerged brood adult in–flight mortality rate (BEAMR), and adult beetle rearrival rate (BRR). A procedure was designed to estimate the values of these rates. This entailed setting the rates at the "best" values, based on knowledge of the biological system and limited data, and initializing the model with known input data. The output of the model (numbers of infested and dead trees) was compared to the observed data, and the three rates adjusted until the output approximated the observed data for the infested spot. This procedure was repeated for eight infested spots. The final values assigned the three rates agreed with our intuitive expected values for the rates (Taha & Stephen 1984). Egg production rate (BER) was estimated using a similar procedure, where number of eggs produced by 100 beetles over time was used as output data to determine the appropriate value for the rate. Egg production rate is expressed as a multiple of egg development rate (BEDR).

Ambient Temperature Estimation

SPBMODEL contains a subsystem that produces a profile of hourly temperatures from mean daily minimum and maximum temperatures (Hines et al. 1980b). The temperature file presently accessed by the model represents mean daily extreme temperatures in south central Arkansas. The user can modify the extreme temperatures by any constant number of degrees to better represent the location of the infested spot or anticipated temperature conditions. Temperature tables for the southeastern United States are presented in the User's Guide, along with a worksheet that can be used to determine whether the default temperature profile is adequate.

Model Input

A primary goal of the model–building process was to produce a decision–making aid for forest pest managers. Thus, both model input and model output are designed to satisfy that objective. Input primarily consists of easily–acquired forest stand variables, such as mean tree diameter at breast height (d.b.h.) and tree species present in the stand. The only entomological inputs required are counts of currently infested trees and trees that were previously infested but from which beetles have emerged. In addition, it is highly recommended that the user identify and input southern pine beetle brood stage that predominates at breast height in infested trees, particularly for spots with less than 20 infested trees. This requires examination of each tree, but this process can be completed rapidly by an experienced crew. As entomological expertise varies considerably among foresters, a generous default input system is included.

Model Output

Model output consists of a daily or weekly report of the predicted ranges of currently infested trees and cumulative dead trees. A range is presented rather than a single value for each output variable in order to account for some of the uncertainty involved in estimation of model parameters. If model parameter values were randomly selected from assumed probability distributions, then 90% of simulations would fall within the given range. The model is structured around the maturation and reproduction of beetles, but beetle numbers are converted to area of bark presently occupied by beetle life forms, and then to numbers of trees infested by various life stages, in order to calculate number of currently infested trees. The conversion process involves calculating stage–specific beetle densities present within the bark using initial attack and egg densities modified by pertinent mortality rates. Average infested bark area per tree is calculated from mean tree diameter at breast height (Stephen & Taha 1979). Cumulative number of dead trees represents a count of all trees that have been attacked and killed by beetles during the course of this infestation. Output can be adjusted to accommodate other research objectives, e.g. to follow a specific cohort of beetles through their life processes.

The user also can request an estimate of total timber volume and monetary loss to be expected if the infestation proceeds unchecked. This requires that the user input the d.b.h. distribution of the stand and current market prices for beetle–killed pulpwood and saw timber.

Description of Model Processes

When a user invokes the model, a COMMAND MODE statement appears. The user may then enter any of eleven commands in order to input, edit, display, or run an infested plot data set, or perform other tasks.

If the user enters RUN while in COMMAND MODE, simulation of spot growth begins. The user is asked to input the desired number of days of simulation, whether daily or weekly output is desired, whether an estimate of volume and monetary loss is requested, whether a copy of the input file and/or initial values of system parameters should be displayed with the output, and to specify any temperature modification that is to be in effect for this model run. System variables and file processing routines are initialized. Infested plot data are read and used to calculate input—dependent parameters and numbers of life forms in each developmental cohort. Current temperatures are calculated, and system constants and rates are initialized.

At this point the simulation clock, which keeps track of days and hours into the simulation, is incremented to the next hour and compared to the final simulation time to see if the end—of—simulation time has been reached. Hourly processes are updated, i.e. hourly development is calculated from development rates, life forms develop, and the attack process progresses. Then hourly rates, i.e. temperature—dependent development rates, are re—calculated for the next hour. When the clock reads 0200 hours, a period of minimal daily beetle activity, current status of the spot is recorded for output at the end of simulation. At 0000 hours the daily processes are updated, i.e. mortality rates and life stage densities are recalculated and the next day's temperatures are determined. This sequence is repeated until the end—of—simulation time is reached.

The user also is given an opportunity to modify certain model parameters during simulation. This is primarily a research tool but could also be used to specify changing stand conditions that may occur as the infestation expands. The model assumes uniform stand conditions unless otherwise indicated by the user.

At the end of simulation, output is displayed at the terminal. The model may append some warning messages to the output if extreme conditions were encountered during the simulation run. The user is given the opportunity to store model output in a file for later reference.

Results

SPBMODEL has been tested using infested spot data from 70 infested spots located in five states. Average absolute percentage error was calculated for cumulative number of dead trees and for number of currently infested trees using the following equation:

$$100 \cdot \left[\sum_{i=1}^{n} \left(|A_i - P_i| / A_i \right) \right] / n$$

where A_i is the actual number of dead (or infested) trees, P_i is the predicted number of dead (or infested) trees, and n is the total number of observations (n = 350 for dead trees and n = 338 for infested trees, since each infested spot was observed several times over three months). Average percentage error was calculated as above except the residual rather than the absolute value of the residual was used in the equation. The calculations use the individual value of dead or infested trees that is predicted by the model for a given spot and day; the range of values that is output does not enter into these testing procedures.

Model predictions had a mean absolute error of 16.7%, and a mean error of –2.8%, for cumulative number of dead trees over a 92–day prediction period, not including the initial day of input. For number of currently infested trees the mean absolute error was 66.0%, and mean error was 8.6%. Average absolute percentage error provides the mean percentage error in the predicted values over all infested plots tested and includes all 92 days of the prediction, while average percentage error gives an indication of the tendency of the model to overpredict or underpredict. The model more accurately predicts dead trees than infested trees (Figs. 2 & 3). The error values indicate that many of the observed values would fall within the predicted range that is output.

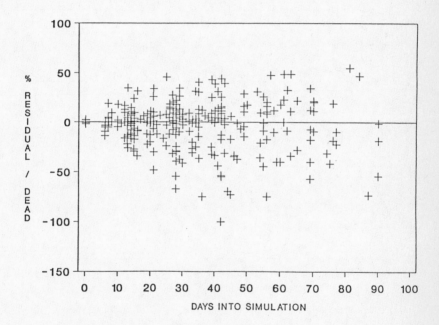

Fig. 2. Deviation of simulated infestations from actual data for cumulative numver of dead trees, expressed as percent residual: 100 X (Actual – Prediction) / Actual.

Discussion

SPBMODEL is a reliable decision–support tool for forest pest managers. The U.S. Forest Service uses the model to evaluate potential spot growth in wilderness and proposed wilderness areas, particularly in situations where private lands or Red–cockaded Woodpecker (*Picoides borealis*) colonies are threatened. SPBMODEL predictions are used to evaluate cost effectiveness of control/non–control decisions, for instance, to assess what monetary losses might have resulted if control measures had not been taken or to justify the decision to let a spot grow or decline on its own. SPBMODEL has been included in a prediction module of an Expert System being developed at Texas A&M University. The model is used by the Texas Forest Service, and private companies are beginning to explore its usefulness.

SPBMODEL is available through the U.S. Forest Service Data General computer, and a Personal Computer version of the model has been widely disseminated to State Forestry personnel and pest management researchers. A <u>User's</u> <u>Guide</u> includes field data collection and data summary forms, temperature tables, a description of the model, and a sample computer session.

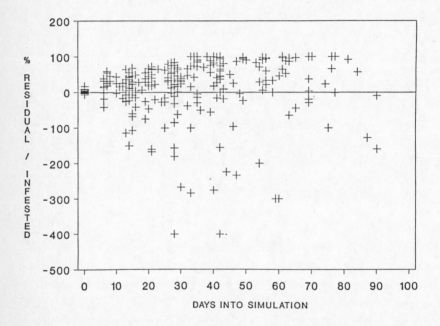

Fig. 3. Deviation of simulated infestations from actual data for number of currently infested trees, expressed as percent residual: 100 X (Actual − Predicted) / Actual.

The model has been used to test research hypotheses, such as the impact of natural enemies on spot growth (Stephen & Lih 1985, Stephen et al. 1988). SPBMODEL was used as an additional check on potential spot growth in a recent study that evaluated effects of the pesticide fenitrothion applied by the hack–and–squirt method on southern pine beetle population dynamics.

Impending Challenges

Our current research is focused on defining mechanisms of host tree resistance and suitability and evaluating their roles on southern pine beetle reproduction and mortality (Stephen & Paine 1985, Paine & Stephen 1987a, b, c). Data have been collected on numbers of unsuccessful attacks, but the study populations were rather atypical, and additional information must be collected before modifications can be implemented.

Regional beetle population levels appear to reflect some measure of population quality and infestation growth potential. A conceptual model has been developed that expresses stand risk (i.e. the potential for an infestation to start and grow) as a functional relationship between beetle population size and physiological condition of trees (Paine et al. 1984). Another area that has received little study is dynamics of overwintering beetle populations. Currently the model may be initialized between May and October, as data on southern pine beetle population dynamics are only available for these months. SPBMODEL may require extensive revisions to accommodate year–round predictions, as it appears that beetle population dynamics differ considerably between seasons, and the model is not currently structured to accommodate these differences. The anticipated restructuring will be examined in more depth soon, as the U.S. Forest Service has expressed an interest in using the model on a year–round basis. The model has been tested for much of the range of southern pine beetle using infestation data from Alabama, Arkansas, Louisiana, Mississippi, and Texas. In order to extend its utility, additional data will be collected from the Piedmont region of Georgia.

There are several difficult problems that are inherent to southern pine beetle population dynamics research. The spot growth data used to test model predictions must reflect natural growth patterns over extended periods of time. Many landowners are reluctant, unwilling, or unable to risk unchecked spot growth. Thus, most data used for testing the model are collected on national forests, which may introduce some biases in stand and age types. Collection of population samples from standing trees is a highly labor intensive and expensive operation, as is intensive interpretation and analysis of these samples in the laboratory. Also, beetle population dynamics vary tremendously from region to region and year to year, necessitating extensive study before significant conclusions can be drawn. SPBMODEL currently performs well under limited, but nonetheless very useful, situations. There is a need to extend its predictive capabilities to other regions and other seasons. Many subtleties of beetle dynamics are only now beginning to receive attention, such as interactions of the beetle with its three species of symbiotic fungi. The model will be improved and be even more useful as additional data become available that help unravel this complex ecological system.

Acknowledgements

The authors gratefully acknowledge the contributions of H. A. Taha, G. S. Hines, M. Motamedi, J. Ghosh, G. W. Wallis and R. C. Sanger. Thanks are extended to K. G. Smith, T. J.

Kring and W. C. Yearian for their reviews of this manuscript. This paper is published with the approval of the Director, Arkansas Agricultural Experiment Station. This study was supported in part by the U. S. Department of Agriculture's Integrated Pest Management Research, Development and Applications Program, Pineville, LA.

References Cited

Connor, M. D., M. Remion, J. Solomon, & D. Ward. 1987. Survey of damage caused by forest insects in the southeast in calendar year 1986. Southern forest insect work conference, Committee on losses caused by forest insects, mimeo. 7 August 1987, San Antonio, TX.

Cooper, M. E., & F. M. Stephen. 1978. Parent adult reemergence in southern pine beetle populations. *Environ. Entomol.* 7: 574–577.

Gagne, J. A. 1980. The effects of temperature on population processes of the southern pine beetle, *Dendroctonus frontalis* Zimmermann. Ph.D. dissertation, Texas A&M University, College Station.

Hines, G. S., H. A. Taha, & F. M. Stephen. 1980a. Model for predicting southern pine beetle population growth and tree mortality, pp. 4–12. In F. M. Stephen, J. L. Searcy, & G. D. Hertel [eds.], Modeling southern pine beetle populations, Symposium proceedings. 20–22 February 1980, Asheville, N.C. U.S. Department of Agriculture, Forest Service, *Tech. Bull.* 1630.

Hines, G. S., F. M. Stephen, & H. A. Taha. 1980b. Uses and structure of a simulation model for investigating the population dynamics of a southern pine beetle, pp. 696–699. In Proceedings of the 1980 summer computer simulation conference. 25–27 August 1980, Seattle, WA.

Paine, T. D., F. M. Stephen, & H. A. Taha. 1984. Conceptual model of infestation probability based on bark beetle abundance and host tree susceptibility. *Environ. Entomol.* 13: 619–624.

Paine, T. D., & F. M. Stephen. 1987a. Fungi associated with the southern pine beetle: avoidance of induced defense response in loblolly pine. *Oecologia* 74: 377–379.

Paine, T. D., & F. M. Stephen. 1987b. Influence of tree stress and site quality on the induced defense system of loblolly pine. *Can. J. For. Res.* 17: 569–571.

Paine, T. D., & F. M. Stephen. 1987c. The relationship of tree height and crown class to the induced plant defenses of loblolly pine. *Can. J. Botany* 65: 2090–2092.

Payne, T. L. 1980. Life history and habits, pp. 7–28. In R. C. Thatcher, J. L. Searcy, J. E. Coster, & G. D. Hertel [eds.], The southern pine beetle. U.S. Department of Agriculture, Forest Service, *Tech. Bull.* 1631.

SAS Institute Inc. 1982. SAS user's guide: statistics, 1982 edition. SAS Institute Inc., Cary, N.C. pp. 101–110.

Stephen, F. M., & M. P. Lih. 1985. A *Dendroctonus frontalis* infestation growth model: Organization, refinement, and utilization, pp. 186–194. In S. J. Branham & R. C. Thatcher [eds.], Integrated pest management research symposium: The proceedings. 15–18 April 1985, Asheville, N.C. U.S. Department of Agriculture, Forest Service, *Gen. Tech. Rep.* SO–56.

Stephen, F. M., & T. D. Paine. 1985. Seasonal patterns of host tree resistance to fungal associates of the southern pine beetle. *Z. ang. Ent.* 99: 113–122.

Stephen, F. M., & H. A. Taha. 1979. Tree mortality, infested bark area, and beetle population measurements as components of treatment evaluation procedures on discrete forest management units, pp. 45–53. In J. E. Coster & J. L. Searcy [eds.], Evaluating control tactics for the southern pine beetle: Symposium proceedings. 30 Jan. – 1 Feb. 1979, Many, La. U.S. Department of Agriculture, Forest Service, *Tech. Bull.* 1613.

Stephen, F. M., M. P. Lih, & G. W. Wallis. 1988. Impact of arthropod natural enemies on *Dendroctonus frontalis* (*Coleoptera: Scolytidae*) mortality and their potential role in infestation growth. In D. L. Kulhavy & M. C. Miller [eds.], Potential for biological control of *Dendroctonus* and *Ips* bark beetles: Conference proceedings. 9 Dec. 1986, Reno, NV. Stephen F. Austin State University Press, Nacogdoches, TX (in press).

Taha, H. A., F. M. Stephen, & M. Motamedi. 1980. Sensitivity analysis and uncertainty in estimation of rates for a southern pine beetle model, pp. 13–19. In F. M. Stephen, J. L. Searcy, & G. D. Hertel [eds.], Modeling southern pine beetle populations, Symposium proceedings. 20–22 February 1980, Asheville, N.C. U.S. Department of Agriculture, Forest Service, *Tech. Bull.* 1630.

Taha, H. A., & F. M. Stephen. 1984. Modeling with imperfect data: A case study simulating a biological system. *Simulation* 42(3): 109–115.

Wagner, T. L., H. Wu, P. J. H. Sharpe, R. M. Schoolfield, & R. N. Coulson. 1984a. Modeling insect development rates: a literature review and application of a biophysical model. *Ann. Entomol. Soc. Am.* 77: 208–225.

Wagner, T. L., H. Wu, P. J. H. Sharpe, & R. N. Coulson. 1984b. Modeling distributions of insect development time: a literature review and application of the Weibull function. *Ann. Entomol. Soc. Am.* 77: 475–487.

Wagner, T. L., H. Wu, R. M. Feldman, P. J. H. Sharpe, & R. N. Coulson. 1985. Multiple–cohort approach for simulating development of insect populations under variable temperatures. *Ann. Entomol. Soc. Am.* 78: 691–704.

Application of Catastrophe Theory to Population Dynamics of Rangeland Grasshoppers

Dale R. Lockwood[1] and Jeffrey A. Lockwood[2]

ABSTRACT Catastrophe theory is a form of non–linear mathematics which allows predictions of discontinuities within systems that are governed by a relatively small number of control variables and are derivable from smooth potentials. Grasshopper population dynamics appear to possess properties which can be described by a cusp catastrophe. Previous work has shown that weather conditions at the time of hatching and early development appear to mediate outbreaks and crashes of grasshopper populations. Using several measures of temperature and precipitation and various expressions of grasshopper infestations in Wyoming, we developed a series of models using Catastrophe Theory. The cusp catastrophe based on outbreak levels of infestation (> 9.6 grasshoppers per m²) and monthly (April and May) weather data generally provided the most appropriate descriptions of grasshopper population dynamics; the best models were up to 76% accurate based on *a priori* criteria. Catastrophic changes in grasshopper populations occur on a very large scale (state–wide), and it appears that different weather parameters affect high and low density outbreaks.

In all of the natural sciences a recent trend has been to re–examine phenomena using techniques in non–linear mathematics. This research has shed new light on the unexplainable perturbations in real world models. Physics, chemistry and engineering have seen most of the impact of these new approaches, because the phenomena in these fields are easily described by exact mathematical equations. The biological sciences have also been greatly impacted by non–linear mathematics. The study of population dynamics has been one of the most affected fields. It was the intent of this research to develop a model for grasshopper populations in Wyoming.

Catastrophe Theory can describe the qualitatively different discontinuities in a given system of n state variables and m control variables. It can model events that are usually continuous but exhibit discontinuities and describe singularities which depend on the number of control variables. The inherent assumptions of Catastrophe Theory include the following. The system to be modeled is under control of few (usually no more than five) control variables. Specifically, if there are four or fewer control variables, only seven possible catastrophes can occur. As Saunders (1980) points out, this assumption is less of a restriction than it might seem since we can leave out any independent variables that do not have a significant effect on the discontinuity that we are studying. Moreover, if a system is behaving in a discontinuous manner with more than five independent variables critically involved, then it is obviously going to be extremely difficult to

---

[1]International Business Machines, Sunnyvale, California 94086

[2]Department of Plant, Soil and Insect Sciences, University of Wyoming, Laramie, Wyoming, 82071.

make sense of it by any means. The only other assumption is that the dynamic of the system is derivable from a smooth potential, although the requirement of a potential is much more rigorous than we really need. The potential does not necessarily imply that the system possesses gradient dynamics. So, the population does not have to follow an ordinary differential equation, but could be governed by partial differential equations or limit cycles.

For reasons which will become clear, we were concerned with one catastrophe in relation to the grasshopper populations, viz, the cusp catastrophe. The cusp catastrophe occurs in a response surface of a state variable in relation to two control variables. The potential is described by the polynomial:

$$V(x) = x^4 + ux^2 + vx,$$

where x is the state variable and u and v are the control variables. As the control variables change, the state variable transverses the equilibrium surface. If it crosses the cusp edge, it will suddenly jump to the other leaf of the surface (Fig. 1).

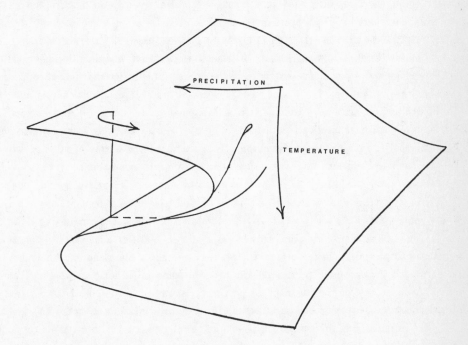

Fig. 1. The equilibrium surface of a cusp catastrohpe for rangeland grasshopper outbreaks showing how a sudden jump may occur.

It is this bifurcation that is most important in the application of the cusp to the development of a grasshopper population model. Since the potential is fourth order, the phase space is three—dimensional and the cusp surface is described by

$$4x^3 + 2ux + v = 0.$$

The coefficients u and v are the normal and splitting factors, respectively (Saunders 1980). As v changes from negative to positive and $u \leq 0$, bifurcation in the surface can occur; while if u remains positive, the changes in x are only smooth (Cobb 1980).

There are five flags that are inherent in a cusp catastrophe (Zeeman 1976). It is the existence of these flags in the grasshopper data that suggests that the grasshopper population follows a cusp model. The first flag, sudden jumps, has been explained above. The second is hysteresis; the jumps up do not occur in the same location as the jumps down. Inaccessibility is the third flag, exhibited by areas of the equilibrium surface where the state variable is unstable and will not stay but jumps one way or the other. Divergence occurs when two nearby trajectories eventually produce drastically different results in the position of the state variable. The final flag is bimodality, where a given set of values in the control variables can result in two stable positions of the state variable.

The ability to predict grasshopper outbreaks is generally recognized as a fundamental component of an effective, integrated pest management strategy. Entomologists agree that fluctuations in numbers of grasshoppers and weather conditions are in some way intrinsically related, notwithstanding the fact that all attempts to show a consistent, direct correlation have been largely unsuccessful (Watts et al. 1982). In general, the weather conditions during egg hatch and early development of grasshoppers seem to be the primary abiotic determinants of population dynamics. Gage & Mukerji (1977) examined 32 years of grasshopper annual survey data in context of weather variables to determine trends in population dynamics. Heat accumulation above 1600 $days_{50}$ and precipitation less than 254 cm during April–June were found to favor high grasshopper populations. Nerney & Hamilton (1969) examined the effects of fall and winter precipitation and subsequent range conditions on grasshopper populations over 15 years. Abundant range forage following above average precipitation preceded grasshopper population increases. In dry years, plant growth was retarded and grasshopper populations decreased drastically and remained low for two to four years regardless of subsequent conditions. Edwards (1960) also reported a high correlation between grasshopper densities and weather conditions (temperature and precipitation). Watts et al. (1982) noted that a cause and effect relationship between weather and grasshopper populations can not be defended; it is not clear whether there is a direct action of weather on the grasshopper's physiology, an indirect action on the food plants, a differential effect on their predators, parasites and diseases, or a measure of each. This lack of understanding of underlying mechanisms of population regulation has led to attempts to model grasshopper population dynamics without regard to specific external variables using methods such as autocorrelation (Gage & Mukerji 1977) and Markov chain models (Lockwood & Kemp 1987, Kemp 1987). However, the traditional forecasting technique used in the U.S. and Canada is not based on assessment of environmental parameters but relies on summer adult and fall egg surveys, and this technique has never provided the desired degree of accuracy (Edwards 1960).

Materials and Methods

Data Acquisition

Climatic data were determined from the Water Resources Data System (Wyoming Water Research Center, Laramie, WY). The mean temperature and total precipitation for April and May, the mean annual temperature, and the total annual precipitation for the 81 stations in Wyoming which had complete weather records from 1960 to 1984 were averaged for each year over all stations. Annual U. S. Department of Agriculture–Animal and Plant Health Inspection Service adult grasshopper survey maps from 1960 to 1984 were used to quantify infestations. These surveys classified infestations as noneconomic, < 3.6 grasshoppers/m^2, threatening, ≥ 3.6 grasshoppers/m^2 but < 9.6 grasshoppers/m^2, or outbreak, ≥ 9.6 grasshoppers/m^2. The state of Wyoming was divided into 414, 6,950 km^2 blocks, and for each year of available data, each block was classified as either infested (having any area with a given level of infestation) or uninfested (having no area with a given level of infestation).

Model Development

Since the data were biological, fluctuating and nonrepeatable, the model was developed statistically instead of being based purely on the canonical model. The cusp estimation program of Cobb (1980) was modified for this purpose. Using this approach, the surface equation was redefined as:

$$0 = a + b(y-x)/s - [(y-x)/s]^3,$$

where x is the sample mean and s is the sample standard deviation. If $z = (y-x)/s$, then the equation becomes:

$$0 = a + bz - z^3,$$

with z being the standardized state variable. In the program, the canonical factors, a and b, are approximated by:

$$a = a_0 + a_1x_1 + \ldots + a_nx_n,$$
$$b = b_0 + b_1x_1 + \ldots + b_nx_n,$$

where the variables $x_1, \ldots, x_n$ are the actual measured control variables. The program calculates the splitting and normal factors, both standardized and nonstandardized, as well as the statistics. It then plots a representation of the control surface and plots all data in the surface. The method for estimating the cusp parameters is based on the method of moments (Cobb 1980). The program can be obtained by contacting the authors.

The model was executed using the number of infested blocks as the state variable and temperature and precipitation as control variables. Within this framework, three criteria of infestation and four criteria of weather were used, so 12 runs of the model were executed. The infestation criteria consisted of the number of blocks with a threatening infestation, the number of blocks with an outbreak infestation, and the number of blocks with threatening or outbreak infestations. The weather criteria included the mean temperature and total precipitation for April, May, April and May combined, and the entire year. These time periods were chosen based on the timing of egg hatch in Wyoming and the susceptibility of early instars to climatic conditions. The catastrophes predicted by the model (i.e., a prediction of an upper and lower surface in context of the sign of the bifurcation [Stewart & Peregoy 1983]) were compared with the observed population dynamics. A catastrophe in the actual grasshopper populations was defined *a priori* as a 50% change in the number of infested blocks from one year to the next.

Results

When the program was executed with the 12 data sets, all of the analyses resulted in two or more catastrophes. When infestations were expressed as the number of blocks with outbreak infestations there were seven to nine predicted catastrophes, with threatening infestations there were three to five catastrophes, and with threatening/outbreak infestations there were two to seven catastrophes, depending on the expression of weather data.

To determine the accuracy of the various analyses with regard to real events, the catastrophes predicted by the model were compared to the actual occurrence of catastrophic events (Table 1). On the average, analyses which used annual weather data resulted in the fewest predicted catastrophes, followed by analyses which incorporated weather in April, April/May, and May, respectively. Analyses which were based on the number of blocks with outbreak infestations predicted more catastrophes than analyses which used threatening or threatening/outbreak infestations, respectively. No form of the model accurately predicted all observed catastrophic changes in the grasshopper population.

An ideal model would accurately predict events (catastrophes) both when they occur and when they do not occur. Thus, one measure of a model's accuracy is the proportion of correct predictions (including both correct occurrences and nonoccurrences). More specific measures of a model's accuracy include the proportion of correctly predicted occurrences and the proportion of correctly predicted nonoccurrences (Table 2). The best analyses with regard to correct prediction of catastrophes used outbreak densities and monthly weather data. Indeed, models using these parameters were the most accurate predictors of sudden jumps in the area infested, regardless of the density criterion used for the catastrophe. All models were generally less accurate in predicting catastrophes based on threatening/outbreak densities than catastrophes based on either level of infestation alone. No form of the model predicted more than 40% of the catastrophes based on the combination of outbreak and threatening densities or more than 57% of the

Table 1. Number of cases in which Catastrophe Theory models agree and disagree with observed grasshopper population dynamics.

Model		Infestation criterion [a]											
Infestation	Weather	Threatening				Outbreak				Threatening/Outbreak			
		A	B	C	D	A	B	C	D	A	B	C	D
Threatening	April/May	4	1.	14	6	2	3	12	8	1	4	14	6
	April	1	2	13	9	2	1	14	8	1	3	15	6
	May	1	2	13	9	3	0	15	7	1	2	16	6
	Annual	3	2	13	7	4	1	14	6	3	2	16	6
Outbreak	April/May	4	3	12	6	5	2	13	5	4	3	15	3
	April	5	2	13	5	5	2	13	5	4	3	15	3
	May	5	4	11	5	5	4	11	5	4	5	13	3
	Annual	4	3	12	6	4	3	12	6	3	4	14	4
Outbreak or Threatening	April/May	4	1	14	6	2	3	12	8	1	4	14	6
	April	3	2	13	7	3	2	13	7	1	4	14	6
	May	4	3	12	6	4	3	12	6	2	5	13	5
	Annual	2	0	15	8	1	1	14	9	1	1	17	6

[a]
A = Catastrophe predicted and one occurred (correct)
B = Catastrophe predicted but none occurred (incorrect)
C = Catastrophe not predicted and none occurred (correct)
D = Catastrophe not predicted but one occurred (incorrect)

catastrophes based on outbreak or threatening densities. Several of the analyses more reliably predicted noncatastrophic population dynamics; all but one of the analyses were at least 80% accurate in this regard with respect to threatening or outbreak infestations, and six of the analyses were at least 80% accurate in this regard with respect to infestations based on threatening/outbreak conditions. However, based on the measure of total reliability (including accurate predictions of both occurrences and nonoccurrences of catastrophes) only eleven forms of the model fared significantly better than chance, as determined by a binomial test. These analyses included: prediction of threatening infestations by using weather data from April/May with threatening or threatening/outbreak levels of infestation, or weather data from April with outbreak infestations, prediction of outbreak infestations using weather from May or annual records with threatening infestation or weather data from April/May or April with outbreak infestations, and prediction of threatening/outbreak infestations by using annual weather data with threatening or threatening/outbreak infestations or weather data from April or April/May with outbreak infestations (e.g., Fig. 2).

Table 2. Percentage of cases in which Catastrophe Theory models correctly predict grasshopper population dynamics.

Model		Infestation criterion [a]								
Infestation	Weather	Threatening			Outbreak			Threatening/ Outbreak		
		A	B	C	A	B	C	A	B	C
Threatening	April/May	72	40	93	56	20	80	60	14	78
	April	56	10	87	64	20	93	64	14	83
	May	56	10	87	72	30	100	68	14	89
	Annual	60	30	87	72	40	93	76	43	89
Outbreak	April/May	64	40	80	72	50	87	76	57	83
	April	72	50	87	72	50	87	76	57	83
	May	64	50	73	64	50	73	68	57	72
	Annual	64	40	80	64	40	80	68	43	78
Outbreak or Threatening	April/May	72	40	93	56	20	80	60	14	78
	April	64	30	87	60	30	87	60	14	78
	May	64	40	80	64	40	80	60	29	72
	Annual	68	20	100	60	10	93	72	14	94

[a]
$$A = \frac{\text{\# of correctly predicted occurrences and nonoccurrences}}{\text{Total number of events}}$$

$$B = \frac{\text{\# of correctly predicted catastrophic events}}{\text{\# of catastrophic events}}$$

$$C = \frac{\text{\# of correctly predicted noncatastrophic events}}{\text{\# of noncatastrophic events}}$$

Discussion

The success of the cusp catastrophe in predicting outbreaks and crashes with selected data sets provides considerable insight as to the properties of grasshopper population dynamics. With regard to weather, early spring conditions are clearly important in mediating populations. Catastrophe Theory is consistent with previous models and observations (Gage & Mukerji 1977, Smith 1969, Dempster 1963, Edwards 1960) in that warm, dry conditions during hatching and early development apparently promote outbreaks while cold, wet conditions may induce crashes. Annual weather conditions do not appear to govern population dynamics. This may be due to the apparent detrimental effects of relatively hot, dry conditions in the Summer and the beneficial effects of these conditions in the Spring (Nerney & Hamilton 1969).

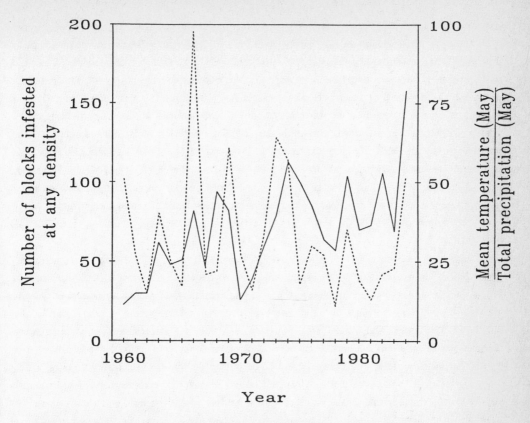

Fig. 2. Relationship between Wyoming rangeland grasshopper outbreaks (————————)
and a temperature/precipitation index (————).

The intensity of an outbreak (i.e., outbreak versus threatening population densities) was an
important factor in the analyses. Outbreak densities produced more accurate predictions of
catastrophes than threatening densities when incorporated into models. That is, a sudden increase
or decrease in the area infested can be effectively modelled using outbreak densities and Spring
climatic factors as the controlling variables in a cusp catastrophe. Population dynamics based on
outbreak densities do not appear to differ substantially from threatening densities with regard to
the expected properties of a cusp catastrophe (i.e., sudden jumps, hysteresis, divergence,
bimodality and inaccessibility). Thus, one explanation of the difference between models of
population dynamics using threatening and outbreak densities is that the former is not as clearly
mediated by weather factors, or at least tne weather factors that mediate the latter (Spring
temperature and precipitation). Perhaps catastrophes defined by jumps in the area infested by
low densities can be modelled with different weather variables or the same weather parameters for
different times of year. Summer weather patterns may be important in this regard, since evidence
indicates that conditions which promote good forage growth may be necessary to support high
population densities (Capinera 1987, White 1976, Nerney 1961).

Another approach to understanding the application of the cusp catastrophe to grasshopper population dynamics is to develop *a posteriori* criteria of catastrophes. That is, we can attempt to determine what criterion of a catastrophe will maximize the accuracy of the model. Since the most accurate predictions based on the *a priori* criterion incorporated monthly weather data and used outbreak densities in the model, it is reasonable to examine this case more carefully. In particular, when weather data for April/May and outbreak densities are used as the basis of the model, we find that the predicted catastrophes are 100% accurate if a catastrophe is defined as the point at which the area of outbreak crosses a threshold of 60 blocks. If the other weather parameters are used with outbreak densities and the above criterion of a catastrophe, accuracies range from 84 to 92%. Thus, some forms of the model may provide insight into the mathematical nature of population dynamics. We are continuing to pursue assessments of Catastrophe Theory from the perspectives of both *a priori* and *a posteriori* criteria of grasshopper population outbreaks.

Catastrophe Theory offers several potential advantages over existing, linear models. Catastrophe Theory is designed to model sudden perturbations in a system, and it is sudden grasshopper outbreaks and their collapse that are fundamentally of interest to pest management efforts. Indeed, the occurrence of population explosions and crashes seems to typify grasshopper population dynamics (Capinera 1987). Given the assumptions of Catastrophe Theory, the number of qualitatively different configurations of discontinuities that can occur depends not on the number of state variables, which may be very large, but on the number of control variables, which is generally small (Saunders 1980). To model large and complex systems such as grasshopper populations by conventional means would require us to develop n differential equations (where n is the number of variables needed to completely specify the system), supply initial conditions, solve the equations and then try to comprehend the solutions. On the basis of very few assumptions, Catastrophe Theory makes it possible to predict much of the qualitative behavior of the system without even knowing what the differential equations are, much less solving them (Saunders 1980).

Acknowledgements

We thank Dr. W. Hoffman, Department of Mathematics, New Mexico State University for his help in developing the model and E. Miller, College of Agriculture, University of Wyoming for her conversion of the program to an IBM—PC compatible system.

References Cited

Capinera, J. L. 1987. Population ecology of rangeland grasshoppers. In Integrated pest management of rangeland, J. L. Capinera [ed.], pp. 196–204. Westview Press, Boulder, Colorado.

Cobb, L. 1980. Cusp surface estimation: programs and examples. University of South Carolina Press, Greeneville, South Carolina.

Dempster, J. P. 1963. The population dynamics of grasshoppers and locusts. *Biol. Rev.* 38: 490–529.

Edwards, R. L. 1960. Relationship between grasshopper abundance and weather conditions in Saskatchewan, 1930–1958. *Can. Entomol.* 92: 619–624.

Gage, S. H. & M. K. Mukerji. 1977. A perspective of grasshopper population distribution in Saskatchewan and interrelationships with weather. *Environ. Entomol.* 6: 469–479.

Kemp, W. P. 1987. Probability of outbreak for rangeland grasshoppers in Montana: application of Markovian principles. *J. Econ. Entomol.* 80: 1100–1105.

Lockwood, J. A. & W. P. Kemp. 1988. Probabilities of rangeland grasshopper outbreaks in Wyoming counties. Wyoming Agricultural Experiment Station, Bulletin B–896.

Nerney, N. J. 1961. Effects of seasonal rainfall on range condition and grasshopper population, San Carlos Apache Indian Reservation, Arizona. *J. Econ. Entomol.* 54: 382–385.

Nerney, N. J. & A. G. Hamilton. 1969. Effects of rainfall on range forage and populations of grasshoppers, San Carlos Apache Indian Reservation, Arizona. *J. Econ. Entomol.* 62: 329–333.

Saunders, P. T. 1980. An introduction to catastrophe theory. Cambridge University Press, Cambridge, Massachusetts.

Smith, L. B. 1969. Possible effects of changes in the environment on grasshopper populations. *Manitoba Entomol.* 3: 51–55.

Stewart, I. N. & P. L. Peregoy. 1983. Catastrophe theory modeling in psychology. Psychol. Bull. 94: 336–362.

Watts, J. G., E. W. Huddleston, & J. C. Capinera. 1982. Rangeland entomology. *Ann. Rev. Entomol.* 27: 283–311.

White, J. 1977. Weather, food, and plagues of locusts. *Oecologia* 22: 119–134.

Zeeman, E. C. 1976. Catastrophe theory. *Sci. Amer.* 234: 65–83.

Derivation and Analysis of Composite Models
for Insect Populations

Jesse A. Logan[1]

ABSTRACT In this paper I discuss the concept of a composite modeling approach as applied to insect population analysis. A composite model is a term that has been applied to mathematically tractable models that have parameters which were derived in an obvious fashion from more complex, and ecologically meaningful, simulation models. This approach to modeling was motivated by the difficulty in producing models that simultaneously satisfy the requirements of ecological fidelity and mathematical tractability. The modeling paradigm discussed in this paper is to first develop a model that satisfies the requirement of ecological realism. Then, through application of various mathematical procedures and ecological intuition, successively more abstract, simplified models are developed. The final objective is an ecologically credible model amenable to mathematical analysis. I first discuss the problem in general terms and then illustrate its application by a specific problem.

A paradox that has been apparent from the first applications of mathematical analysis of ecological systems is the dichotomy between models useful for representing a specific biological association and those capable of producing general insights. In a classic paper, Levins (1966) convincingly argued that, at most, two of three potential attributes (generality, precision, and realism) of ecological models can be achieved simultaneously. At least part of the reason for Levins' paradox may lie in the dichotomy of the world view between mathematician and biologists (Fig. 1). On one hand, biologists are confronted daily with complexity of real world, ecological systems. On the other, the tools of applied mathematics have traditionally required relatively simple systems of equations amenable to mathematical analysis. In historical perspective, Levins' contentions seem to have been supported, and two relatively distinct approaches to modeling have evolved. The first attempts to capture empirical complexity and realism (simulation approaches) whereas the second attempts to capture generality (typically the analytically tractable Lotka/Volterra–type models of theoretical ecology). Levins' dilemma may be responsible for the occasionally discordant differences between systems ecologists and theoretical ecologists. In fact, it is becoming increasingly apparent that neither approach by itself is capable of adequately addressing the important ecological issues and questions facing us today. What is needed is a systematic modeling paradigm that accommodates both empirical detail and the power of mathematical abstraction.

The production of mathematical models that realistically represent specific associations, and that are simultaneously capable of producing general insights, remains an elusive goal. One possible paradigm for accomplishing this goal is development of composite models (Logan 1982,

---

[1]Department of Entomology and Forestry, Virginia Polytechnic Institute and State University, Blacksburg, Virginia 24061.

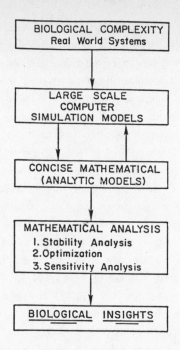

Fig. 1. Relationship between the complexity of the real world and model descriptions of that complexity.

Plant & Mangle 1987). In this approach, a simulation model is first developed that captures ecological complexity with sufficient resolution to satisfy the goals of precision and realism. Then, through a systematic process of simplification and aggregation, progressively more analytically tractable models are developed. This finally results in a model amenable to mathematical analysis. Throughout this process, procedures are followed that serve to maintain the link between simplified models and the more complex models from which they were derived; and, thus, to the actual ecological system of interest. Derivation of a composite model requires three steps, (1) development of the original simulation model (s), (2) model simplification, (3) analysis of the derived analytic model.

Development of a Composite Model

The first step in composite model development is to produce a model description of sufficient detail to satisfy the biologists perception of the complex, real world system. The description must be stated in terms that are readily understood by biologists. Historically, these essential attributes are best captured in complex simulation models. Simulation modeling and system analysis are essentially synonymous terms for an approach to modeling that was formalized by

Forrester in his influential book Industrial Dynamics (1961). Primarily due to Forrester's work, it was recognized early in the history of simulation modeling that techniques were readily generalized, and several computer languages have been developed to facilitate model building based on these generalizations. The philosophy of a generalized modeling paradigm can effectively be applied to building a simulation of a stage specific model of insect populations. The two most basic requisites for such a modeling system are adequate representation of phenology and demography. Recently, expert systems concepts have been employed to facilitate development of insect population models (Logan 1988).

Once a sufficient simulation exists, the next step in composite model development is to produce the most concise simulation model possible (Logan 1982, Logan & Hilbert 1983). The objective of parsimonious model development is to produce a simulation model that captures the essence of the targeted system within the stated research objectives. In other words, to produce as economical a model as possible while retaining ecological fidelity. Specific tactics that have proven useful in attaining this goal include sensitivity analysis, parameter, and variable aggregation, and phenomenological modeling. Sensitivity analysis is a standard method of analysis for simulation models (e. g. Miller et al. 1973). The objective of sensitivity analysis is to identify those variables, processes, and parameters that are most important in determining critical model responses. Through this procedure, those factors determined not to be of critical importance can either be ignored (removed from the model), set to constants, or aggregated into composite factors.

Parameter and variable aggregation is primarily based in ecological intuition, although important new insights into the aggregation process are beginning to emerge from hierarchical theory (O'Neill et al. 1986). Finally, phenomenological model representation is a procedure borrowed from the physical sciences in which an appropriate parameterized function (canonical form) is used to represent a more complex mechanistic process (Wollkind et al. 1979, Logan 1988).

The meta–model (model of a model) produced by procedures in the previous paragraphs will most likely still be too complex for classical mathematical analysis. A technique that has proven useful for derivation of analytically tractable models from parsimonious simulation models is termed "elaboration". In this procedure, a holistic viewpoint is adopted, and the question is asked: What is the simplest model representation possible that captures the essence of the system? After a candidate model is selected, an attempt is made to establish direct relationships between the concrete parameters of the parsimonious simulation model and the abstract parameters in the analytic model. If rigorous relationships can be established between the two model representations, then one may reasonably conclude that the analytic model is a sufficient representation of the parsimonious simulation model. This conjecture can be strengthened by two procedures: (1) use of the simulation model to generate data for parameter estimation in the analytic model, and (2) similarity in responses of the two models over a reasonable range of parameter perturbations and/or variations in driving variables. The objective in this stage of model simplification is to establish the relationship between a simulation model that has intuitive ecological meaning and a highly abstracted mathematical model representation.

The final step in development of a composite model is simplification of the analytical model. Simplification in this context is the same as that used in applied mathematics, and the most useful techniques are those borrowed from applied mathematics, namely scaling and dimensional analysis (Segel 1972). Although these techniques are routinely applied in the physical sciences, they are infrequently applied for analysis of ecological models. It is usually well worth the effort to scale equations and proceed with analysis of nondimensional variables and parameters. Not only do such procedures typically reduce the numbers of parameters in a model; but also, through a judicious choice of scaling factors (Wollkind et al. 1988), dimensionless parameters may actually enhance ecological interpretation.

The objective of the procedures outlined above is to produce a composite model amenable to mathematical analysis. A central theme of theoretical ecology has been stability analysis (May 1981). Many ecologically meaningful concepts, such as domains of attraction (Holling 1973), require establishment of global stability criteria. Determining the global stability properties of even relatively simple models is a difficult problem; however, powerful new tools have recently become available for the numerical analysis of dynamical systems. With these new tools, it is possible to perform analyses that only a short time ago were unthinkable. Among the most powerful of these new capabilities is the ability to perform a bifurcation analysis of higher order, nonlinear systems (e. g., AUTO, Doedel, and Kernevez 1986). The procedure followed in this type of analysis is to plot the stability characteristics of a system's fixed points as a parameter is continuously varied over a range of values. This results in characterization of the model stability properties over an entire region of parameter space. Through this process, it is finally possible to analytically determine ecological properties such as domains of attraction (Holling 1973) for a wide range of ecological meaningful models.

Development of composite models has provided a useful concept for facilitating the mathematical analysis of ecological systems (Plant & Mangle 1987). As interesting new attributes of dynamical systems (such as deterministic chaos) are discovered, the importance of developing mathematically tractable models that are ecological credible will become even more apparent. In the next section I will briefly describe development of a composite model describing the interaction between an economically important mite species and its biological control agent.

A Case History

Among the most widespread and economically important arthropods in agricultural systems are the spider mites. In many crop systems, spider mite populations are maintained at low densities by the action of predacious mites belonging to the family Phytoseiidae. The population control exhibited by phytoseiid predators is often so efficient that it is difficult to sample populations of either species. However, if conditions become exceptionally favorable for spider mites or if the prey/predator interaction is disrupted, spider mites are capable of explosive and spectacular population irruptions. Outbreak densities may be high enough to result in total defoliation.

During the mid 1960's spider mites constituted a serious economic threat to apple production in the Pacific North West region of the United States. By far the most important species in Washington orchards was the McDaniel spider mite, *Tetranychus mcdanieli* McGregor. Mite problems were exacerbated (and apparently precipitated) by the disruptive effects of pesticides applied to control codling moth. In 1967 S. C. Hoyt discovered populations of the phytoseeid mite, *Metaseiulus occidentalis* (Nesbitt), that had survived a codling moth control program (Hoyt 1969). He was able to culture this pesticide–resistant predator strain in the laboratory and subsequently reintroduce it into other orchards. What followed was truly a success story in the annals of modern biological control (Hoyt & Caltagirone 1971). However, during the early phases of establishing integrated mite control, it was often observed that high temperatures led to breakdown of predator effectiveness. Motivated by these observations, a research project was initiated with the objective of elucidating the mechanistic role that temperature played in the prey–predator interaction between *T. mcdanieli* and *M. occidentalis* (Tanigoshi, et al. 1975 a, b).

As a component of this research program, I developed a computer simulation model that provided a detailed description of the prey–predator interaction. Temperature dependence was included in model representation of phenology, demography, and the predation process (Logan 1977). Although the simulation model provided a heuristically satisfying description of the interaction between *M. occidentalis* and *T. mcdanieli*, and its performance was judged to be both quantitatively and qualitatively acceptable (Logan 1982, Logan & Hilbert 1983); the simulation model also had definite limitations. First, like most simulation models, the model was relatively expensive to run in terms of both computer time and computer space. Secondly, the simulation included more detail than was absolutely necessary for description of the basic predation process. Finally, and perhaps most importantly, the complexity and detail of the simulation model tended to obscure important ecological relationships (Plant & Mangle 1987). For these reasons, the simplest biologically reasonable representation of the system was considered in conjunction with the simulation model.

The simplest representation of a spider mite/phytoseiid interaction that is likely to maintain biological credibility is provided by the Tanner–May model:

$$\frac{dH}{dt} = r_1H(1. - \frac{H}{K}) - \frac{\alpha PH}{H + B} \tag{1a}$$

$$\frac{dP}{dt} = r_2P(1. - \frac{P}{\gamma H}) \tag{1b}$$

where H is prey density; P is predator density; r_1 is the maximum intrinsic population growth rate for the prey; K is the prey carrying capacity; α corresponds to the maximum predation rate; B is the prey density at which the functional response curve reaches its "half–maximum" point; r_2 is the maximum predator intrinsic population growth rate; and γ is a measure of conversion efficiency of prey biomass to predator biomass.

Estimating parameters in Eq. (1) is an interesting, and nontrivial, problem. Logan (1977, 1982) employed the detailed simulation model to facilitate parameter estimation, and to provide the link between empiricism and mathematical tractability. Fig. 2a illustrates such a process. The "data" in this Fig. were generated from a simulation model that was based on empirical information describing developmental periods for each individual instar, fecundity schedules, and longevity (Tanigoshi et al. 1975 a, b). Other parameters in the model were estimated by similar procedures. The empirical effect of temperature on the prey/predator interaction was, therefore, included in the estimation of important model parameters (Logan 1977). A comparative plot of temperature effect on population growth rates for the prey and predator is illustrated in Fig. 2b. Temperature effects on the stability properties of the prey/predator interaction were determined by performing a structural stability analysis on model Eq. (1). The stability of Eq. (1) as a function of temperature is shown in Fig. 3. This criterion was obtained by first performing a linear stability analysis of model (May 1981), which resulted in a relationship between model parameters and model stability. The temperature dependency was then included in the parameter representation, resulting in a stability criterion as a function of temperature. As seen from Fig. 3, model instability occurs at approximately 32°C.

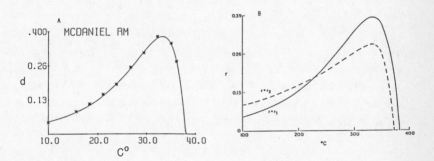

Fig. 2.   (a) Functional relationship describing r_m for *T. mcdanieli*. The curve was determined by least squares fit of a generalized rate–temperature model (Logan et al. 1976) to simulated data points.   (b) Comparative r_m values for *T. mcdanieli* (solid line) and *M. occidentalis* (broken line) as a function of temperature. Point of crossover is approximately 32°C.

An appreciation for the critical importance of temperature can be gained from time plots of the populations when perturbed from steady state. Two such plots are provided in Fig. 4. Note that the density scale and basic system behavior are quite different between the two plots shown in Fig. 4. Even at a high temperature slightly below the stability threshold, perturbations damp to steady state. However, at temperatures only slightly above the threshold, model behavior is qualitatively much different; populations no longer damp to equilibrium but continue to rebound in a limit cycle mode.

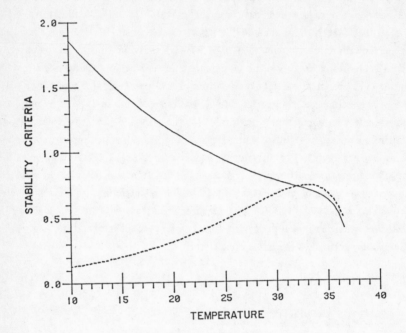

Fig. 3. Structural stability of Eq. (1) as a function of temperature. As long as the solid line remains above the broken line, the model is stable. When the broken line is below the solid line, the model is characterized by limit cycle behavior (May 1973).

Outbreak Dynamics In A Simple Model

Powerful new tools have recently become available for the numerical analysis of dynamical systems. Among the more useful of these tools is the bifurcation code developed by Doedel (Doedel & Kernevez 1986). Wollkind et al. (1988) applied this code in analysis of Eq. (1), a system for which the global stability properties were thought to be completely understood. Contrary to a previous analysis of this model that had indicated a relatively simple dynamic (May 1971; see previous section), Wollkind et al. (1988) found that over a range of high temperatures, the global stability properties are characterized by a fixed point surrounded by an <u>unstable</u> limit cycle, which in turn is surrounded by a <u>stable</u> limit cycle (Fig. 5). Trajectories initiated within the inner cycle damp back to the fixed point. Trajectories initiated outside the inner but within the outer cycle, will increase to the stable limit of the outer cycle. Finally, trajectories initiated outside of the outer cycle will damp to the stable limit of the outer cycle, but may do so in a complex fashion. Examples of three trajectories (T=30⁰C) are shown in Fig. 6. Most startling of these is Fig. 6c, in which the initial densities of both species were set below the nadir of the outer cycle. Note in particular the violent oscillation this leads to. Prey density abruptly increases to almost its carrying capacity.

The work by Wollkind et al. (1988) led to several additional insights into model behavior, including an evaluation of sensitivity of model stability to various model parameters. However, the most important insight is that simple prey/predator models are capable of dynamic behavior that is much more complex than had previously been appreciated. A corollary to this general result is that the specific interaction between *T. mcdanieli* and *M. occidentalis* is even more susceptible to pesticide disruption than had previously been appreciated. These insights have largely resulted from the interaction between abstract mathematical analysis and empiricism. This interaction was vastly facilitated through the concept of composite model development.

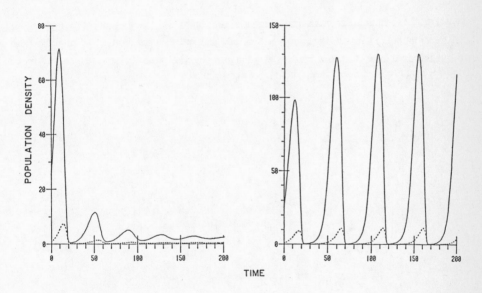

Fig. 4. Plots of population density vs. time for prey (solid line) and predator (broken line) densities for Eq. (1). (a) Model results at T=31.5⁰C; and initial conditions of H=25; P=2. (b) model results at T=32.5⁰C; and initial conditions of H=25; P=2.

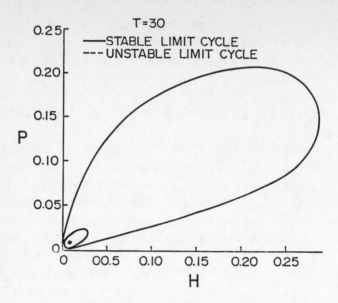

Fig. 5. The domain of attraction for a fixed point in the region of complex, high temperature dynamics. Trajectories initiated within the inner cycle damp to the fixed point, while those initiated outside the inner cycle will approach the outer cycle (see text for further explanation).

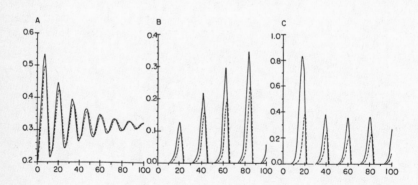

Fig. 6. Plots of population density vs. time for scaled prey (h) (solid line) and predator (p) (broken line). Solutions are for parameter values evaluated at T=30ºC. (a) Trajectory initiated within the inner, unstable limit cycle (b) Trajectory initiated outside of the unstable inner cycle but within the stabe outer limit cycle (c) Trajectory initiated outside and below the nadir of the outer stable limit cycle.

Acknowledgements

Drs. David M. Swift and H. William Hunt provided useful reviews of an earlier version of this manuscript. Research described herein was supported through a Cooperative Research Agreement between the Environmental Protection Agency and Colorado State University (CR813688–01–0), and a National Science Foundation Research Grant (BSR8418049).

References Cited

Doedel, E. J., & J. P. Kernevez. 1986. AUTO: Software for continuation and bifurcation problems in ordinary differential equations. Technical Report in Applied Mathematics. California Institute of Technology, Pasadena.

Forrester, J. W. 1961. Industrial dynamics. MIT Press, Cambridge, 464 pp.

Holling, C. S. 1973. Resiliency and stability of ecological systems. *Ann Rev. Ecol. Syst.* 4: 1–24.

Hoyt, S. C. 1969. Integrated chemical control of insects and biological control of mites on apple in Washington. *J. Econ. Entomol.* 62: 74–86.

Hoyt, S. C., & L. E. Caltagirone. 1971. The developing programs of integrated control of pests of apples in Washington and peaches in California, pp. 395–421. In C. B. Huffaker [ed.], Biological control. Plenum Press, New York.

Levins, R. 1966. Strategy of model building in population biology. *Am. Sci.* 54: 421–431.

Logan, J. A. 1977. Population model of the association of *Tetranychus mcdanieli* (Acarina: Tetranychidae) with *Metaseiulus occidentalis* (Acarina: Phytoseiidae) in the apple ecosystem. Ph. D. dissertation Washington State University, Pullman.

Logan, J. A. 1982. Recent advances and new directions in Phytoseiid population models, pp. 49–71. In M. A. Hoy [ed.], Recent advances in knowledge of the Phytoseiidae. *Agric. Sci. Publ.* 3284, University of California, Berkeley.

Logan, J. A. 1988. Toward an expert system for development of pest simulation models. *Environ. Entomol.* (in press).

Logan, J. A., & D. W. Hilbert. 1983. Modeling the effect of temperature on arthropod population systems, pp. 113–122. In W. K. Lauenroth, G. V. Skogerboe, and M. Flug [eds.], Analysis of ecological systems: state–of–the–art in ecological modeling. Elsevier Sci. Publ. Co., Amsterdam.

May R. M. 1981. Models for two interacting populations, pp. 8–104. In R. M. May [ed.], Theoretical ecology: principles and applications (second edition). Blackwell Sci. Publ. Co., London.

Miller, D. R., D. E. Weidhass, & R. C. Hall. 1973. Parameter sensitivity in insect population modeling. *J. Theor. Biol.* 42: 263–274.

O'Neill, R. V., D. L. DeAngelis, J. B. Waide, & T. F. H. Allen. 1986. A hierarchical concept of ecosystems. Princeton University Press, Princeton, NJ.

Plant, R. E., & M. Mangle. 1987. Modeling and simulation in agriculture pest management. *SIAM Rev.* 29: 235–261.

Segel, L. A. 1972. Simplification and scaling. *SIAM Rev.* 14: 547–571.

Tanigoshi, L. K., S. C. Hoyt, R. W. Browne, & J. A. Logan. 1975a. Influence of temperature on population increase of *Tetranychus mcdanieli* (Acarina: Phytoseiidae). *Ann. Entomol. Soc. Am.* 68: 972–978.

Tanigoshi, L. K., S. C. Hoyt, R. W. Browne, & J. A. Logan. 1975b. Influence of temperature on population increase of *Metaseiulus occidentalis* (Acarina: Phytoseiidae). *Ann. Entomol. Soc. Am.* 68: 979–986.

Wollkind, D. J., J. A. Logan, & A. A. Berryman. 1979. Asymptotic methods for modeling biological processes. *Res. Popul. Ecol.* 20: 49–59.

Wollkind, D. J., J. B. Collins, & J. A. Logan. 1988. Metastability in a temperature–dependent model system for predator–prey mite outbreak interactions on fruit trees. *Bull. Math. Biol.* (in press).

Leslie Matrix Models for Insect Populations
With Overlapping Generations

Erik V. Nordheim[1,3], David B. Hogg[2], and Shun–Yi Chen[1]

ABSTRACT Leslie matrix models appear a useful tool for studying population dynamics for insects with overlapping generations. Our studies have focused on aphids. Analysis using deterministic Leslie matrices has demonstrated the importance of relative length of the reproductive period and also the starting age distribution. Using the length of reproductive period from field pea aphid data, we conclude that aphid populations have the potential to achieve a stable age distribution in a typical temperate growing season. We suggest use of stochastic Leslie matrices for more realistic modelling. We point out that simulation studies using these stochastic models require careful interpretation. Using the beta and gamma distributions to describe survival and natality, respectively, we demonstrate the performance of a simulated aphid population. The realized growth rate is lower than would be suggested from the deterministic model.

Our main objective is to gain an improved understanding of population dynamics for insects with overlapping generations with particular interest focused on aphid populations. The primary analytical device we have considered is the Leslie matrix (Leslie, 1945). We have studied some uses of the deterministic Leslie matrix in describing population dynamics and have also explored the benefits of allowing the matrix to be random. In this paper we describe some of our results to date. As our work in this area is rather recent, this manuscript should best be viewed as a report on preliminary findings. In time we intend to present more complete results.

The starting point for our efforts came from a study of the conclusions from Hughes (1962). Hughes utilized an assumption of stable age distribution (SAD) in the development of a life table analysis for use with aphid populations. This procedure has been widely used (e.g., Mackauer & Way, 1976; see also Hutchison & Hogg (1985) for further discussion). Carter et al. (1978) provided some cautionary comments about the Hughes procedures, particularly concerning a chi–squared test for stable age structure. They conducted a fairly extensive numerical study to analyze the dynamics of aphid populations. They utilized a deterministic Leslie matrix approach with seven age classes, one representing each of the first 3 aphid instar periods, two for the fourth instar period, and two for reproductive adults. Their analyses were conducted for 30 time periods where the length of a time period is the length of an "instar period" (IP) equal to the duration of each of the first 3 instars. They used a single starting age distribution, a survival rate of 0.8, and a range of reproductive rates from 1.5 to 10 offspring per reproductive aphid per IP. The starting

[1]Department of Statistics, University of Wisconsin, Madison, Wisconsin 53706.

[2]Department of Entomology, University of Wisconsin, Madison, Wisconsin 53706.

[3]Department of Forestry, University of Wisconsin, Madison, Wisconsin 53706.

point in our study of aphid population dynamics was a reanalysis of the Carter et al. numerical results.

Performance of Deterministic Leslie Matrix Models

Although we agree with the cautionary comments of Carter et al. concerning application of the Hughes methodology, we have some questions about certain aspects of their methodology. Herein we address only some of these. One of their conclusions that has been questioned elsewhere (Hutchison & Hogg, 1985), but that we wished to analyze in some detail, was that aphid field populations are unable to achieve an SAD within 30 IP's.

Before we present our findings, we briefly discuss the Leslie matrix formulation. For simplicity of presentation, we show a model with 3 age classes with reproduction in the oldest 2 classes, resulting in a 3 x 3 Leslie matrix. The basic equation can be written

$$
\begin{bmatrix} n_{1,t+1} \\ n_{2,t+1} \\ n_{3,t+1} \end{bmatrix} = \begin{bmatrix} 0 & r_2 & r_3 \\ s_1 & 0 & 0 \\ 0 & s_2 & 0 \end{bmatrix} \begin{bmatrix} n_{1,t} \\ n_{2,t} \\ n_{3,t} \end{bmatrix} = \begin{bmatrix} r_2*n_{2,t} + r_3*n_{3,t} \\ s_1*n_{1,t} \\ s_2*n_{2,t} \end{bmatrix} \tag{1}
$$

where $n_{i,t}$ indicates the number of individuals in the ith age class at time t, s_i is the survival rate for an individual in age class i, and r_i is the reproductive rate for an adult in age class i.

First, we wish to point out the effect on population dynamics of initial values for $n_{i,0}$. Consider the 3 x 3 Leslie matrix of the form given by equation (1) with $r_2 = 1.5$, $r_3 = 2.0$, $s_1 = .5$, and $s_2 = .25$. This matrix has an SAD given by a ratio of 8:4:1 for $n_{1,t}:n_{2,t}:n_{3,t}$ at all time points t. This means that once this ratio has been achieved, it will be maintained. The given matrix also will result in stationary population size for the SAD. The SAD and constant population size are demonstrated by the following matrix equation.

$$
\begin{bmatrix} 0 & 1.5 & 2.0 \\ .5 & 0 & 0 \\ 0 & .25 & 0 \end{bmatrix} \begin{bmatrix} 8 \\ 4 \\ 1 \end{bmatrix} = \begin{bmatrix} 8 \\ 4 \\ 1 \end{bmatrix}
$$

Suppose we consider a numerical study of the time course of a population described by this matrix. Consider as the initial values $n_{1,0} = 0$, $n_{2,0} = 0$, $n_{3,0} = 8000$. Presented below are the numbers of individuals in each age class for the first 8 time periods.

time period	0	1	2	3	4	5	6	7	8
$n_{1,t}$	0	16000	0	12000	4000	9000	6000	7750	6750
$n_{2,t}$	0	0	8000	0	6000	2000	4500	3000	3875
$n_{3,t}$	8000	0	0	2000	0	1500	500	1125	750

Note the effects of the initial values. It requires 5 time periods before individuals can be found in all age classes. This performance is due to the 'mixing' of age classes from the Leslie matrix. Suppose in the current example that $r_2 = 0$. Then there would be no mixing and individuals would be found in only a single age class at every point in time. The fact that r_2 is non—zero ($r_2 = 1.5$) allows individuals in the second age class to lead to offspring in the first age class and older adults in the third age class at the next time period. Over time, this mixing causes the ratios of individuals in the 3 age classes to approach the SAD.

Carter et al. assumed an aphid population with 7 age classes, 2 of the classes for reproductive adults, for their numerical studies. This allows for slow mixing, implying that an initial age structure far from the SAD will approach stability slowly. However, aphid species characteristically exhibit a reproductive period that is longer in duration than the respective pre—reproductive period (see Taylor, 1979). Thus, a reproductive period equivalent to only 2 IP's is unrealistically short for most aphids.

To show the effect of increased reproductive life span, we repeated the Carter numerical studies using 10 age classes with 5 classes for reproductive adults in addition to the 7 age class case. We used a survival rate of 0.8 for all age classes. Our initial population distribution placed 100 individuals in the youngest reproductive age class. We generated data for 60 instar periods. A constant reproductive rate was used for all reproductive age classes; we considered a range of rates from 1.5 to 10. The results for all rates were qualitatively similar. We calculated the proportion of individuals in all age classes and plotted this against time (number of IP's). Fig. 1 shows the results with a reproductive rate of 10, for both the 7 x 7 and 10 x 10 cases. For the 10 x 10 case, substantial mixing has occurred and the age distribution appears quite close to stability by 30 IP's. For the 7 x 7 case, the discreteness effects are still very striking at 60 IP's. Thus the relative length of adult aphid reproduction is very important in population dynamics. Our results are in agreement with Taylor (1979), whose analysis of life history data for 5 aphid species suggests an average convergence time of about 20 IP's, assuming fixed mortality and natality schedules.

Introduction to Stochastic Leslie Matrices

It became evident to us that continued use of the deterministic Leslie matrix in studying insect population dynamics was of limited value. Clearly natality and mortality are subject to stochastic mechanisms in nature. Thus, rather than pursue investigations on the deterministic case, we focused attention on the stochastic.

Sykes (1969) wrote an important paper describing three stochastic formulations for the Leslie model. Writing equation (1) in matrix form, $X_{t+1} = AX_t$, the three approaches are the following: (a) allow an additive error term to this equation, (b) consider the elements of A as probabilities rather than rates, and (c) consider A itself to be random. The last approach seems

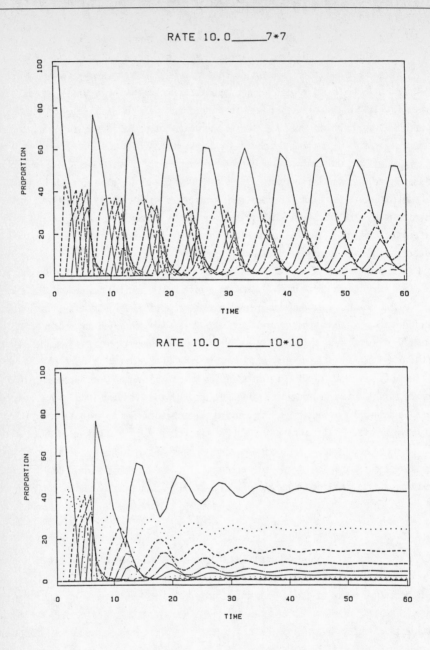

Fig. 1. Proportions of individuals in each age class for a numerical study of aphid populations using 7 x 7 and 10 x 10 deterministic Leslie matrices.

the most realistic for insect population dynamics; it has also received the most attention in the biological literature. Most of our work has used this formulation.

The biological literature on the stochastic Leslie matrix has been of a conceptual or theoretical nature and has focused on the 3 x 3 case. Much attention has been devoted to simulation studies. In this numerical work, an assumption of normality for matrix elements has been seemingly universal. Some of the important papers in this biological development are Boyce (1977), Tuljapurkar & Orzack (1980), Tuljapurkar (1982), and Slade & Levenson (1982). An important issue in this literature is the asymptotic performance of the Leslie model. This can be very useful in the determination of long–term growth rates. In his papers, Tuljapurkar showed that the total population size at time t, N_t, is asymptotically lognormal. Thus, $\log(N_t)$ is asymptotically normal with mean μt and variance $\sigma^2 t$.

We have analyzed some of the issues involved in these asymptotic arguments. Much of this discussion is beyond the scope of the current manuscript; a paper with the relevant details is currently in preparation. One of the major issues is presented here and is of considerable importance in understanding the asymptotic performance of Leslie matrix models. This argument relies on the unusual properties of the lognormal. The mean of the asymptotic lognormal distribution for N_t is

$$\text{mean}(N_t) = \exp(\mu t + 1/2\ \sigma^2 t)$$

where exp( ) denotes exponentiation (see Johnson & Kotz, 1970). Consider the αth quantile of N_t. With Z_α as the αth quantile from a standard normal distribution, the αth quantile of N_t is

$$\alpha\text{th quantile}(N_t) = \exp(\mu t + Z_\alpha\ \sigma\ \sqrt{t}).$$

When t is large enough, mean (N_t) is larger than any quantile of interest due to the stronger dependence of mean (N_t) on t than that of αth quantile (N_t). This demonstrates the unusual asymptotic performance of N_t from the Leslie model. Numerical studies we have conducted have indicated that simulated population sizes (for the means of hundreds of runs) can be lower by orders of magnitude than the expected value for large t.

Stochastic Leslie Matrix Simulation Useful To Aphid Studies

A primary goal of our work on stochastic Leslie matrices has been to develop a model structure useful to insect population dynamics. An important component of this goal is the choice of distributions for the survival and reproductive rate (or natality) terms in the stochastic Leslie matrix. Apparently, rather little attention has been devoted to this issue. For survival, it is essential to consider a distribution that can only range over the values from 0 to 1. A versatile distribution that meets the objective is the beta (Johnson & Kotz, 1970). This distribution has two parameters and allows the mean to vary between 0 to 1 and the variance to range quite broadly. For natality, the values need to be positive. The gamma distribution, also with two parameters, is a versatile choice. Thus, in the 3 x 3 example modeled in equation (1), we suggest that s_1 and s_2 be random variables described by the beta distribution and r_2 and r_3 be randomly distributed according to the gamma distribution.

For each time period there will be a new realization of the Leslie matrix. It is possible that the matrices at different times are independent of one another; similarly, one could consider correlation across time. Autocorrelation in time might be explained by the existence of natural enemies (e.g., lady beetles) or extreme weather conditions that persist over several time periods. Similarly, within a time period the elements of the Leslie matrix can be independent of one another or could be correlated. The existence of positive correlation here could again be explained by natural enemies or weather. From the perspective of simulation studies, this stochastic Leslie matrix formulation provides a rich and flexible modelling framework.

We have completed some simulation studies that attempt to describe in a realistic way, the dynamics of the pea aphid *Acyrthosiphon pisum* (Harris). Much of the data base for selection of parameters of our model comes from Hutchison & Hogg (1984) and unpublished data (W. D. Hutchison, personal communication). These data were obtained from experiments conducted in an outdoor, screened insectary; they thus provide useful information on reproduction but little information on natural mortality. The data indicate that the effective lifespan of the pea aphid equals about 15 instar periods. Reproduction begins in the 6th period, remains high for about 5 instar periods, and is then lower for the last 5 periods. The mean reproductive rate for periods 6 through 10 is about 10 offspring per adult per instar period; the mean rate for periods 11 through 15 is around 6.

We conducted a study of this pea aphid situation with a 15 x 15 stochastic Leslie matrix using the beta and gamma distributions for survival and reproductive rates. To date we have only considered the case where rates are independent within a time period and also across time periods. We report here on simulations with a mean survival rate of 0.8 for all periods and mean reproductive rates given by the field insectary data. Thus, using the notation introduced for equation (1), $r_i = 0$ for $i = 1, ..., 5$; $r_i = 10$ for $i = 6, ..., 10$; and $r_i = 6$ for $i = 11, ..., 15$. We performed simulations allowing the distributions for the r's and s's to have different variances; we present results only for a single case. For each s term, the standard deviation was 0.32; the standard deviations for the r terms were 5 and 3 corresponding to the mean values of 10 and 6, respectively. The standard deviation for the s terms is somewhat arbitrary but may be typical of a high variability situation; the standard deviations for the r terms come from the field insectary data. Our simulations were conducted for a time duration equal to 30 instar periods. We considered three distinct starting conditions: (a) 100 individuals placed initially in age class 6; (b) 34 individuals placed initially in age class 6 with 33 individuals placed in age class 6 after a time interval of one IP and again after two IP's; and (c) an SAD corresponding to a deterministic Leslie matrix with the mean survival and natality. The total number of individuals initially placed in (c), 530, was chosen so that the total population size after 30 instar periods would equal, approximately, the sizes obtained from the other two conditions. (The pattern described by (b) was selected as a potentially realistic case for colonization by winged adults.) The simulation results based on 200 runs for each starting distribution are shown in Fig. 2.

This Fig. shows plots of natural log (total population size) versus time for each of the three starting conditoins. The dashed line is the population curve from the deterministic Leslie matrix.The upper solid line is the log of the arithmetic mean from the 200 runs. The lower solid line is the mean of the log of the total population sizes. The dotted lines around each solid line represent approximate 95% confidence limits for the log of the arithmetic mean and the mean of the logs, respectively.

The upper solid line follows the deterministic line quite closely, as expected. The distinction between the two solid lines deserves some discussion. The distribution of total population sizes is very skewed. (As noted above, for large t the distribution is well approximated by a lognormal, a very skewed distribution.) Thus the arithmetic mean is strongly influenced by the largest simulated population size from the 200 runs. The mean of the logs can be thought of as the log of the geometric mean of the population sizes. Asymptotically, the geometric mean of the lognormal distribution is equivalent to the median; thus the lower line should be viewed as being representative of a 'typical outcome'. Slade & Levenson (1982) reported on simulations performed in the 3x3 case with only s_1 random. Their results indicate that for suitable choices of parameters with finite rate of increase slightly greater than 1.0, the arithmetic mean (upper line) could be made to increase with time while the geometric mean (lower line) would decrease. We agree with their conclusion that the lower line is the better predictor of probable growth.

The comparison among the graphs for the three starting conditions is interesting. The population growth for case (c) (SAD) shows a very steady growth with only slight fluctuations. The growth of case (a) (100 individuals initially in the 6th class) shows substantial unevenness (cyclical behavior) initially. This is due to the highly discrete initial conditions. It takes some time for adequate mixing to occur in this case. Case (b) (34–33–33 individuals introduced into stage 6) is intermediate between the other two. The separation (arithmetic difference) between the upper and lower solid lines is greatest for case (a), intermediate for (b) and smallest for (c). This follows since the variability and also the skewness for the total population sizes are decreasing from (a) to (c). The widths of the confidence intervals show the same pattern.

Discussion

We are optimistic about the potential for stochastic Leslie matrices in studying population dynamics for insects with overlapping generations. Thus far we have gained some preliminary appreciation for the performance of these Leslie models and believe them to be flexible enough to describe real insect populations. The major task confronting us is the estimation of Leslie parameters from field data. We do not believe that brute force estimation of parameters

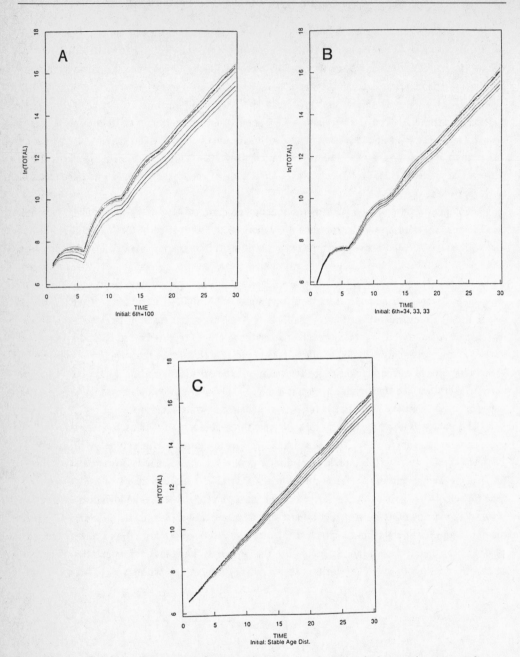

Fig. 2. Total population sizes for a stochastic pea aphid simulation study presented on a natural logarithm scale for three distinct initial conditions. The dashed line is the (deterministic) population curve and the upper and lower solid lines are the arithmetic and geometric means, respectively. The dotted lines are approximate 95% confidence limits.

describing the random Leslie matrix elements is plausible. Initial investigations with maximum likelihood and method of moments estimators have been discouraging. However, if field data can be augmented by screenhouse experiments, i.e., to estimate parameters of reproduction independently, then we believe estimates of parameters of survival from the Leslie matrix is possible. This work is currently in progress.

Acknowledgements

We thank G. Hoffman and H. Triwidodo for discussing with us some of these ideas, M. Clayton and J. Carey for reviewing the manuscript, and W. Hutchison for use of his unpublished data. This research was suppported by the College of Agricultural and Life Sciences, University of Wisconsin—Madison, and by the UW—Madison Graduate School.

References

Boyce, M. S. 1977. Population growth with stochastic fluctuations in the life table. *Theor. Popul. Biol.* 12: 366–373.

Carter, N., D. P. Aikman & A. F. G. Dixon. 1978. An appraisal of Hughes' time–specific life table analysis for determining aphid reproductive and mortality rates. *J. Anim. Ecol.* 47: 677–687.

Hughes, R. D. 1962. A method for estimating the effects of mortality on aphid populations. *J. Anim. Ecol.* 31: 389–396.

Hutchison, W. D. & D. B. Hogg. 1984. Demographic statistics for the pea aphid (Homoptera: Aphididae) in Wisconsin and a comparison with other populations. *Environ. Entomol.* 13: 1173–1181.

Hutchison, W. D. & D. B. Hogg. 1985. Time–specific life tables for the pea aphid, *Acyrthosiphon pisum* (Harris), on alfalfa. *Res. Popul. Ecol.* 27: 231–253.

Johnson, N. L. & S. Kotz. 1970. Distributions in Statistics. Continuous Univariate Statistics 1 and 2. John Wiley and Sons, New York.

Leslie, P. H. 1945. On the use of matrices in certain population mathematics. *Biometrika* 33: 183–212.

Mackauer, M. & M. J. Way. 1976. *Myzus persicae* Sulz., an aphid of world importance. pp. 51–119 In V. L. Delucchi [ed.], Studies in Biological Control. Cambridge Univ. Press, London.

Slade, N. A. & H. Levenson. 1982. Estimating population growth rates from stochastic Leslie matrices. *Theor. Popul. Biol.* 22: 299–308.

Sykes, Z. M. 1969. Some stochastic versions of the matrix model for population dynamics. *J. Am. Stat. Assoc.* 64: 111–130.

Taylor, F. 1979. Convergence to the stable age distribution in populations of insects. *Am. Nat.* 113: 511–530.

Tuljapurkar, S. D. 1982. Population dynamics in variable environments. II. Correlated environments, sensitivity analysis and dynamics. *Theor. Popul. Biol.* 21: 114–140.

Tuljapurkar, S. D. & S. H. Orzack. 1980. Population dynamics in variable environments. I. Long–run growth rates and extinction. *Theor. Pop. Biol.* 18: 314–342.

Relationships Among Recent Models for Insect Population Dynamics with Variable Rates of Development

G. Bruce Schaalje[1] and H. R. van der Vaart[1]

ABSTRACT   A classification scheme for those population models which allow variation in development rates is proposed, based on two ways of modifying standard age structured models.   General formulations for the various classes of models are developed, and specific models from the literature are shown to fit into the scheme. Relationships among the classes are investigated and the use of the scheme in population modelling is discussed.

Over the past decade authors in numerous entomological endeavors have developed mathematical models for population dynamics in which the idea that individuals mature (i.e. pass from one growth stage to another) at different rates is modelled explicitly.   The need for models including this additional level of complexity has arisen in estimating demographic parameters of insect populations (Southwood 1979, Kempton 1979, Shoemaker et al. 1986) from field data since a cohort of insects tends to 'spread out' in developmental status as time passes.   Integrated pest management researchers have also developed such models (Barr et al. 1973, Welch et al. 1978, Plant & Wilson 1986) because their purposes often require prediction of the date at which the first individuals of some growth stage begin to emerge.   Other relevant papers have considered the effects of variable development rates on stability, intrinsic growth rates and other demographic features (Blythe et al. 1984, Bellows 1986).

There have been some helpful partial reviews of these models (Shaffer 1983, Shoemaker et al. 1986, Plant & Wilson 1986, Curry & Feldman 1987), but the purpose of this paper is to outline and discuss a possible synthesis of modelling strategies for insect populations in which individual variation in development rates cannot be ignored.

A Classification of Models

Consider a population in which all individuals advance in age at the same rate, and the age of an individual carries with it all pertinent, vital information.   Here the (time–varying versions of the) models of Leslie (1945), Von Foerster (1959), and some near relatives (Vansickle 1977a) are appropriate for age–specific dynamics.   Growth stage information is not explicitly included in these models but is required implicitly because the age–specific fecundity and death rate of a cohort may change abruptly as the individuals in the cohort advance from one stage to the next.

---

[1]Biomathematics Graduate Program, Department of Statistics, North Carolina State University, Raleigh, NC 27695.

In the above models, as used for insect populations, all individuals in a cohort (a) advance in age at the same rate and (b) change from one growth stage to the next at the same age. In order to modify these models to allow variation among individuals in their rate of maturation, two alternative approaches are possible:

1. Retain property (b) but relax property (a). That is, we would continue to require all individuals to change from one growth stage to the next at the same 'age', but allow them to advance in 'age' at various rates. Since this 'age' will continue to carry with it the vital information, it is an index of the actual developmental status of the individual (where the technology used to measure it will be at present left open). The assumption is implicitly made that this index is a scalar quantity.

2. Retain property (a) but relax property (b). That is, we would continue to require all individuals to advance in 'age' at identical rates, but allow them to change from one growth stage to the next at various 'ages'. Hence, this 'age' communicates little about the individual's developmental status, but does tell us how long it has sojourned in the current stage. (For the remainder of this paper, sojourn time will be measured in chronological time units, and parameters of models using this approach may depend on the temperature regime. A commonly used alternative is to measure sojourn time in physiological time units.)

Models based on the first approach will be called 'development index' models. Models based on the second approach will be called 'sojourn time' models.

To be able to compare behaviors of sojourn time and development index models we divide the development index (r) into series of contiguous intervals corresponding to the growth stages of the insect (Van Straalen 1986). We replace the single development index by a series of development indices (r_h), one for each growth stage interval. By convention we let r_h range from 0 to β_h. Similarly, sojourn time models depend on the quantities time (t) and sojourn time (s_h) in growth stage h. The model does not impose an upper limit on s_h. A complete mathematical description of the dynamics of a population following either model then consists of equations describing the number of individuals attaining the various development index values/sojourn times in each growth stage at any time, various renewal conditions giving the relationships between subsequent growth stages, a driving function describing the number of individuals entering the population from the outside at any time (hatching of diapaused eggs and immigration are examples of this), and initial conditions. The two classes of models have the same mathematical structure except for the equations describing the internal dynamics of each stage. The remainder of this paper discusses and compares models for these internal dynamics. The subscript for growth stage (h) will be left off of s_h and r_h when it will not cause confusion to do so.

Our models are essentially of a bookkeeping nature, where certain variables may be declared to be random in order to allow for individually varying development rates. The bookkeeping for the internal dynamics of each growth stage is done by means of the following balance equations:

1. For development index models: number of individuals of development index r at time t+dt
   = number of individual of development index r at time t.
   − number of individuals who increased in development from r to any r'>r during (t,t+dt)
   + number of individuals who increased in development from any r"<r to r during (t,t+dt)
   − number of individuals of development index r at time t who died during (t,t+dt)
   + number of individuals of development index r at time t+dt who entered the population from the outside during (t,t+dt)

2. For sojourn time models: Number of individuals of sojourn time s+dt at time t+dt
   = number of individuals of sojourn time s at time t
   − number of individuals of sojourn time s at time t who matured to the next stage during (t,t+dt)
   − number of individuals of sojourn time s at time t who died during (t,t+dt)
   + number of individuals of sojourn time s+dt at t+dt who entered the population from outside during (t,t+dt)

For development index models we observe the modelling convention that all individuals recruited from stage h either enter stage h+1 immediately as r_h becomes equal to β_h or die in transition, and those that enter do so with $r_{h+1} = 0$. For all models we assume that age (of any kind) always changes positively, and we consider only state–determined models.

Model Development and Examples from the Literature

Depending on which quantities (time, development index, sojourn time) are taken as continuous and which as discrete, we consider five cases in formulating mathematical models from the balance equations. For more detail see Schaalje (1988).

Model 1.1

When time and development index are discrete quantities, the balance equation for $n_{h,r}(t)$, the number of individuals of development index r (r= 1,2,...,β−1) at time t in growth stage h is

$$n_{h,r}(t+1) = n_{h,r}(t) - \sum_{j=r+1}^{\beta} p_{h,jr}(t)\,[1 - m_{h,r}(t)]\,n_{h,r}(t)$$

$$+ \sum_{i=0}^{r-1} p_{h,ri}(t)\,[1-m_{h,i}(t)]\,n_{h,i}(t) - m_{h,r}(t)\,n_{h,r}(t) + u_{h,r}(t+1) \tag{1}$$

where we adopt the convention that a sum is defined to be zero if its upper limit is less than its lower limit, and

$p_{h,ij}(t)$ = proportion of individuals of growth stage h who advance from development index j to development index i between t and t+1 given that they survive,

$m_{h,i}(t)$ = proportion of individuals of growth stage h with development index i who die between t and t+1 given that they are alive at time t,

$u_{h,i}(t+1)$ = number of individuals of growth stage h with development index i who enter the growth stage from the outside between t and t+1.

The initial conditions are given by $n_{h,r}(t_0)$. For the description of the internal dynamics of stage h the case $r=\beta$ is not needed but choosing $m_{h,\beta}(t)=0$ will make $n_{h,\beta}(t)$ the cumulative number of individuals who before or at time t have become candidates for transition to stage h+1.

Equations (1) for the internal dynamics within growth stage h could be written in matrix form reminiscent of the Leslie (1945) model as

$$\underline{n}_h(t+1) = P_h(t)L_h(t)\underline{n}_h(t) + \underline{u}_h(t+1) \qquad (2)$$

where $L_h(t) = I - \text{diag}\,(m_r(t))$, and

$P_h(t)$ = a lower triangular column–stochastic matrix in which $p_{h,ij}$ gives the proportion of survivors advancing in development index from j to i (j<i) and

$$p_{h,ii} = 1 - \sum_{i=j+1}^{\beta} p_{h,ij} \text{ for } i<\beta, \text{ and } p_{h,\beta\beta}=1.$$

The solution to equation (2) is

$$\underline{n}_h(t) = \sum_{\tau=0}^{t-2-t_o} \left[\prod_{\sigma=0}^{\tau} P_h(t-1-\sigma)L_h(t-1-\sigma) \right] \underline{u}_h(t-1-\tau)$$

$$+ \left[\prod_{\sigma=0}^{t-1-t_o} P_h(t-1-\sigma)L_h(t-1-\sigma) \right] \underline{n}_h(t_0) + \underline{u}_h(t) \qquad (3)$$

The discrete model proposed by Plant & Wilson (1986) for use when developmental variance is small relative to life span is a special case of equation (2) with

$$
P_h \, L_h \;=\;
\begin{bmatrix}
e & & & & & & \\
p_1-2e & e & & & & 0 & \\
e & p_2-2e & e & & & & \\
0 & e & p_3-2e & e & & & \\
0 & 0 & e & p_4-2e & e & & \\
\cdot & \cdot & \cdot & \cdot & \cdot & \cdot & \\
\cdot & \cdot & \cdot & \cdot & \cdot & \cdot & \cdot \\
\cdot & \cdot & \cdot & \cdot & \cdot & \cdot & \cdot & \cdot
\end{bmatrix}
$$

where e and p_i are constants and $m_r = 1 - p_r$. Plant and Wilson did not worry about the $\beta\beta$ element because they presented the matrix as being of infinite order.

Schneider & Ferris (1986) used a special case of equation (1) as the basis of their procedure for estimating development and mortality rates of insects from field data. They allowed surviving individuals to either stay at their current substage (development index value) or advance to the next.

Model 1.2

With continuous time and discrete development index, we again define $n_{h,r}(t)$ as in model 1.1. The balance equation eventually yields the differential equation

$$
\dot{n}_{h,r}(t) = -l_{h,r}(t)\, n_{h,r}(t) + \sum_{i=1}^{r-1} q_{h,ri}(t)\, n_{h,i}(t)
$$

$$
- w_{h,r}(t)\, n_{h,r}(t) + v_{h,r}(t) \tag{4}
$$

where $q_{h,ij}(t)dt$, $w_{h,i}(t)dt$ and $v_{h,i}(t)dt$ are the continuous counterparts of $p_{h,ij}(t)$, $m_{h,i}(t)$, and $u_{h,i}(t)$, respectively, used in equation (1) and

$$
l_{h,j}(t) = \sum_{i=j+1}^{\beta} q_{h,ij}(t)\,.
$$

Equations (4) for all values of the development index within growth stage h can be arranged in matrix form as

$$\dot{\underline{n}}_h(t) = [Q_h(t) - M_h(t)]\,\underline{n}_h(t) + \underline{v}_h(t) \tag{5}$$

where $Q_h(t)$ = a lower triangular matrix with entries $q_{h,ij}(t)$ for $i>j$, $q_{h,ii}(t) = -l_{h,i}$ for $i<\beta$ and $q_{h,\beta\beta} = 0$, and

$M_h(t) = \mathrm{diag}[w_{h,i}(t)]$ with $w_{h,\beta} = 0$.

The solution to equation (5) cannot be written in terms of integrals of $Q_h(t)$ and $M_h(t)$ unless $[Q_h(t) - M_h(t)]$ is communtative with its integral. In the case of constant Q_h and M_h, the solution is

$$\underline{n}_h(t) = \int_0^{t-t_0} \exp[(Q_h - M_h)\tau]\underline{v}_h(t-\tau)d\tau + \exp[(Q_h - M_h)(t-t_0)]\underline{n}_h(t_0) \tag{6}$$

where $\underline{n}_h(t_0)$ gives the initial numbers of individuals.

Manetsch (1976) and Vansickle (1977b) developed the 'linear distributed delay system' in the spirit of a sojourn time model, but they worked analytically and numerically with a system of ordinary differential equations whose solution was equivalent. This system of equations is a special case of equation (5) with

$$Q_h = \begin{bmatrix}
k/del & & & & & 0 \\
-k/del & k/del & & & & \\
0 & -k/del & k/del & & & \\
0 & 0 & -k/del & k/del & & \\
\cdot & \cdot & \cdot & \cdot & \cdot & \\
\cdot & \cdot & \cdot & \cdot & \cdot & \cdot \\
\cdot & \cdot & \cdot & \cdot & \cdot & \cdot
\end{bmatrix}$$

and $M_h = ar\,I$, where k, del, and ar are constants.

Manetsch's work has been the basis for numerous pest management programs (Welch et al. 1978, Carruthers et al. 1986). Although many users appear to simply regard the equations as an algorithm for a sojourn time model (Vansickle 1977b), others (Curry & Feldman 1987 p. 171, Plant & Wilson 1986) advocate that index values in the system of equations be interpreted as ages or stages (development index values).

Model 1.3

A mathematical formulation of the development index model using continuous time and continuous development index was developed by Weiss (1968). Slightly modified, his balance equation is

$$\frac{\partial n_h(r,t)}{\partial t} = - \left[\int_r^\beta R_h(r,z,t)dz \right] n_h(r,t)$$

$$+ \int_0^r n_h(z,t)R_h(z,r,t)dz - m_h(r,t)n_h(r,t) \tag{7}$$

where $n_h(r,t)dr$ = number of individuals in growth stage h, of development index between r and r + dr at time t,

$R_h(r,z,t)dz$ = instantaneous rate of transitions in development index from r to (z,z+dz) for z $\geq$ r at time t for individuals in stage h, and

$m_h(r,t)$ = instantaneous death rate for individuals of development index r in stage h at time t.

This equation does not have a closed form solution for any but the simplest of situations, but Weiss argued that if the distribution of transitions in development index for individuals were such that all moments of order $\geq$ 3 could be ignored, he could use a Taylor series expansion to obtain an equation closely resembling the forward Kolmogorov equation (FKE):

$$\frac{\partial n_h(r,t)}{\partial t} = -m_h(r,t)n_h(r,t) - \frac{\partial}{\partial r}[g_{h,1}(r,t)n_h(r,t)]$$

$$+ \frac{1}{2}\frac{\partial^2}{\partial r^2}[g_{h,2}(r,t)n_h(r,t)] \tag{8}$$

where $g_{h,1}(r,t)$ and $g_{h,2}(r,t)$ are the mean and variance functions of the development index transition process.

The FKE as a model for population dynamics was discussed or used by Barr et al. (1973), Auslander et al. (1974), and Welch et al. (1978)(who also applied the Manetsch equations and found both gave similar results). Closely related to these models are those discussed by Munholland et al. and Kemp et al. (both in these proceedings).

Although they did not use our bookkeeping formulation, the work of Curry et al. (1978) is also relevant in connection with completely continuous development index models. The simplest and most widely used case considered by them involved rates of progression through the whole range of the development index which were determined by a single random quantity (the level of a controlling enzyme) multiplied by a temperature dependent constant.

Model 2.1

The sojourn time model with discrete time and discrete sojourn time can be formulated from the balance equation as

$$n_h(s+1,\,t+1) = n_h(s,t)[1 - p_h(s,t)][1 - m_h(s,t)] + u_h(s+1,\,t+1) \tag{9}$$

where $n_h(s,t)$ = the number of individuals in stage h at time t of sojourn time s,

         $p_h(s,t)$ = proportion of individuals of sojourn time s who mature out of stage h between t and t+1 given they survive,

         $m_h(s,t)$ = proportion of individuals in stage h of sojourn time s who die between t and t+1 and,

         $u_h(s,t)$ = net number of individuals of sojourn time s at time t who enter stage h from the outside between t−1 and t.

The proportions $p_h(s,t)$ and $m_h(s,t)$ can be taken as conditional probabilities of maturing or dying. Initial conditions are $n_h(s,t_0)$.

The solution to the difference equation (9) is, for $s = 1,2,...,$

$$n_h(s,t) = \begin{cases} n_h(s-t-t_0,t_0)\left\{ \prod_{j=t_0}^{t-1} [1 - p_h(s-t+j,j)][1 - m_h(s-t+j,j)] \right\} \\[2ex] \quad + \sum_{i=s-t+t_0+1}^{s-1} u_h(i,t-s+i)\left\{ \prod_{j=i}^{s-1} [1 - p_h(j,t-s+j)][1 - m_h(j,t-s+j)] \right\} \\[2ex] \quad + u_h(s,t) \qquad\qquad\qquad\qquad\qquad\qquad\qquad \text{if } s > t-t_0 \\[2ex] \sum_{i=0}^{s-1} u_h(i,t-s+1)\left\{ \prod_{j=1}^{s-1} [1 - p_h(j,t-s+j)][1 - m_h(j,t-s+j)] \right\} \\[2ex] \quad + u_h(s,t) \qquad\qquad\qquad\qquad\qquad\qquad\qquad \text{otherwise} \end{cases} \tag{10}$$

Lefkovitch (1965) used a simple model of this type. He took the $p_h(s,t)$ and $m_h(s,t)$ to be constant over both s and t. He organized the difference equation (9) and the renewal conditions into a single generalized Leslie matrix equation and used regression to estimate the parameters (see Kempton (1979) for more details).

Bellows and Birley (1981) and Soendgerath (1987) independently developed a more general case than the Lefkovitch model. In these papers the assumption was made that the conditional probabilities $p_h(s,t)$ and $m_h(s,t)$ varied with sojourn time but were not time dependent. If the 'latent exit times' (Prentice et al. 1978) due to death and maturation are independent, distribution functions $F_h(s)$ and $G_h(s)$ (with corresponding density functions $f_h(s)$ and $g_h(s)$) for the latent maturation and death times, respectively, can be defined in terms of the $p_h(s)$ and $m_h(s)$ such that

$$n_h(s+1,t+1) = n_h(s,t)\left[1 - \frac{g_h(s+1)}{1-G_h(s)}\right]\left[1 - \frac{f_h(s+1)}{1-F_h(s)}\right] + u_h(s+1,t+1). \tag{11}$$

The commonly used iterative cohort approach (Curry & Feldman 1987, p. 187–191) to modelling insect population dynamics is also of the form of model 2.1. In this approach, the $p_h(s,t)$ and $m_h(s,t)$ are time dependent, but are computed on the basis of the temperature regime during the sojourn time of the individuals in each cohort.

Model 2.2

For a sojourn time model with continuous time and continuous sojourn time, the balance equation can be manipulated to produce the partial differential equation

$$\frac{\partial n_h(s,t)}{\partial t} + \frac{\partial n_h(s,t)}{\partial s} = -q_h(s,t)n_h(s,t) - w_h(s,t)n_h(s,t) + v_h(s,t) \tag{12}$$

which is similar to the Von Foerster equation mentioned earlier. Here $n_h(s,t)dt$, $q_h(s,t)dt$, $w_h(s,t)dt$, and $v_h(s,t)dt$ are the continuous counterparts of $n_h(s,t)$, $p_h(s,t)$, $m_h(s,t)$ and $u_h(s,t)$ of equation (9). If the initial conditions are denoted by $n_h(s,t_0)$, the solution is

$$
\left[\begin{array}{l}
n_h(s-t+t_0,t_0)\left[\exp \int_{t_0}^{t} \{-q_h(s-t+\tau,\tau) - w_h(s-t+\tau,\tau)\}d\tau\right] \\[2mm]
+ \int_{s-t+t_0}^{s} v_h(\tau,t-s+\tau)\left[\exp \int_{\tau}^{s}\{-q_h(\sigma,t-s+\sigma) - w_h(\sigma,t-s+\sigma)\}d\sigma\right]d\tau \\[4mm]
\hspace{4cm} \text{if } s \geq t-t_0 \\[2mm]
\int_{0}^{s} v_h(\tau,t-s+\tau)\left[\exp \int_{\tau}^{s}\{-q_h(\sigma,t-s+\sigma) - w_h(\sigma,t-s+\sigma)\}d\sigma\right]d\tau \\[2mm]
\hspace{4cm} \text{otherwise.}
\end{array}\right. \tag{13}
$$

Helpful discussions of models of this form for insect population dynamics are given by Curry & Feldman (1987, p. 191–196) and Gurney et al. (1986). In modelling velvetbean caterpillar populations, Wilkerson et al. (1986) described their population model using equation (12). Blythe et al. (1984), specializing to the case of an organism with two growth stages and time–invariance for all parameters except the death rate of 'adults', were able to use model 2.2 to examine questions of stability in variable maturation models.

Analogously to the discrete case, $q_h(s,t)ds$ and $w_h(s,t)ds$ can be viewed as conditional probabilities. If these probabilities are time dependent, and if the latent exit times from the stage due to death or maturation are independent, then

$$\frac{f_h(s)}{1\text{-}F_h(s)} \, ds \text{ and } \frac{g_h(s)}{1\text{–}G_h(s)} \, ds$$

where $F_h(s)$, $f_h(s)$, $G_h(s)$, and $g_h(s)$ have definitions similar to those in model 2.1, give the conditional probabilities of maturing and dying during a small interval of sojourn time ds and can be substituted into equations (12) and (13). In this form, model 2.2 can be seen to include models developed by Manetsch (1977) and Lewis (1977) using Laplace transforms, and Read & Ashford (1968), Kempton (1979), Kalbfleisch et al. (1983) and Curry & Feldman (1987 p. 85–91) using probabilistic language.

Relationship Between the Classes of Models

Since we have two classes of models, each of which seems to be a logical expression of reality, it is natural to expect them to be equivalent. If a model in one class could not be matched by a model in the other, the assumptions implicit in the second class would warrant examination. Therefore, the extent to which the two classes of models are equivalent is an important subject for further research. Much of this work is yet to be done, but some results are available.

In their most general form, sojourn time models are just a matter of bookkeeping. Development index models, however, involve the modelling assumption that a scalar development index exists.

The sojourn time distributions generated by the class of development index models in which variability in rates of progression is determined by variation in levels of a certain controlling enzyme were investigated by Curry et al. (1978). Also, it is well known that the distributed delay model (Vansickle 1977b), which we have shown to have the form of a specific development index model (see the matrix following equation (6)), is equivalent to a sojourn time model with gamma distributed sojourn times and exponentially distributed survival times (Curry & Feldman 1987 p. 88).

Using the bookkeeping approach, the authors have considered the general question of equivalence of the classes of models when the parameters are time–invariant. This research is important because the time–invariant versions of these models are the most commonly used, often with time measured on some physiological scale (Curry & Feldman 1987 p. 62). In order to study 'equivalence' the assumptions were made that all individuals which become candidates for maturation to the next growth stage successfully make the transition, that initial conditions are zero for the stage in question, and there is no immigration (see Schaalje 1988).

Under the continuous sojourn time model (2.2) the rate of maturation of individuals from stage h to stage h+1 is

$$v_{h+1}(0,t) = \int_0^t v_h(0,t-s)[1 - G_h(s)] \, f_h(s) \, ds \, . \tag{14}$$

Under the development index model with continuous time and discrete development index (model 1.2) the maturation rate can be shown to be (Schaalje 1988)

$$v_{h+1,0}(t) = \int_0^t v_{h,0}(t-s)\{\exp[(Q-M)s](Q-M)\}_{\beta,1} ds \, . \tag{15}$$

For the simple but widely used case in which the instantaneous rate of mortality is constant over the development index values, equation (15) becomes

$$v_{h+1,0}(t) = \int_0^t v_{h,0}(t-s) \, e^{-ms} \, [\exp(Qs) \, Q]_{\beta,1} ds \, . \tag{16}$$

It has been shown (Schaalje 1988) that if $G_h(s)$ in equation (14) is the cumulative exponential distribution function with parameter m, there exists a probability distribution with density function $f_h(s)$ such that equation (14) is identical to equation (16) for any choice of Q satisfying the conditions following equation (5). A similar result holds for discrete time development index and sojourn time models. Thus, all (discrete) development index models with time–invariant parameters and constant death rates over the values of the development index have equivalent sojourn–time counterparts. However, the converse does not hold. There are some sojourn time models, for example, those in which the distribution of sojourn times has finite support, for which no counterpart time–invariant development index model exist.

On the other hand, it appears that whenever death rates vary with the development index, there is no possibility that the maturation and mortality terms in equation (15) could be disentangled. It would be interesting to investigate whether such time–invariant development index models have sojourn time counterparts. If such development index models had no exact time–invariant sojourn time counterparts, it would be of interest to investigate how well one could approximate them with sojourn time models.

Discussion and Summary

None of the variable development models discussed in this paper are new, but the relationships among them and their relative advantages in various applications have not been fully understood. Our classification scheme is intended to be a starting point in investigating the similarities among these models and identifying their unique characteristics. This information will help the researcher to develop an appropriate model for his application, identify the properties of his model, generalize his model, and consider its relationship to alternative models.

Some issues the researcher might consider in selecting the model or class of models for his application are:

1. Since a development index model with time–invariant parameters and constant death rates can be replaced by an equivalent sojourn time model, the choice of models in this case can be based on considerations of computational and parameter estimation ease.

2. The data necessary to implement sojourn time models are often relatively easy to obtain. Distributions of sojourn times and survival times can be readily determined in the labratory.

3. Sojourn time models are easy to modify if their behavior is not as desired. For example, Gamma distributed sojourn times can be easily replaced by, say, Weibull distributed sojourn times.

4. Sojourn time models are easily modified to allow for nonconstant death rates.

5. If either the death rates are known to depend on the development index or the intended use of the model requires that the distribution of development index values (ages) in the population be predicted, development index models are a natural choice. For example, larvae of some pest may burrow into the fruit at some point in their development and thus become impervious to insecticide–induced mortality for the rest of the growth stage.

6. Sojourn times of individuals within a growth stage are difficult to determine in the field but sometimes development index values (or something highly correlated with them such as length) can be observed. In such cases a development index model seems natural.

References Cited

Auslander, D. M., G. F. Oster & C. B. Huffaker. 1974. Dynamics of interacting populations. *J. Franklin Inst.* 297: 345–376.

Barr, R. O., P. C. Cota, S. H. Gage, D. L. Haynes, A. N. Karkar, H. E. Koenig, K. Y, Lee, W. J. Ruesnik & R. L. Tummala. 1973. Ecologically and environmentally compatible pest control P. 241–264. In Geier, P. W., L. R. Clark, D. J. Anderson & H. A. Nix. Insects: studies in population management. Ecol. Soc. Aust. Canberra.

Bellows, T. S. Jr. & M. H. Birley. 1981. Estimating developmental and mortality rates and stage recruitment from insect stage–frequency data. *Res. Popul. Ecol.* 23: 232–244.

Bellows, T. S. Jr. 1986. Impact of developmental variance on behavior of models for insect populations I. Models for populations with unrestricted growth. *Res. Popul. Ecol.* 28: 53–62.

Blythe, S. P., R. M. Nisbet & W. S. C. Gurney. 1984. The dynamics of population models with distributed maturation periods. *Theor. Pop. Biol.* 25: 289–311.

Carruthers, R. I., G. H. Whitfield, R. L. Tummala & D. L. Haynes. 1986. A systems approach to research and simulation of insect pest dynamics in the onion agro–ecosystem. *Ecol. Mod.* 33: 101–121.

Curry, G. L., R. M. Feldman & P. J. H. Sharpe. 1978. Foundations of stochastic development. *J. Theor. Biol.* 74: 397–410.

Curry, G. L. & R. M. Feldman. 1987. Mathmatical Foundations of Population Dynamics. Texas A&M Press. College Station.

Gurney, W. S. C., R. M. Nisbet & S. P. Blythe. 1986. The systematic formulation of models of stage structured populations. P. 474–494. In Metz, J.A.J. & O. Diekmann. The Dynamics of Physiologically Structured Populations. Lecture Notes in Biomath. No. 68. Springer–Verlag, New York.

Kalbfleisch, J. D., J. F. Lawless & W. M. Vollmer. 1983. Estimation in Markov models from aggregate data. *Biometrics* 39: 907–919.

Kempton, R. A. 1979. Statistical analysis of frequency data obtained from sampling an insect population grouped by stages. P. 401–418. In Ord, J. K., G. P. Pail & C. Taille. Statistical distributions in ecological work. Int. Coop. Publishing House. Fairland, Maryland.

Lefkovitch, L. P. 1965. The study of population growth in organisms grouped by stages. *Biometrics* 21: 1–18.

Leslie, P. H. 1945. On the use of matrices in certain population mathmatics. *Biometrika* 35: 183–212.

Lewis, E. R. 1977. Linear population models with stochastic time delays. *Ecology* 58: 738–749.

Manetsch, T. J. 1976. Time–varying distributed delays and their use in aggregative models of large systems. *IEEE Trans. Systems Man Cybernet.* 6: 547–553.

Plant, R. E. & L. T. Wilson. 1986. Models for age structured populations with distributed maturation rates. *J. Math. Biology* 23: 242–262.

Prentice, R. L., J. D. Kalbfleisch, A. V Peterson Jr., N. Flournoy, V. T. Farewell & N. E. Breslow. 1978. The analysis of failure times in the presence of competing risks. *Biometrics* 34: 541–554.

Read, K. L. Q. & J. R. Ashford. 1968. A system of models for the life–cycle of a biological organism. *Biometrika* 55: 211–221.

Schaalje, G. B. 1988. Models for stage structured insect populations affected by pesticides with applications to pesticide efficacy trials. Ph.D. Thesis. North Carolina State University.

Schneider, S. M. & H. Ferris. 1986. Estimation of stage–specific developmental times and survivorship from stage frequency data. *Res. Popul. Ecol.* 28: 267–280.

Shaffer, P. L. 1983. Mathematical structures for modelling pest populations. Institute of Statistics Mimeo Series. No. 1643. North Carolina State University.

Shoemaker, C. A., G. E. Smith & R. G. Helgesen. 1986. Estimation of recruitment rates and survival from field census data with application to poikilotherm populations. *Agric. Sys.* 22: 1–21.

Soendgerath, D. 1987. Eine Erweiterung des Lesliemodells fuer die Beschreibung populations– dynamischer Prozesse bei Spezies mit mehreren Entwicklungsstadien. Ph.D. Thesis. University of Dortmund.

Southwood, T. R. E. 1979. Ecological Methods. Chapman and Hall. London.

Vansickle, J. 1977a. Analysis of a distributed parameter population model based on physiological age. *J. Theor. Biol.* 64: 571–586.

Vansickle, J. 1977b. Attrition in distributed delay models. IEEE Trans. *Systems Man Cybernet.* 7: 635–638.

Van Straalen, N. M. 1982. Demographic analysis of arthropod populations using a continuous stage–variable. *J. Anim. Ecol.* 51: 769–783.

Von Foerster, H. 1959. Some remarks on changing populations. P. 382–407. In Stohlman, F. Jr. The Kinetics of cellular proliferation. Grune & Stratton. New York.

Weiss, G. H. 1968. Equations for the age structure of growing populations. *Bull. Math. Biophys.* 30: 427–435.

Welch, S. M., B. A. Croft, J. F. Brunner & M. F. Michels. 1978. PETE: an extension phenology modeling system for management of multi–species pest complex. *Environ. Entomol.* 7: 487–494.

Wilkerson, G. C., J. W. Mishoe & J. L. Stimac. 1986. Modelling velvetbean caterpillar (*Lepidoptera: Noctuidae*) populations in soybeans. *Environ. Entomol.* 15: 809–816.

Models of Development in Insect Populations

Glen H. Smerage[1]

ABSTRACT The features of development essential to analyses of insect population dynamics, development time and mortality, are discussed in the context of convolution and network models. Whether representing one or several successive stages of a life cycle, a model in either class is designed from relevant statistics of development time and mortality. Network models are emphasized, and Erlang and Bessel networks are discussed in particular. Use of networks in models of population systems is discussed briefly with examples.

Although the experimentation is time and labor intensive, entomologists now develop quantitative information about life cycles of pest and beneficial insects in integrated pest management programs. Typical data include development times and physiological mortalities of life stages and reproductive properties (Caswell 1972, Curry & Feldman 1987, Sharpe et al. 1977, Curry, Feldman & Sharpe 1978, Johnson et al., 1983). Complete studies include variations of those properties with significant environmental factors, such as temperature and nutrition. Additional studies provide quantitative information about predation, parasitism, and migration. All those data correspond to information needed for modeling insect populations.

Several approaches are used to model populations (Tummala et al. 1975, Gutierrez et al. 1980, Curry & Feldman 1987, Shoemaker 1980, Ruesink 1976). Beginning with Leslie (1945, 1948), contemporary approaches generally attempt to model in terms of the biological processes. This is a significant improvement over the dynamic curve fitting employed in early models by Verhulst (1838), Pearl & Reed (1920), Lotka (1925), Volterra (1926), and Nicholson & Bailey (1935).

This chapter discusses two intrinsic properties of insect life cycles − − development and physiological mortality − − relative to formulation of population models. Convolution and network models are preferred for representing individual and successive stages of life cycles. The complexity of a model of a life stage is set by the statistics of development time and mortality in the stage represented. Network models consist of three genera of elementary processes − − storage, transformation, and inertance. Whole life cycles, consisting of egg, larval (possibly partitioned), pupal, and adult stages, are represented by cascades of stage models.

---

[1]Department of Agricultural Engineering, University of Florida, Gainesville, FL 32611.

Statistics of Insect Development

The statistics of development in an insect life stage are determined from the following general experiment. A large cohort corresponding to the beginning of a specified life stage is established in a constant temperature environment. The time interval for each survivor of the cohort to emerge from the stage is recorded as its development time, which also is known as residence or transit time. The relative frequency distribution, mean, and variance of cohort survivor times are calculated from the data as estimates of the true probability density function, mean, and variance. Repeating the experiment at several temperatures over a range (e.g., 10–35 C) establishes the effect of temperature on development. Analogous experiments are performed when dependence on nutrition or another factor is to be established.

Information about physiological mortality is obtained from the experiment as an adjunct to development time. Since time of death in an immature stage is difficult or impossible to establish, physiological mortality for those stages usually is accounted for by survivorship at emergence from the stage. The relative frequency of cohort survivors provides an estimate of the probability of stage survivorship, q, at the stated temperature; the corresponding probabillity of death $p = 1-q$.

The adult stage is a somewhat different matter, where emergence is by death, so development time is better viewed as residence time or adult longevity. The relative frequency distribution, mean, and variance of adult longevity are calculated from data from an experiment, of the sort described above, with an adult cohort under stated conditions. Of course, $p = 1$, and $q = 0$ for adults.

A hypothetical family of isothermal probability density functions of development time is presented in Fig. 1. Corresponding graphs of mean development time, t_d, and its reciprocal, development rate r_d, as functions of temperature are given in Fig. 2. The general assumption is that isothermal development occurs at uniform rate over the whole time interval. Both graphs in Fig. 2 exhibit threshold temperature, T_0 (e.g., 10–12 C). Wagner et al. (1985) successfully fitted relative frequency distributions of development time with the Weibull density function. Otherwise there appear to be no attempts in the literature at generic characterization of the probability density function of development time on theoretical or empirical bases. Nevertheless, it generally is recognized that density functions of isothermal development for a given species are similar in shape and asymmetrical about their means.

Linear temperature dependent development and the degree–day concept have received considerable attention in models and the literature. It applies to poikilothermic organisms whose development time and rate at high temperature follow the dashed curves in Fig. 2. However, most insects appear to exhibit the peak and subsequent decline in development rate at high temperature shown by the solid line in Fig. 2, and the degree–day concept applies only at lower temperature. It follows that degree–day duration of a life stage as a species specific parameter looses meaning at high temperature.

Population Models

The objective of models presented here is to represent a population of an insect species in a stage of its life cycle as affected by development and physiological mortality. The population resides in a region of spatial extent E (area or volume). By appropriate interconnection of stage models, development and mortality in populations of the species in several stages or over the whole life cycle may be represented. The models are discussed in the context of systems theory, and they are based on two related concepts of that theory, convolution and networks. Convolution models have theoretical value, while network models are preferred in practice. The parameters of a model in either class are determined from statistical data on development time and survivorship of the stage. For simplicity, all variables are considered to be continuous functions of time.

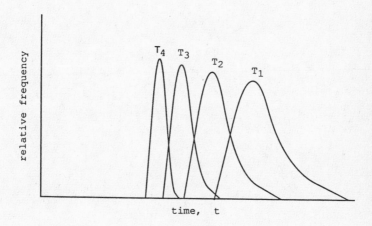

Fig. 1. Relative Frequency of insect development time t at four temperatures.

Convolution Models.

The experiment outlined above is interpretable in the concepts of systems analysis. The box in Fig. 3(a) depicts a stage of an insect life cycle (e.g., egg, larval, pupal, adult) as a system; it may similarly represent a sequence of stages. The system input, γ_i, pertains to a flow of individuals into the stage; output, γ_0, corresponds to a flow of individuals out of the stage. A cohort of M individuals entering the stage is equivalent to an impulse (depicted by a vertical arrow in Fig. 2) of value M into the system (Papoulis 1980, Director & Rohrer 1972, Cooper & McGillem 1967, Lathi 1965). In a constant temperature environment, survivors of the cohort

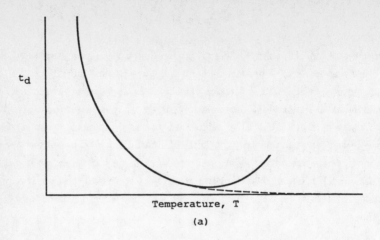

(a)

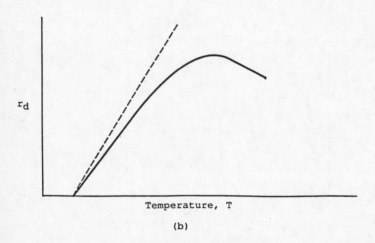

(b)

Fig. 2 a,b. Mean devlopment time, t_d, and mean development rate, r_d, as functions of temperature.

subsequently leave the stage at random times, t, producing a delayed, finite pulse flow of individuals at the system output. With q denoting survivorship, the area of the output pulse equals qM, the total survivors in the cohort. Scaling the output pulse by $1/qM$ yields the impulse response, h, of the system (ibidem), it corresponds to the isothermal probability density function of survivor development time. The conclusions are: (1) the appropriate system representation of the temporal aspect of development in a life stage is dispersive delay and (2) experimental data on development in the stage are directly applicable to formulation of a system model of that stage. General, continuous input, γ_i, of individuals to the life stage in an isothermal environment establishes a resident population and evokes continuous output of individuals, γ_0, given by the convolution of γ_i with h (ibidem);

$$\gamma_0(t) = \gamma_i * h = \int h(t-x)\ \gamma_i(x)\ dx. \tag{1}$$

Network Models.

In theory, network models of a life stage are derivable from the probability density function of survivor development time by methods of network synthesis. In practice, mostly empirical methods have been used. Whatever the approach, the objective is to synthesize a network with a dispersive delay impulse response that approximates the density function of survivor development time and the survivorship of the stage represented. Whereas a convolution model provides an external or black box view of a life stage (Fig. 3), a network provides internal structure to that box. It accepts an input flow of individuals; it retains them in a resident population until, after random delay (development time), it releases survivors at its output. Individuals dying in the stage are transferred to a sink. Synthesis procedures operate on the system transfer funciton, which is the ratio of Laplace transformed system output to input (ibidem). Values of parameters of components in a network relate to the mean and variance of the density function of development time and the survivorshiop of the stage.

The Components

Preliminary to presenting two network models of development, some elementary components of population networks are defined (Smerage 1985, Martens & Allen 1969, Shearer et al. 1967). Components of network population models pertain to elementary processes; they are defined by relations between flows and densities, denoted by γ and n, respectively, of members of relevant populations. A terminal labeled T in the graph of any component denotes dependence of that process on temperature.

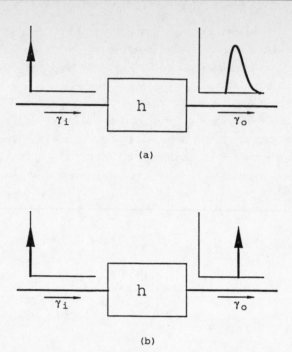

Fig. 3. A stage of an insect life cycle viewed as a system with (a) dispersive and (b) ideal delay impulse responses, h.

The first component, store C in Fig. 4(a), is used to represent a population. It is defined by the following relation between density, n, of its population and flow, γ, of individuals incident to it,

$$C: \quad \gamma = C \frac{dn}{dt} \ . \tag{2}$$

Capacitance parameter $C = E$, the area or volume of the region occupied by the population.

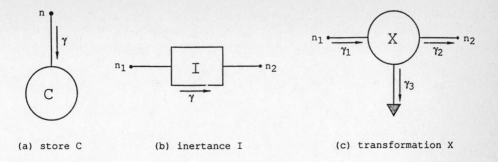

(a) store C   (b) inertance I   (c) transformation X

Fig. 4. Elementary components of population networks.

Inertance I, Fig. 4(b), pertains to inertia of flow. It is defined by the following relation between flow, γ, though a path and difference in densities, $\Delta n \equiv n_1 - n_2$, between ends of the path,

$$\text{I:} \quad \Delta n = I \frac{d\gamma}{dt} \tag{3}$$

where I is the inertance parameter.

Transformation X in Fig. 4(c) has an input flow γ_1 in response to density n_1; its outputs are flows γ_2 of live individuals and γ_3 of dead individuals. The latter go to a sink denoted by the triangle in the Fig.. The following equations describe X, with q and R being its survivorship and input resistance, respectively,

$$\text{X:} \quad \begin{aligned} \gamma_1 &= n_1/R \\ \gamma_2 &= q\, \gamma_1 \\ \gamma_3 &= (1-q)\, \gamma_1 \ . \end{aligned} \tag{4}$$

Erlang Networks

The probability density function of survivor development time, t, often is approximated by the Erlang function with parameters m, the order of approximation, and t_d, The mean development time, (Gross & Harris 1974, Cooper 1972, Manetsch 1976);

$$h_E(t) = \frac{(t_d/m)^{-m}}{(m-1)} t^{m-1} e^{-m \, t/t_d} \ , \ t \geq 0. \tag{5}$$

The mean, $<t>$, variance, σ^2, and transfer function, σ_E, of this function (ibidem) are given, respectively, by

$$<t>_E = t_d \tag{6}$$

$$\sigma_E^2 = t_d^2/m \tag{7}$$

$$H_E(s) = \left[\frac{1}{1 + t_d \, s/m}\right]^m \tag{8}$$

where s is the complex variable of Laplace Transforms (Papoulis 1980, Director & Rohrer 1972, Cooper & McGillem 1967, Lathi 1965). Graphs of several Erlang functions are shown in Fig. 5. To employ this function, parameter t_d is equated to the mean of development time data and m is set to approximate the variance of that data by σ_E^2. Then a development network for the stage is synthesized from H_E for those parameter values.

A discrete network described by the Erlang approximation is obtained as follows. The rational function within the parenthesis of H_E in (8) is the transfer function of the elementary store—transformation network, the first order Erlang network, in Fig. 6. It realizes $1/m$ of overall mean devleopment time t_d and variance t_d^2. Capacitance $C = E/m$. Transformation X has no mortality ($q = 1$), and its input resistance $R = t_d/E$, which incorporates in X, through t_d, the dependence of development time on temperature. A network model for the life stage is obtianed by cascading m of these elementary networks, as in Fig. 7 for $m = 5$. It realizes the m—th order Erlang density function with mean time t_d and variance t_d^2/m. All stores and all transformations except the last are identical, and the sum of all stores equals E. Mortality for the whole stage is accounted for by end transformation X_0 with $q =$ survivorship of the stage.

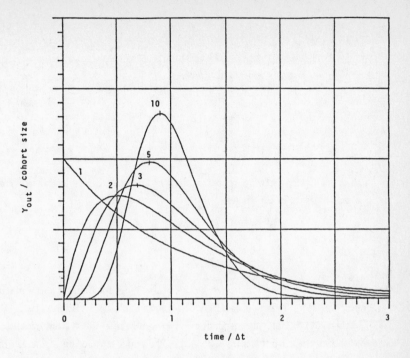

Fig. 5. Impulse responses of Erlang networks normalized by the size of the input impulse. Numbers on curves indicate network order.

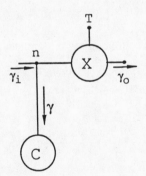

Fig. 6. Elementary store–transformation network is a first order Erlang network. $C = E/m$; for X, $R = t_d/E$ and $q = 1$.

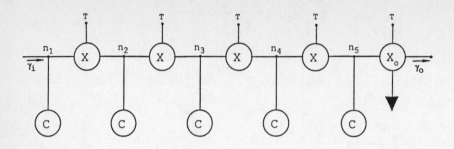

Fig. 7. Fifth order Erlang network representation of an insect life stage. $C = E/m$; for X_o, $R = t_d/E$ and $q =$ survivorship of the stage.

Bessel Networks

An alternative class of population networks is obtained by optimally approximating ideal, finite delay (Smerage 1985). Ideal delay (Papoulis 1980, Director & Rohrer 1972, Cooper & McGillem 1967, Lathi 1965) exhibits no dispersion—an impulse into the system evokes a delayed impulse of equal value at the output, as in Fig. 3(b). The transfer function of ideal, finite delay t_d,

$$H_D(s) = e^{-t_d s}, \tag{9}$$

is not realizable by finite networks. However, H_D is optimally approximated by the following function (Smerage 1985, Storch 1954, Van Valkenburg 1960, Balabanian 1958), with parameters m, the order of approximation, and t_d, the mean delay,

$$H_B(s) \equiv \frac{b_0}{B_m(t_d\ s)} . \tag{10}$$

It is realized by a finite network. Denominator B_m in (10) is the m–th degree Bessel polynomial (ibidem),

$$B_m(x) = b_0 + b_1 x + \cdots + b_m x^m .$$ (11)

Unfortunately, a closed form expression for impulse response h_B (density function) corresponding to H_B does not appear to exist. It may be determined by numerical methods, and several impulse responses are shown in Fig. 8. However, the mean and variance of h_B are known (Smerage, 1985).

$$<t>_B = t_d .$$ (12)

$$\sigma_B^2 = t_d^2/(2m - 1) .$$ (13)

As with the Erlang approximation, the Bessel approximation is employed by equating t_d to the mean of development time data and setting m to approximate the variance of that data by σ_B^2. Then a development network for the stage is synthesized from H_B for those parameter values.

The discrete network for a life stage based on (9), the m–th order Bessel delay network, consists of stores, inertances, and an output transformation (Smerage, 1985). The stores are unequal fractions of E, but the sum of all stores equals E. The inertances also are not identical; each is proportional to t_d^2/E. Output transformation X_0 incorporates stage mortality, through survivorship q, and its input resistance $R = t_d/E$, which incorporates in X_0 the dependence of development time on temperature. Fig. 9 illustrates the fifth order Bessel network model of development in a life stage. Synthesis procedures for networks of other orders are given in references (Smerage 1985, Storch 1954, Van Valkenburg 1960, Balabanian 1958).

Variable Temperature Regimes

To this point, the discussion has been for a constant temperature environment. In real systems, temperature vaires, often widely. All preceding concepts extend, with complications, to variable temperature environments. The impulse response now varies with time in a manner that must be known over the interval of convolution, and convolution will be difficult to calculate in all but simple cases.

Network models are better able to incorporate variable temperature for low to moderate rates of variation. As the examples above indicate, inertance parameters and input resistances of transformations depend on mean development time, t_d, which, in turn, is the function of temperature determined in development experiments. Variation of those parameters in a network model of a life stage according to a temperature scenario and the t_d–temperature relationship approximates the dynamics of the population in that stage.

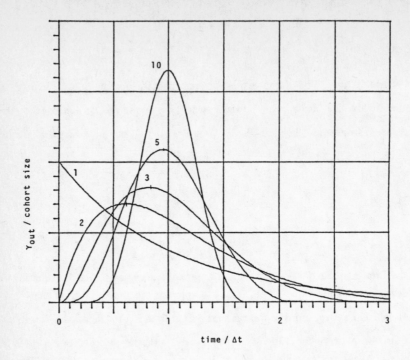

Fig. 8. Impulse responses of Bessel networks normalized by the size of the input impulse. Numbers on curves indicate network order.

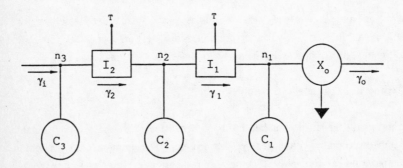

Fig. 9. Fifth order Bessel network representation of an insect life stage. $C_1 = 0.067$ E; $C_2 = 0.310$ E; $C_3 = 0.623$ E; $I_1 = 0.195$ t_d^2/E; $I_2 = 0.42$ t_d^2/E. For X_o, R $= t_d/E$ and q $=$ survivorship of the stage.

Applications

Convolution Models

Convolution models are useful in studies of the dynamics of single and multiple cohorts in a single life stage, in multiple, successive stages, or over a whole life cycle. They are employed using established methods of convolution (Papoulis 1980, McGillem & Cooper 1974, Director & Rohrer 1972, Cooper & McGillem 1967, Lathi 1965). When augmented by an appropriate treatment of reproduction, population dynamics throughout a life cycle over several generations may be studied by these models. However, the same information may be obtained from network models, and those models may seem more "natural" to users. Convolution models tend to be abstract, while network models correspond more to processes in and passage through a life cycle. Furthermore, networks are more interpretable and easier and more intuitive to implement, expecially for comprehensive systems involving multiple, interacting species.

Network Models

The choice between Erlang and Bessel networks for models of life stages is based on the following considerations. Both networks achieve mean delay t_d, but their capabilities to achieve variance differ. Comparing (7) and (13) for equal variance,

$$m_E = 2m_B - 1 \ . \tag{14}$$

Alternatively, for specified network order m,

$$\sigma_E^2 = (2 - 1/m) \ \sigma_B^2 \ . \tag{15}$$

Thus, to obtain an impulse response approximting a probability density function of development time of specified variance, an Erlang network nearly twice the size of a Bessel network may be required. Stated differently, an Erlang network of given order produces an impulse response with nearly twice the variance of a Bessel network of the same order. This difference is readily seen by comparing Fig.s 5 and 8. Beyond this criterion, some analysts may prefer the storage–transformation composition of Erlang networks as more "natural" representations of the biology. They may consider the inertance unbiological. However, if efficiency is the criterion, choose Bessel networks.

Analysis of the population dynamics of a community of insects with network models generally proceeds in four major steps. First, a block diagram is developed showing all stages of each species as well as reproductive, interactive (predation, parasitism), and migratory processes. This is illustrated in Fig. 10 for two species interacting through predation in a case where migration is negligible. Boxes represent the four major life stages of each species, with output flow, γ_0, of one stage equaling input flow, γ_i, of the succeeding stage. Circles X_p, X_{r1} and X_{r2} represent predation and reproduction, respectively.

The second step is to select and substitute for each box in the block diagram a development network, like Fig. 7 or 9, appropriate to the statistics of development for the stage represented (Smerage 1985; Storch 1954; Van Valkenburg 1960; Balabanian, 1958). The result is a network diagram of the total system consisting of stores, transformations, and (possibly) inertances.

The third step is formulation of a mathematical model of the system from its network model. Several alternative methods for formulation exist (Martens & Allen 1969, Athans et al. 1974, Director & Rohrer 1972, Koenig et al. 1967). The state variables method may be preferred. Stores and inertances are the only processes exhibiting state. The state variable of a store is its density, n; the state variable of an inertance is its flow, γ. The state variables mathematical model of an mth order network model of development in a life stage is formulated (ibidem) as a set of m equations in m state variables, one variable, and therefore one equation, for each store and one for each inertance in the network. Each equation expresses the rate of change of a state variable.

Since an mth order Erlang network model of development in a life stage contains only m stores and m transformations, its m state variables are identified only with the densities of the m stores. The corresponding Bessel network contains both stores and inertances (total of m; about m/2 of each), and it follows that its m state variables are identified with the densities of all stores and the flows in all inertances in the network. Models of fifth order Bessel and Erlang networks in the Appendix illustrate these points.

The fourth and final step is analysis of the dynamics of the real population system by analysis of the behavior of the model. Simulation and other mathematical analyses of the model are used to predict, interpret, and understand those dynamics.

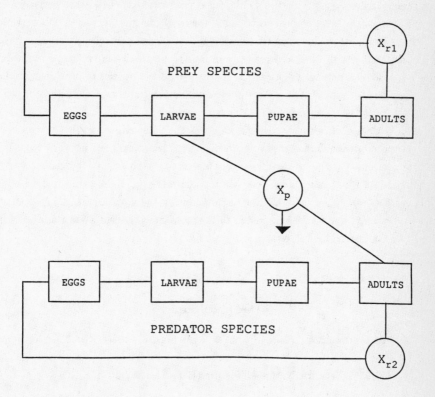

Fig. 10. Block diagram of an insect population system consisting of two species interacting through predation, X_p, in the absence of migration. Processes X_{r1} and X_{r2} represent reproduction by the species.

Conclusion

It was shown that process—based networks may be used in population models to represent development in a life stage. This is possible when one focuses on the essence of development in population dynamics—development time and physiological mortality. A network receives individuals at its input, retains them for an interval of time, and subsequently releases them at its output in conformity with flows and residence of individuals developing in a real life stage. The impulse response of a network is shaped to approximate the statistics of development time and mortality in the stage represented. These properties and associated ease of analysis of population dynamics are justifications for using network models.

The two classes of networks, Erlang and Bessel, discussed as specific models consist of storage, transformation, and inertance components. The size (order) of a network depends on the variance of development time of the life stage represented— —the smaller the variance, the larger the network. Both Bessel and Erlang networks reproduce the mean development time, but Bessel networks were shown to be more efficient in reproducing the variance. In applied analyses of population dynamics, state variables models of networks in both classes are readily formulated, analyzed, and understood by well established procedures.

References Cited

Athans, M., M. L. Dertouzos, R. N. Spann & S. J. Mason. 1974. Systems, Networks, and Computation. McGraw—Hill, New York.

Balabanian, N. 1958. Network Synthesis. Prentice Hall, Englewood Cliffs, N.J.

Caswell, H. 1972. A simulation study of a time lag population model. *J. Theor. Biol.* 34: 419–439.

Cooper, G. R. & C. D. McGillem. 1967. Methods of Signal and System Analysis. Holt, Rinehart, and Winston, New York.

Cooper, R. B. 1972. Introduction to Queueing Theory. MacMillan, New York.

Curry, G. L. & R. M. Feldman. 1987. Mathematical Foundations of Population Dynamics. Texas A&M University Press, College Station.

Curry, G. L., R. M. Feldman & P. J. H. Sharpe. 1978. Foundations of stochastic development. *J. Theor. Biol.* 74: 397–410.

Director, S. W. & R. A. Rohrer. 1972. Introduction to Systems Theory. McGraw—Hill, New York.

Gross, D. & C. M. Harris. 1974. Fundamentals of Queueing Theory. J. Wiley & Sons, New York.

Gutierrez, A. P., D. W. DeMichele, Y. Wang, G. L. Curry, R. Skeith & L. G. Brown. 1980. The systems approach to research and decision making for cotton pest control. In C. B. Huffaker [ed.] New Technology of Pest Control. J. Wiley & Sons, New York.

Johnson, D. W., C. S. Barfield & G. E. Allen. 1983. Temperature—dependent development model for the velvetbean caterpillar (*Lepidoptera: Noctuidae*). *Environ. Ent.* 12(6): 1657–1663.

Koenig, H. E., Y. Tokad & H. K. Kesavan. 1967. Analysis of Discrete Physical Systems. McGraw—Hill, New York.

Lathi, B. P. 1965. Signals, Systems, and Communication. J. Wiley & Sons, New York.

Leslie, P. H. 1945. On the use of matrices in certain population mathematics. *Biometrika* 33: 183–212.

Leslie, P H. 1948. Some further notes on the use of matrices in population mathematics. *Biometrika* 35: 213–245.

Lotka, A. J. 1925. Elements of Physical Biology. Williams and Wilkins, Baltimore.

Manetsch, T.J. 1976. Time—varying distributed delays and their use in aggregative models of large systems. IEEE *Trans. Sys. Man. Cybern.* 6: 547–553.

Martens, H. R. & D. R. Allen. 1969. Introduction to System Theory. C. E. Merrill, Columbus, OH.

McGillem, C. D. & G. R. Cooper. 1974. Continuous and Discrete Signal and System Analysis. Holt, Rinehart and Winston, New York.

Nicholson, A. J. & V. A. Bailey. 1935. The balance of animal populations. *Proc. Zool. Soc. Lond.* 1: 551–598.

Papoulis, A. 1980. Circuits and Systems. Holt, Rinehart and Winston, New York.

Pearl, R. & L. J. Reed. 1920. On the rate of growth of the population of the United States since 1790 and its mathematical representation. *Proc. Natl. Acad. Sci.* U.S.A. 6:275–288.

Ruesink, W. G. 1976. Status of the Systems Approach to Pest Management. *Ann. Rev. Entomol.* 21: 27–44.

Sharpe, P. J. H., G. L. Curry, D. W. DeMichele & C. L. Cole. 1977. Distribution model of organism development times. *J. Theor. Biol.* 66: 21–38.

Shearer, J. L., A. T. Murphy & H. H Richardson. 1967. Introduction to System Dynamics. Addison—Wesley, Reading, MA.

Shoemaker, C. A. 1980. The role of systems analysis in integrated pest management. In C. B. Huffaker [ed.], New Technology of Pest CJontrol. J. Wiley & Sons, New York.

Smerage, G. H. 1985. Bessel Delay Networks for Population Models. *Trans.* ASAE 28: 1269–1278.

Storch, L. 1954. Synthesis of constant—time—delay ladder networks using Bessel polynomials. Proc. IRE 42: 1666–1675.

Tummala, R. L., W. G. Ruesink & D. L. Haynes. 1975. A discrete component approach to the management of the cereal leaf beetle ecosystem. *Environ. Entomol.* 4: 175–86.

Van Valkenburg, M. E. 1960. Introduction to Modern Network Synthesis. J. Wiley & Sons, New York.

Verhulst, P. F. 1838. Notice sur la loi que la population suits dons son accroissement. *Corresp. Math. Phys.* 10: 113–121.

Volterra, V. 1926. Variazione e fluttuazioni de numero d'individui in specie animali conviventi. *Mem. Accad. Lincei.* 2: 31–113.

Wagner, T. L., H. Wu, R. M. Feldman, P. J. H. Sharpe & R. N. Coulson. 1985. Multiple–Cohort Approach for Simulating Development of Insect Populations Under Variable Temperatures. Forum: *Ann. Ent. Soc. Am.* 78: 691–704.

Appendix

State Variables Models for Fifth Order Networks

Bessel Network

$$\frac{dn_1}{dt} = \frac{1}{C_1}\gamma_1 - \frac{1}{R\,C_1}n_1$$

$$\frac{dn_2}{dt} = \frac{1}{C_2}\gamma_2 - \frac{1}{C_2}\gamma_1$$

$$\frac{dn_3}{dt} = \frac{1}{C_3}\gamma_i - \frac{1}{C_3}\gamma_2$$

$$\frac{d\gamma_1}{dt} = \frac{1}{I_1}n_2 - \frac{1}{I_1}n_1$$

$$\frac{d\gamma_2}{dt} = \frac{1}{I_2}n_3 - \frac{1}{I_1}n_2$$

$$\gamma_0 = (q/R)\,n_1$$

Erlang Network

$$\frac{dn_1}{dt} = \frac{1}{C}\gamma_i - \frac{1}{RC}n_1$$

$$\frac{dn_2}{dt} = \frac{1}{RC}n_1 - \frac{1}{RC}n_2$$

$$\frac{dn_3}{dt} = \frac{1}{RC}n_2 - \frac{1}{RC}n_3$$

$$\frac{dn_4}{dt} = \frac{1}{RC}n_3 - \frac{1}{RC}n_4$$

$$\frac{dn_5}{dt} = \frac{1}{RC}n_4 - \frac{1}{RC}n_5$$

$$\gamma_0 = (q/R)\,n_5$$

SECTION III

ANALYSIS OF SPATIAL DATE

A Significance Test for Morisita'a Index of Dispersion and the Moments when the Population is Negative Binomial and Poisson

K. Hutcheson[1] and N. I. Lyons[2]

ABSTRACT The moments of Morisita's index of dispersion are derived assuming the observed counts follow negative binomial and Poisson distributions. The moments are expressed as truncated infinite series. Bounds are placed on the truncation error. The rate of convergence to normality of the index appears to be slow for populations with low mean density. A significance test for the comparison of the dispersions of two populations is suggested. Data from a census of the presence of the southern green stinkbug (*Nezara Viridula*) on three crops are used to examine the effect of sample size and quadrat size on the power of this test and on confidence interval coverages.

A common measure of dispersion of individuals in a population is Morisita's Index (1959),

$$I = q \sum_{i=1}^{q} \frac{x_i(x_i - 1)}{T(T-1)} \ ,$$

where x_i is the number of individuals in the ith sample unit ($i = 1, ..., q$), q is the number of sample units, and $T = \Sigma x_i$. Often the sampling unit is a quadrat and the individuals are plants or animals dispersed in a habitat.

The most important property of this index is that if each sample unit is randomly taken from each of q groups within which an infinite population is randomly dispersed, then it is not influenced by T. Methods for analyzing the pattern of distribution of individuals and estimating the clump sizes in an area have been developed through use of this index. In order to develop useful statistical estimation and hypothesis testing procedures based on I, it is desirable that the relationships between I and the parameters of the statistical distributions commonly applied to the natural and experimental populations be clarified. In particular it is useful to determine the moments of the distribution of I under certain assumptions about the distribution of the x_i's.

When the distribution of individuals per sampling unit is Poisson, it can be shown that $E(I) = 1$. If the distribution is contagious or uniform, $E(I)$ will be larger or smaller than unity, respectively. Most biological populations are clumped, or follow a contagious distribution. A commonly used distribution to exhibit this clumping is the negative binomial.

[1]Department of Statistics, University of Georgia, Athens, GA 30602.
[2]Institute of Ecology, University of Georgia, Athens, GA 30602.

In the following sections the first four moments of I will be given when the support is negative binomial and Poisson. The approach in each case will be to condition on the value of T, and then take expectations with respect to the distribution of T. Both of the distributions considered satisfy the reproductive property, i.e., T has a distribution belonging to the same family as the original observations. In each case the distribution of the observations conditional on T has a well–known distribution as pointed out by Stiteler & Patil (1969). The large sample $(q \to \infty)$ asymptotic distribution of I is shown to be normal in each case. The moments of I are used to study the rate of convergence and adequacy of standard error estimates. Based on these results large sample confidence interval estimation and hypothesis testing procedures are proposed.

Moments of I when the Support is Negative Binomial

Let the x_i's by independent identically distributed negative binomial random variables with parameters P and N. The density function is

$$f(x) = \binom{N+x-1}{N-1} P^x (1+P)^{-N-x}, \quad P > 0, \quad N > 0 \text{ an integer}$$

and $x = 0, 1, 2, \ldots$.

In this case T has a negative binomial distribution with parameters P and NP. The density function of T is

$$f(T) = \binom{qN+T-1}{qN-1} P^T (1+P)^{-qN-T} .$$

Let $N^{[r]} = N(N+1) \ldots (N+r-1)$, with $N^{[1]} = N$ and $N^{[0]} = 1$ and $N^{(r)} = N(N-1) \ldots (N-r+1)$ with $N^{(1)} = N$ and $N^{(0)} = 1$. The conditional density of the x_i's is

$$f(x_1, x_2, \ldots, x_q \mid T) = \frac{T!}{(qN)^{[T]}} \prod_{i=1}^{q} \frac{N^{[x_i]}}{x_i!} ,$$

the negative multivariate hypergeometric distribution. The joint factorial moments of the x_i's are given by

$$\mu(r_1, r_2, \ldots r_q) = T^{(\Sigma r_i)} \Pi \frac{N^{[r_i]}}{(qN)^{[\Sigma x_i]}} .$$

If $z_i = x_i(x_i - 1)$, then

$$E(\Sigma z_i \,|\, T) = q \frac{T^{(2)} N^{[2]}}{(qN)^{[2]}} \; .$$

For evaluation of $E((\Sigma z_i)^k \,|\, T)$ for $k = 2, 3, 4$, see Chiang (1980).

Since I is not defined for $T = 0$ and $T = 1$, the truncated negative binomial is used for T. Therefore,

$$E(I^k) = \frac{1}{c} \sum_{T=2}^{\infty} \begin{bmatrix} qN+T-1 \\ qN-1 \end{bmatrix} P^T (1+P)^{-qN-T} \left[\frac{q}{T(T-1)} \right]^k E(z^k \,|\, T) \; , \qquad (3)$$

where $c = 1 - \Pr(T=0) - \Pr(T-1) = 1 - (1+P)^{-qN} - \begin{bmatrix} qN \\ qN-1 \end{bmatrix} P(1+P)^{-qN-1}$. For $k = 1$, Morisita gives

$$E(I) = \frac{N+1}{N+\frac{1}{q}} \; .$$

The series given in (3) are convergent for all four moments and the error involved in truncating these series can be bounded by using properties of a probability distribution. For $k = 2$ the error in truncating the series at $T = T_0$ is bounded by $\frac{1}{c} A \Pr(T \geq T_0)$ where

$$A = q^2 \left[\frac{qN^{[4]}}{(qN)^{[4]}} + \frac{4 T_0^{(3)} N^{[3]}}{T_0(T-1))^2} (qN)^{[3]} + \frac{2 T_0^{(2)} N^{[2]}}{(T_0(T_0-1))^2} (qN)^{[2]} + q\frac{(q-1) T_0^{(2)} N^{[2]}}{(qN)^{[4]}} \right] \; .$$

Moments of I when the Support is Poisson

The conditional distribution of $x_1, x_2, ..., x_q$ given T fixed is multinomial $\text{Mult}(T, \Pi_1, \Pi_2, ..., \Pi_q)$ with equal probabilities, $\Pi_i = 1/q$, $i = 1, 2, ..., q$.

I can be expressed as

$$I = -\frac{qT}{T}(d-1+1/T) \; ,$$

where d is the complement of Simpson's index of diversity,

$$d = 1 - \Sigma x_i^2 / T^2 \ ,$$

(see Pielou 1969).

The first and second moments of d are given by Simpson (1949) and the third and fourth moments by Lyons & Hutcheson (1979) in the case T is assumed fixed. Expressing I as a function of d

$$I = -\frac{qT}{T-1}d + q \ ,$$

it follows that

$$E(I \mid T) = -\frac{qT}{T-1}E(d \mid T) + q$$

$$= -\frac{qT}{T-1} \frac{q-1}{q} \frac{T-1}{T} + q$$

$$= 1 \ .$$

Since $E(I \mid T)$ is independent of T, $E(I \mid T) = E(I)$. The moments of I can be written in terms of the conditional moments of d using the iterative property of conditional expectation. This approach is taken by Stiteler & Patil (1969) in obtaining the variance of I.

The moments of I can be obtained from

$$E_T[(I - E(I \mid T))^k \mid T] = E_T\{\left[-\frac{qT}{T-1}\right]^k [d - E(d \mid T)]^k \mid T\}$$

$$= \left[-\frac{qT}{T-1}\right]^k \mu_k(d \mid T)$$

where

$$\mu_k(d \mid T) = E_T[(d - E(d \mid T))^k \mid T], k = 1, 2, 3, 4.$$

From Lyons & Hutcheson (1979), the moments of d given T are (note correction in the third moment)

$$\mu_1(d \mid T) = \frac{q-1}{q}\left[1 - \frac{1}{T}\right]$$

$$\mu_2(d \mid T) = (q-1)\left[\frac{1}{T^2} - \frac{1}{T^3}\right]$$

$$\mu_3(d \mid T) = -\frac{4(q-1)}{3}\left[\frac{2}{T^3} + \frac{q-8}{T^4} - \frac{q-6}{T^5}\right]$$

$$\mu_4(d \mid T) = \frac{4(q-1)}{q^4}\left[\frac{3(q+3)}{T^4} + \frac{6(3q-19)}{T^5} + \frac{2q^2-81q+285}{T^6} - \frac{2(q^2-30q+90)}{T^7}\right].$$

Thus the kth central moment of I is in general

$$\mu_k(I) = \frac{1}{c}\sum_{T=2}^{\infty} \frac{e^{-q^\lambda}(q\lambda)^t}{T!}\mu_k(d \mid T).$$

Expressions for the moments when k = 2, 3, 4 are given in Vormongkul (1982).

Stiteler & Patil point out that by a simple rearrangement of terms $\mu_2(I)$ can be expressed as a product of Π_i and expectations of simple functions of T. The Π_i's are probabilities of the general multinomial where i = 1, 2, ..., q. Due to the usefulness of these expectations, and the fact that there was an error in their simplification of $\mu_2(I)$, a corrected table of values is presented. The correct version of $\mu_2(I)$ is

$$\mu_2(I) = 4E_T\left[\frac{1}{T}\right]\left[\Sigma\Pi_i^3 - \left[\Sigma\Pi_i^2\right]^2\right]$$

$$+ 2E\left[\frac{1}{T(T-1)}\right]\left[\Sigma\Pi_i^2 + \left[\Sigma\Pi_i^2\right]^2 - 2\Sigma\Pi_i^3\right]$$

which reduces to

$$2\frac{q-1}{q}E\left[\frac{1}{T(T-1)}\right]$$

in the equiprobable case. If $E\left[\dfrac{1}{T(T-1)}\right]$ is approximated by truncating the series at N,

$$E\left[\frac{1}{T(T-1)}\right] = \frac{1}{c}\sum_{T=2}^{N}\frac{1}{T(T-1)}e^{-q\lambda}(q\lambda)^t/T! \quad ,$$

where $c = 1 - e^{-q\lambda} - q\lambda e^{-q\lambda}$, the error due to truncation is bounded by

$$e_n = \frac{1}{N(N+1)}[1 - P(T \leq N)].$$

Some values of $E\left[\frac{1}{T}\right]$ and $E\left[\frac{1}{T(T-1)}\right]$ are given in Table 1, accurate to six significant digits. Limitation on most computers is in the calculation of $e^{-q\lambda}$.

Table 1. expectations of 1/T and 1/(T(T−1)) for zero—one—truncated Poisson

$q\lambda$	$E\left[\dfrac{1}{T}\right]$	$E\left[\dfrac{1}{T(T-1)}\right]$
.01	.499444	.498889
.05	.497216	.494450
.10	.494422	.488913
.15	.491616	.483389
.20	.488800	.477878
.25	.485973	.472382
.30	.483136	.466902
.35	.480290	.461437
.40	.477435	.455990
.45	.474570	.450560
.50	.471697	.445148
.60	.465927	.434383
.70	.460128	.423701
.80	.454303	.413108
.90	.448454	.402609
1.00	.442586	.392211
1.50	.413084	.341937
2.00	.383652	.295010
2.50	.354766	.251989
3.00	.326877	.213264
3.50	.300375	.179029
4.00	.275562	.149279
4.50	.252640	.123836
5.00	.231710	.102387
5.50	.212784	.084529
6.00	.195800	.069811
7.00	.167165	.048002
8.00	.144592	.033683
9.00	.126787	.024310
10.00	.122618	.018109
15.00	.071868	.006225
20.00	.052797	.003158
25.00	.041746	.001915

The values of the skewness, $\sqrt{\beta_1} = \sqrt{\mu_3^2/\mu_2^3}$ and kurtosis, $\beta_2 = \mu_4/\mu_2^2$ along with the approximate variance of I and the percentage error in using the asymptotic results are given in Table 2 for $\lambda = .5$ and q increasing. For moderately large values of λ the normal approximation appears to do well for moderate values of q. However, for small values of λ, q must be very high.

Table 2. The Variance, Skewness and Kurtosis of I for $\lambda = .5$ and increasing q.

q	Variance	Skewness	Kurtosis	Error in Variance
2	.784422	.000000	1.242603	$.304679 \ (10)^{-16}$
3	1.367749	.849842	2.165580	$.781647 \ (10)^{-17}$
4	1.770060	1.372815	3.670968	$.146901 \ (10)^{-15}$
5	2.015914	1.807943	5.567622	$.111677 \ (10)^{-15}$
10	1.842975	3.486627	21.578876	$.911241 \ (10)^{-16}$
15	1.122360	4.160798	43.915080	$.251681 \ (10)^{-15}$
20	.688140	3.463631	49.287558	$.440484 \ (10)^{-16}$
25	.474391	2.374620	34.068507	$.932505 \ (10)^{-16}$
30	.361051	1.671010	18.942138	$.118905 \ (10)^{-15}$
35	.292463	1.317979	11.154115	$.110634 \ (10)^{-15}$
40	.246328	1.132822	7.939227	$.834891 \ (10)^{-16}$
45	.212998	1.018756	6.574966	$.544329 \ (10)^{-16}$
50	.187715	.937949	5.879400	$.319175 \ (10)^{-16}$
60	.151811	.825118	5.139923	$.725884 \ (10)^{-16}$
70	.127493	.746824	4.720762	$.408993 \ (10)^{-16}$
80	.109912	.687892	4.442881	$.494621 \ (10)^{-16}$
90	.096604	.641301	4.243738	$.489401 \ (10)^{-16}$
100	.086176	.603202	4.093581	$.419381 \ (10)^{-16}$
120	.070882	.543962	3.881669	$.509440 \ (10)^{-16}$
130	.065107	.520270	3.803999	$.330698 \ (10)^{-16}$
140	.060203	.499450	3.738983	$.434868 \ (10)^{-16}$
150	.055986	.480960	3.683747	$.221232 \ (10)^{-15}$

Asymptotic Normality of I

Stiteler & Patil (1969) state that a disadvantage in using Morisita's index of dispersion, I, is the lack of a test of significance. They also note that under the Poisson assumption the test statistic proposed by Morisita (1959) is equal to the variance–to–mean ratio. A test for significance for the variance–to–mean ratio from one, when the population mean is large, is based on the work by Hoel (1943). In this section the asymptotic distribution of I is derived assuming the negative binomial, binomial and Poisson. Based on these results a procedure for determining confidence interval estimation of the population value of I are discussed. A large sample test of the difference between two population dispersions is also suggested.

Assuming that $x_1, x_2, ..., x_q$ have underlying distributions with mean μ and variance σ^2 then $\mu_t = q\mu$ and $\sigma_t^2 = q\sigma^2$. For large q we have approximately,

$$I = \frac{s^2 + \overline{x}^2 - \overline{x}}{\overline{x}^2}$$

and

$$\sqrt{q}(\overline{x}, s^2) \xrightarrow{D} N\left[(\mu_1, \mu_2), \begin{bmatrix} \mu_2 & \mu_3 \\ \mu_3 & \mu_4 - \mu_2^2 \end{bmatrix} \right].$$

By the transformation of variables theorem, the ratio is asymptotically normal with (Lloyd 1967)

$$\text{mean} = \frac{\mu_2 + \mu_1^2 - \mu_1}{\mu_1^2}$$

and

$$\sigma^2 = \frac{1}{q\mu_1^4}\left[\left[\frac{-2\mu_2 + \mu_1}{\mu_1}\right]^2 \mu_2 + 2\left[\frac{-2\mu_2 + \mu_1}{\mu_1}\right]\mu_3 + \mu_4 - \mu_2^2 \right].$$

For the negative binomial,

$$I \sim AN(1 + 1/N, \quad 2(N+1)Q^2/(qN^3P^2)), \quad Q = 1 + P$$

binomial $X \sim b(N, p)$,

$$I \sim AN(1 - 1/N, \quad 2(N-1)Q^2/(qN^3p^2)), \quad Q = 1 - p$$

and Poisson,

$$I \sim AN(1, \quad 2/(q\lambda^2)).$$

Thus, an approximate confidence interval on I is,

$$I \pm Z_{\alpha/2} \sqrt{\text{vâr}(I)}$$

and a test for comparing two population densities is based on the statistic

$$Z = (I_1 - I_2)/\sqrt{\text{vâr}(I_1) + \text{vâr}(I_2)}. \tag{4}$$

If the population is Poisson, a test for randomness is based on

$$Z = (I - 1)/\sqrt{2/(q\lambda^2)}.$$

An Application

Three crops (mustard, tobacco and sorghum) were censused in the 1980 season. Counts of adult and nymph stinkbugs (*Nezara Viridula*) were taken in one square yard quadrats. The data below are the frequencies of adult stinkbugs per quadrat for the three crops,

Frequency	0	1	2	3	4	5	6	7	8
Mustard	2740	93	31	17	5	4			
Tobacco	1054	36	22	10	3	2	2	1	
Sorghum	680	198	76	27	9	6	2	1	1 .

Preliminary results indicate an adequate fit of the negative binomial distribution to all three crops. Estimates of k, P and the values of I and $s^2/\bar{x}$ are,

	$\hat{K}$	$\hat{P}$	I	$s^2/\bar{x}$
Mustard	.074	1.15	14.58	2.15
Tobacco	.077	1.73	14.08	2.73
Sorghum	.645	0.81	2.55	1.81

Pairwise comparison of crops with respect to dispersion result in Z values according to equation (4) as follows,

	I	$s^2/\overline{x}$
Mustard vs Tobacco z =	0.19	−1.77
Mustard vs Sorghum z =	5.50*	1.55
Tobacco vs Sorghum z =	4.90*	2.50*

The results indicate significant differences between tobacco and sorghum, and no difference between mustard and tobacco for both indices. However, a significant difference between mustard and sorghum was indicated for the I index but not with the variance–to–mean ratio.

It would be useful to determine the effect on the results of sampling the population. A simulation was conducted using various size samples. The percentage of rejections out of 500 iterations are,

			Sample Size			
		50	100	200	300	500
Mustard vs tobacco	I	0.0	0.0	0.2	0.4	1.2
	$s^2/\overline{x}$	0.0	0.0	2.4	2.8	5.2
Mustard vs sorghum		0.0	0.0	25.8	73.4	98.8
		0.0	1.4	3.1	2.2	1.0
Tobacco vs sorghum		0.0	3.5	65.8	98.0	100.0
		0.0	0.2	1.0	4.8	21.2

These results support the conclusions based on the entire population provided the sample size is large, say 200. Sample sizes below 100 indicated no significant differences. Confidence estimation of I was simulated and the percent coverage of population values are,

		Sample Size				
		10	20	30	50	100
Mustard	I	76.7	86.5	92.1	92.9	96.0
	$s^2/\overline{x}$	100.0	100.0	98.7	98.4	97.2
Tobacco		69.8	85.5	91.8	87.5	95.5
		98.7	94.5	97.0	92.0	92.6
Sorghum		97.5	95.7	92.0	90.2	93.2
		100.0	94.4	91.3	89.0	89.6

The results of the census of the sorghum field was used to determine the effect of quadrat size on the confidence interval estimation. Adjacent quadrats were combined to form quadrats of size 1 × 2, 2 × 1 and 2 × 2 respectively. Size 1 × 2 and 2 × 1 were both used because some rows of the sorghum field were destroyed by placement of irrigation systems. The percent coverage of the population values are given below for 500 iterations,

		Sample Size			
quadrat size		10	20	30	50
1 × 1	I	85.3	86.0	90.0	93.0
	$s^2/\overline{x}$	81.7	85.2	86.4	90.6
1 × 2		55.9	52.4	48.8	42.6
		86.6	91.0	93.0	96.8
2 × 1		68.2	64.8	64.4	63.6
		89.3	95.0	99.2	99.4
2 × 2		31.1	27.0	20.8	7.6
		92.4	98.2	99.2	99.8

The variance–to–mean ratio has significantly superior coverage, however the estimated variance of $s^2/\overline{x}$ increased as quadrat size increased, while the estimate of the variance of I did not. All confidence intervals for both indices tended to underestimate the population values. The results indicate that I is not independent of quadrat size when the population is negative binomial. Although Reed (1983) suggests use of the jackknife estimator of variance for dispersion indices, we found in simulation studies that the jackknife underestimates the variance of I and $s^2/\overline{x}$ more often than the moment estimators for the negative binomial distribution. Based on efficiency Lloyd (1967) prefers the maximum likelihood estimator, however, Shenton (1987) notes that in estimating parameters of the negative binomial the maximum likelihood estimators often are not the best and they are likely to have more singularities. The maximum likelihood estimators will be investigated in a later report.

Acknowledgments

We thank Dr. Glenn Ware, Station Statistician, University of Georgia for providing the data. This research was partially supported by the Air Force Office of Scientific research under grant number AFOSR–85–0161.

References Cited

Boutwell, J. R. 1982. Computation of moments of Morisita's index of dispersion under binomial sampling. MAMS Thesis, University of Georgia, Athens, Georgia.

Chiang, H. C. 1981. Moments of Morisita's index of dispersion assuming negative binomial frequency counts. MAMS Thesis, University of Georgia, Athens, Georgia.

Hoel, P. G. 1943. On indices of dispersion. *Ann. Math. Stat.* 14: 155–62.

Johnson, N. L. & S. Kotz. 1969. Discrete Distributions. Houghton Miffin Co., Boston.

Lloyd, M. 1967. Mean Crowding. *J. Anim. Ecol.* 36: 1–30.

Lyons, N. I. & K. Hutcheson. 1979. Distributional properties of Simpson's index of diversity. *Commun. Statist. – Theor. Meth.* A8: 569–74.

Morisita, M. 1959. Measuring of the dispersion of individuals and analysis of the distributional patterns. *Mem. Fac. Sci. Kyusha Univ.*, Ser. E, 2: 215–35.

Morisita, M. 1962. I_δ index as a measure of dispersion of individuals. *Res. Pop. Ecol.* 4: 1–7.

Pielou, E. C. 1969. An Introduction to Mathematical Ecology. Wiley–Interscience, New York.

Reed, W. J. 1983. Confidence estimation of ecological aggregation indices based on counts – a robust procedure. *Biometrika* 39: 987–998.

Simpson, E. H. 1949. Measurement of diversity. *Nature* 163: 688.

Stiteler, W. M. & G. P. Patil. 1969. Variance–to–mean ratio and Morisita's index as measures of spatial patterns in ecological populations. *Stat. Ecol.* 1: 423–59, Penn State Press.

Voramongkol, A. 1982. Moments of Morisita's index of dispersion assuming a Poisson distribution of Counts. MAMS Thesis, University of Georgia, Athens, Georgia.

Spatial Analysis of the Relationship of Grasshopper
Outbreaks to Soil Classification

Daniel L. Johnson[1]

ABSTRACT Two spatial modeling methods, based on analysis of either area or point data, were applied to test hypotheses regarding the relationship of grasshopper population density to soil type and texture in southern Alberta. Grasshopper abundance over the last 10 years was higher in certain soil zones than in others. This difference has been attributed in some previous studies to soil surface texture ("intrinsic hypothesis"). Alternatively, the effect may actually be caused by geographical covariables such as weather, vegetation and farming practices ("extrinsic hypothesis"). A geographic information system was utilized to test these hypotheses in two ways. First, area modeling was employed by overlaying maps of grasshopper population density, previous year's population density, soil type, and soil texture. The resulting unique conditions were subjected to analysis of covariance, with the map intersection area as the weight. The second method relied on analysis of point data from the grasshopper survey database, using the same statistical model. The results of both analytical methods indicated that grasshopper abundance was related to soil type ($\underline{P} < 0.001$), but not to soil texture ($\underline{P} > 0.1$), and the intrinsic hypothesis was rejected.

Much of the science of field ecology is concerned with the description and analysis of the relationships of populations to the geographical patterns of abiotic variables. For example, the distribution and abundance of insect species depend to varying extents on the regional patterns of food, habitat, weather, soil, and vegetation. Danks (1979) summarized the factors that affect the Canadian insect fauna. In practice, spatial demographic analysis is conducted as a search for the factors responsible for regional differences in observed population size and structure, and usually consists of a statistical comparison of species and numbers among geographical regions. The recent development of microcomputer–based geographic information systems (GIS) with analytical capabilities provides a means for rigorous tests of biogeographical relationships. For the present study, I applied GIS techniques in combination with standard covariance analysis methods to investigate the relationship of grasshopper outbreaks to soil zones.

Grasshoppers (*Orthoptera: Acrididae*), like a large number of other insects, have close ecological connections with soil attributes and even inhabit the soil environment during part of their life cycle. These insects have one generation per year in Canada's prairie grasslands, and spend the winter in the egg stage, buried 2–5 cm in the soil. It has been noted that some soil zones tend to have grasshopper outbreaks of greater frequency and intensity than do others. Isely (1937) studied the relationship between Texas soils and the kinds and numbers of grasshoppers, and asserted that, "the primary controlling environmental factor in local distribution is doubtless the soil." He found that north–south differences in the Texas acridid fauna were slight, but that east–west differences, i.e., among soil zones, were striking. His monograph emphasized the

---

[1]Agriculture Canada Research Station, Lethbridge, Alberta, T1J 4B1, Canada

importance of soil type in determining acridid distribution, but he could not determine whether the effect was due to soil quality itself or to the related vegetation. In a more detailed study, Isely (1938) again noted a "clear—cut correlation between grasshoppers and different soil types", and asked whether this relationship might be due to soil qualities such as structure, texture, moisture, or pH. For example, one rangeland grasshopper species reached its greatest numbers only in light sandy soil. He concluded from field surveys and soil preference experiments that while the distribution and abundance of some grasshopper species may be determined by the presence of required food plants, "in certain cases soil texture, such as sandy, sandy loam, clay loam, or clay ... appears to determine the soil choices". I will call this the "intrinsic hypothesis", since the expected relationship between grasshoppers and soil type is determined by the intrinsic physical properties of the soil, primarily particle size. The most likely mechanisms behind this relationship would be through the effects of soil drainage, friability, moisture capacity, and thermal conductivity on grasshopper egg—laying, embryo development and hatching success (Hewitt 1985, Johnson et al. 1986). An uncompacted loam or sandy soil would presumably allow a higher reproductive rate than clay and would promote the rapid population growth that results in outbreaks.

I will call the other general hypothesis for the relationship between grasshoppers and soil the "extrinsic hypothesis". This hypothesis would ascribe the greater abundance of grasshoppers in certain soil zones to geographic covariables and not to the particular physical qualities of the soil. For example, grasshopper populations may respond more to the vegetation type, farming practices, and weather in the different soil zones than to the attributes (e.g., sandiness) of the soils themselves (Grace & Johnson 1985).

Hypotheses of this type may be tested in two simple ways: by examination and analysis of map areas, or through analysis of point data. For area analysis the observations on all variables (in this case, insect population samples and soil survey classes) are converted into maps. Tables describing the intersections of the map variables are constructed by overlaying the maps to find all unique conditions, calculate their areas, and determine the correlation or dependence among maps. The other method, more conventional and more likely to meet standard assumptions is based on analysis of the population sample data recorded at points (survey locations). With the points modeling method, the soil attributes must be determined and recorded for each survey site and used as independent variables in the regression model. I used SPANS (Spatial Analysis System, TYDAC Technologies 1987) to conduct and compare the area modeling and the points modeling methods of analysis.

Methods

Location of the Study Area

South—central Alberta provides an excellent setting in which to test the intrinsic and extrinsic hypotheses of the relationship between grasshoppers and soil. Broad bands of

Chernozemic soils divide the region of agricultural land infested during grasshopper outbreaks. Chernozemic soils are usually well drained and have surface horizons that contain a dark accumulation of organic matter originating from the grassland vegetation of the cool, subarid to subhumid climatic zones of western Canada (Agriculture Canada Expert Committee on Soil Survey 1987). Soils in the Black group have surface horizons that are thicker and darker than the other Chernozemic soils. Black, Dark Brown and Brown Chernozemic soils are usually associated with subhumid, semiarid and subarid regions, respectively. Typical vegetation associations found in these three soil types are mixed grasses, forbs, and shrubs; semiarid grasses and forbs; and dryland grasses, respectively. Within these zones are large pockets of Solonetzic soils and a range of soil textures. Solonetzic soils have hard subsurface layers and are typically high in sodium salt content. Their color also varies with aridity and associated vegetation. The common soil textures within the study region are classified as clay, clay loam, loam, loamy sand, sandy loam, and silt loam.

The grasshopper outbreaks and the prairie soils of interest occur primarily in the southern part of Alberta, so this study was confined to the counties in the region bounded by 49–53°N latitude and 110–115°W longitude.

Grasshopper Population Data

The population database was generated from the annual Alberta grasshopper survey. The survey is the basis for management decisions regarding grasshopper control strategy in Alberta. Approximately 50 permanent field personnel and assistants from counties, municipal districts, and improvement districts are involved in the sampling. They are instructed in sampling methods and taxonomic identification in a training course, presently instructed by the author, offered every 2–3 years. Survey sites are chosen at random each year (stratified random sampling), usually about 10 km apart in an area of cropland and rangeland covering approximately 190,000 km². During the survey period, typically 25 July – 10 August, the surveyors visit each site and record the average number of grasshoppers per m² along 100 m of roadside and field margin.

The survey data, collected at typically 1,500–2,000 sites per year, are assembled at the Lethbridge Research Station and used to produce a summary of grasshopper infestation of agricultural land in the current year. Until recently, the grasshopper forecast consisted of a map of the unadjusted adult population density (e.g., Johnson & Andrews 1986), based on a system of combining roadside and field counts to produce risk ratings (Riegert 1968). The method of forecasting has changed because of the availability of GIS technology (Johnson & Worobec 1988), but the annual survey still provides a map of the fundamental variable, population density of adult grasshoppers.

The present study is based on an analysis of the roadside grasshopper counts for the period 1978–87. This period has the most complete database, and represents five years of low grasshopper population density followed by five years of severe outbreak.

The grasshopper survey counts were coded for analysis with SAS (SAS Institute 1982) on a VAX 11/750, and with SPANS on an IBM PC AT equipped for color map production. Each data record consisted of the year, the legal land description of the survey site, and the roadside grasshopper density estimate (number per m²) at that site.

Spatial Autocorrelation

In order to characterize the degree of geographical dependence of grasshopper densities, measures of spatial autocorrelation (Moran 1950) were calculated for each year. Moran's I coefficient is essentially a ratio of the spatial covariance to the variance of the observations. It lies between −1 and +1, and can be interpreted as indicating negative or positive correlation in the same way as simple correlation coefficients (Sokal & Oden 1978). Significant spatial autocorrelation indicates that the value of the mapped variable (population density in this case) is dependent on the values at neighboring sites. Independent correlation indices were calculated for neighborhood distances of 0–5, 5–10, ..., 75–80 km from the points and were plotted to produce correlograms showing the autocorrelation as a function of distance for each year.

Map Construction

The correlograms also provided estimates of reasonable weighting function parameters required for contouring the population data. All contouring of the population data was accomplished with the POTMAP routine of SPANS. Averaging circles of specified radius were used to produce an interpolated map from the weighted averages of the values that fell within the circles.

Mapping by SPANS is based on quadtrees, hierarchical data structures that provide efficient execution time and rapid graphics generation (Samet 1984, Samet et al. 1986). Map regions are repeatedly subdivided into quarters (recursive decomposition) until areas of homogeneous value are attained. The method is referred to as a variable resolution data structure, since small quads are utilized only when necessary, for example at the curved interface of two map colors (Hunter & Steiglitz 1979). Consequently, SPANS is well suited to using the hierarchical structure of the Canada Land Information System. Morton numbers (Morton 1966) are used to manage the quadtrees and keep maps comparable.

Maps of soil type ("type" refers to group and subgroup) and texture were constructed by TYDAC Technologies Inc. from the Canada Soil Information System (CanSIS), compiled and maintained by Agriculture Canada, Ottawa. A vector file delineating the soil classification areas was used to produce a map of unique conditions, and these were recolored to produce maps of the soil variables of interest, in this case soil type and soil texture. The soil type names follow CanSIS classification (Agriculture Canada Expert Committee on Soil Survey 1987). Soil types and textures that either accounted for less than 3% of the study area or fell outside of typical grasshopper habitat were excluded from the analysis.

All population and soils maps were based on Universal Transverse Mercator (UTM) projection, which has proven useful for local and global quadtrees (Mark & Lauzon 1985, Tobler & Chen 1986).

Modeling

To construct a table of map intersections for each of the ten years, SPANS was used to overlay maps of 1) grasshopper density in the year of interest, 2) grasshopper density the year before, 3) soil type, and 4) soil texture. The table of unique conditions and the corresponding areas (in km²) was exported to SAS. The grasshopper density categories (map classes, $1/m^2$, $2/m^2$, $3/m^2$, ..., $30/m^2$) were log–transformed, and the general linear model (GLM) procedure of SAS was used to fit a weighted analysis of covariance to the entire 10–year dataset, with the previous year's grasshopper density as the concomitant variable and the intersection area as the weight. GLM was used because of its model flexibility and reliability for estimation of effects in unbalanced linear models. Year, soil type, and soil texture were the main effects in the model, and the appropriate F–tests were constructed with mean squares for residual, year X type interaction and year X texture interaction, respectively. The two–way interactions were tested with the three–way interaction mean square. Year was considered to be a random effect, and soil type and texture were fixed effects.

The points modeling approach used the same statistical model, but the analysis was unweighted since survey data points and not map areas were used. The matrix was constructed by finding the soil attributes for each of the 12,247 survey sites that fell within the soils study area. SPANS was used to reference the soil maps, determine the soil type and texture, and attach these values to the vector of grasshopper survey counts (number per m²). For each year, the grasshopper counts from the previous year were found by construction of a tessellated map (the Voronoi procedure of SPANS) of polygons that include the area closest to the survey site of each polygon (see Dirichlet tessalation, Diggle 1983). The grasshopper survey observation from the nearest point the year before was then found and added to the matrix. The same model that was applied to the area data was fitted to the point data.

Results and Discussion

Grasshopper Density Maps

The grasshopper species present at the surveyed sites were primarily the major species known to attack crops. For example, identification of 5,324 grasshoppers collected with sweepnets at southern survey sites in 1985 by research station personnel indicated a composition of 40% *Melanoplus sanguinipes* (F.), 34% *Camnula pellucida* (Scudder), 12% *M. bivittatus* (Say), and 6.5% *M. packardii* Scudder. Since these species have similar life histories and destructive potential, the analyses were performed without respect to species. The population maps show the gross density of all grasshoppers counted by surveyors.

Spatial autocorrelation of grasshopper densities was significant and positive out to approximately 50 km in most years, and remained greater than 0.2 (significantly greater than zero, $\underline{P} < 0.01$) within 20 km in most years. Grasshopper population density is "mounded" around peaks of high density, probably indicating its dependence on large–scale phenomena. Johnson (1988) discusses interpretation of the correlograms.

Based on the shape of the correlograms and on known patterns of grasshopper field densities and migration, I chose a POTMAP exponential weighting function that averaged counts within circles of 25–km radius, with weights of 1.0, 0.5 and 0.0 at 0, 5 and 25 km distance, respectively. Computer simulation with a range of these parameters resulted in similar conclusions as those below; the results are not highly sensitive to the choice of weighting function parameters. The contoured SPANS maps were constructed with quads, resulting in 4096 X 4096 equivalent raster resolution. The study region includes an area of 114,150 km², after subtraction of rivers, military bases, cities, parks and minor soil types and texture classes that were excluded from the study. The tesselated (Voronoi) maps required to find the previous year's estimate of population density had polygons averaging 60 km² in size.

Presentation maps (collapsed to 6 density categories) for the last 4 years of the study period are shown in Fig. 1a – 1d. The map areas are listed in Table 1.

Table 1. Land area of grasshopper population density classes (maps of the last four years are shown in Fig. 1). The areas were mapped and summed using SPANS.

Grasshoppers/m²	Year[a]				
	1978	1979	1980	1981	1982
0 to 1	41.1	47.6	32.1	29.7	21.8
1+ to 2	39.8	29.8	31.8	33.4	37.5
2+ to 4	17.4	20.3	29.6	33.1	30.9
4+ to 6	1.6	2.3	4.9	3.2	7.0
6+ to 10	0.0	0.0	1.6	0.6	1.9
> 10	0.0	0.0	0.1	0.0	0.9
Grasshoppers/m²	1983	1984	1985	1986	1987
0 to 1	15.1	9.5	4.4	11.8	12.6
1+ to 2	28.5	16.2	7.3	6.0	17.5
2+ to 4	31.3	28.6	18.7	15.4	28.1
4+ to 6	17.0	17.7	17.7	22.2	21.9
6+ to 10	6.1	16.5	28.1	28.5	12.5
> 10	2.0	11.6	23.7	16.1	7.4

[a] The percentage of the total area is given for each year and category. The total study area is 130,302 square kilometers.

Soil Maps

The soil types used for the analysis were the Chernozemic soils (Black, Dark Brown and Brown great groups) and the Solonetzic soils (Black, Dark Brown and Brown subgroups)(Fig. 2a). The Chernozemic and Solonetzic soils accounted for 82% and 18% of the study area, respectively. The majority of the study area fell into the loam soil texture class, which was well represented in

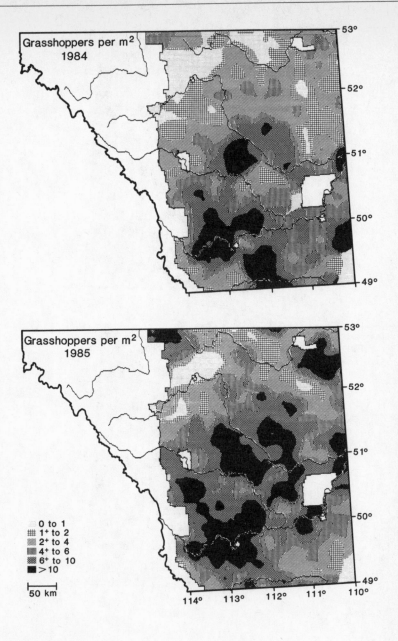

Fig. 1a, 1b. Grasshopper population density in Alberta, 1984–87, based on contoured roadside survey records. The areas in the density classes are shown in Table 1.

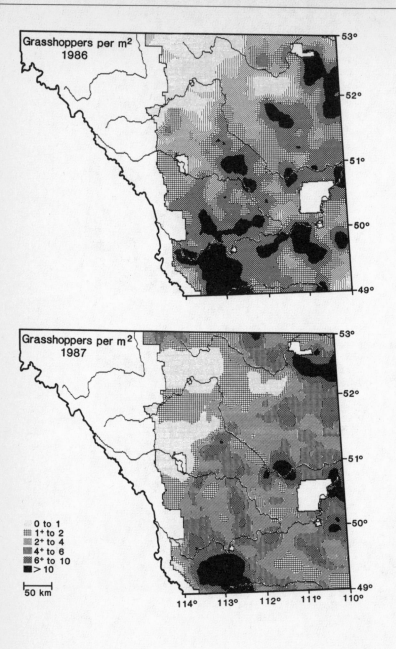

Fig. 1c, 1d. Grasshopper population density in Alberta, 1984–87, based on contoured roadside survey records. The areas in the density classes are shown in Table 1.

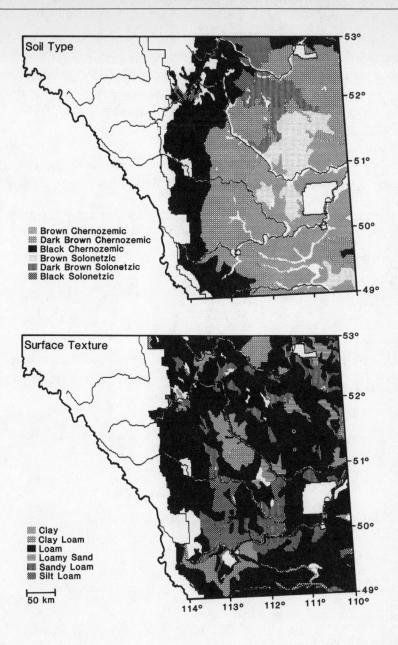

Fig. 2. The soil classification and soil surface texture maps used in the overlay area analysis. The unclassified areas represent less common soil features or types that were omitted from the study. The area of each class is shown in Table 2.

all soil types. The remaining area was roughly divided among the other five soil texture classes (Fig. 2b; Table 2). Soil type and soil texture are slightly correlated, since soil genesis and environment affect both texture and classification. The statistical correlation is significant but not large: a cross–tabulation of the three Chernozemic soil types and the six soil textures yields Cramer's measure of association $V = 0.22$. Maximum association would result in $V = 1.0$ (Bishop et al. 1980).

Table 2. Cross–tabulation of the land area in soil type and soil texture map classes. The percentages of the total land area in the study area, 49 to 53°N latitude in Alberta, are shown.

	Clay	Clay Loam	Loam	Loamy Sand	Sandy Loam	Silt Loam	Total
Brown	516[a]	146	18,251	1,384	3,245	2,966	26,508
Chernozemic	1.9[b]	0.6	68.9	5.2	12.2	11.2	23.2
	12.5[c]	2.4	22.5	33.4	31.7	35.8	
Dark Brown	3,109	1,820	20,577	1,903	4,698	4,909	37,016
Chernozemic	8.4	4.9	55.6	5.1	12.7	13.3	32.4
	75.0	29.8	25.3	46.0	45.8	59.2	
Black	518	2,174	24,288	853	1,630	411	29,874
Chernozemic	1.7	7.3	81.3	2.9	5.5	1.4	26.2
	12.5	35.6	29.9	20.6	15.9	5.0	
Brown	0	0	11,049	0	0	0	11,049
Solonetzic	0	0	100.0	0	0	0	9.7
	0	0	13.6	0	0	0	
Dark Brown	0	0	5,309	0	679	0	5,988
Solonetzic	0	0	88.7	0	11.3	0	5.3
	0	0	6.5	0	6.6	0	
Black	0	1,966	1,752	0	0	0	3,718
Solonetzic	0	52.9	47.1	0	0	0	3.3
	0	32.2	2.2	0	0	0	
Total	4,143	6,106	81,226	4,140	10,252	8,286	114,153
%	3.6	5.3	71.2	3.6	9.0	7.3	

[a] Area in square kilometers.
[b] Row %.
[c] Column %.

Analysis of the Effect of Soil Type and Texture

The results of the analysis of covariance are shown in Table 3. Both models accounted for significant proportions of the total variance of log–transformed densities (56% by area modeling, and 41% by points modeling). In both cases, differences among years were highly significant, and the previous year's population density accounted for more of the total variance than any other factor.

Table 3. Analysis of covariance of ln(grasshopper denisty), as a function of the previous year's density, the soil type, and the soil texture. The results of both area modeling and points modeling are shown.

Source	Area modeling			Points modeling		
	DF	F	P	DF	F	P
Year	9	32.70	<0.001	9	20.19	<0.001
Soil type	5	8.48	<0.001	5	5.04	<0.001
Soil texture	5	1.66	0.165	5	0.55	0.734
Type X texture	12	1.37	0.189	12	1.52	0.130
Type X year	45	5.33	<0.001	45	5.54	<0.001
Texture X year	45	3.26	<0.001	45	3.81	<0.001
Type X text. X yr	108	3.07	<0.001	97	3.03	<0.001
Last year's popn.	1	1,610	<0.001	1	1,149	<0.001
		SS (III)			SS (III)	
Residual	8,414	500,553		10,949	4,286	
Total	8,644	1,154,024		11,168	7,272	

Both statistical analyses suggest that soil type is a highly significant factor ($\underline{P} < 0.001$), but that soil texture and type X texture interaction are not ($\underline{P} > 0.1$). This result allows rejection of the intrinsic hypothesis. The geographical distribution of grasshoppers in Alberta does indeed vary with soil type, but does not vary significantly with the sand, silt, or clay content of the soil, either across the study area as a whole or within the major soil zones. It should be noted that this result pertains to wide–scale population differences only. Small–scale distribution of grasshoppers within a particular habitat is probably affected by soil texture, as well as by local differences in topography, erosion, surface structure, and rockiness.

The interpretation of the results may be questioned in light of the non–orthogonality apparent in Table 2, the soil type, and soil texture cross–tabulation (Gilbert 1972), so I also analyzed the data after excluding the Solonetzic soils. The basic results and interpretation did not change: differences in grasshopper population density among Chernozemic soil types are significant, while differences among the soil textures are not.

The areas of greatest infestation are in the Dark Brown and Brown Chernozemic soil zones. Orthogonal comparisons indicated that these two great groups did not differ in mean numbers of grasshoppers ($\underline{P} > 0.2$), but that Black Chernozemic soil had significantly lower infestations than did the Brown and Dark Brown Chernozemic soils ($\underline{P} < 0.001$).

Area Versus Points Modeling

This comparative analysis indicates that area modeling and points modeling can be used reliably to compare variables over large geographical areas. The example in this paper relied on detailed maps with large numbers of observations and unique conditions, but in most cases area modeling is faster than points modeling methods because of the considerable reduction in data handling. Computation time in area modeling is often insignificant. For example, over 2,000 unique conditions required for an analysis of the 1987 data over the entire survey area were identified, measured, condensed, summed, and exported by SPANS running on an IBM PC AT in under 15 minutes.

Also area modeling is often more appropriate than points modeling to GIS application. For example, when maps produced from a variety of sources and data types are used, point data may be incomplete or lacking on the map of the dependent variable.

Acknowledgements

The CanSIS soil database was provided by Agriculture Canada, and was converted to SPANS maps by Wendy Saxton and associates of TYDAC Technologies Inc., Ottawa, Canada. Production of the grasshopper maps and analysis of the impact of soil variables were performed at the Lethbridge Research Station.

Craig Andrews, Mike Dolinski, Mike Hardman, Tom Kveder, Dave Pang, Karen Wilde and Lynn Wright assisted with compilation of the grasshopper population database. Lyndon Kok assisted in data management and revision of the manuscript. Annual surveys of grasshopper numbers were conducted by the Agriculture Service Board, under the supervision of the author. This research was supported in part by the Agricultural Research Council of Alberta, Farming for the Future Grant No. 870025.

References Cited

Agriculture Canada Expert Committee on Soil Survey. 1987. The Canadian system of soil classification, 2nd ed. Agric. Can. Pub. No. 1646.

Bishop, Y. M. M., S. E. Fienberg & P. W. Holland. 1980. Discrete multivariate analysis: theory and practice. The MIT Press, Cambridge, Massachusetts.

Danks, H. V. 1979. Canada and its insect fauna. *Mem. Entomol. Soc. Can.* 108: 1–573.

Diggle, P. J. 1983. Statistical analysis of spatial point patterns. Academic Press.

Gilbert, N. E. 1972. Biometrical interpretation. Clarendon Press, Oxford, U.K.

Grace, B. D. & D. L. Johnson. 1985. The drought of 1984 in southern Alberta: its severity and effects. *Can. Wat. Res. J.* 10: 28–38.

Hewitt, G. B. 1985. Review of the factors affecting fecundity, oviposition, and egg survival of grasshoppers in North America. U.S.D.A. ARS–36.

Hunter, G. M. & K. Steiglitz. 1979. Operations on images using quad trees. I.E.E.E. Trans. Pattern Analysis and Machine Intelligence. PAMI–1: 145–153.

Isely, F. B. 1937. Seasonal succession, soil relations, numbers, and regional distribution of north–eastern Texas acridians. *Ecol. Monog.* 8: 553–604.

Isely, F. B. 1938. The relations of Texas acrididae to plants and soils. *Ecol. Monog.* 8: 553–604.

Johnson, D. L. & R. C. Andrews. 1986. 1986 Grasshopper Forecast. 35 x 62 cm color map sheet (1:2,000,000), Alberta Bureau of Surveying and Mapping, Edmonton, Alberta.

Johnson, D. L., R. C. Andrews, M. G. Dolinski & J. W. Jones. 1986. High numbers but low reproduction of grasshoppers in 1985. *Can. Agric. Insect Pest Rev.* 63: 8–10.

Johnson, D. L. & A. Worobec. 1988. Spatial and temporal computer analysis of insects and weather: grasshoppers and rainfall in Alberta. In Paths from a viewpoint: the Wellington Festschrift on insect ecology. *Mem. Entomol. Soc. Can.* (in press).

Mark, D. M. & J. P. Lauzon. 1985. Approaches for quadtree–based geographic information systems at continental or global scales. In AUTO–CARTO 7 Proc., pp. 355–364. Washington, D.C.

Moran, P. A. P. 1950. Notes on continuous stochastic phenomena. *Biometrika* 37: 17–23.

Morton, G. 1966. A computer–oriented geodetic data base, and a new technique in file sequencing. Internal memorandum, March, 1966. IBM Canada, Ltd.

Riegert, P. W. 1968. A history of grasshopper abundance surveys and forecasts of outbreaks in Saskatchewan. *Mem. Entomol. Soc. Can.* 52: 1–99.

Samet, H. 1984. The quadtree and related hierarchical data structures. Computing Surveys 16: 187–260.

Samet, H., C. A. Schaffer, R. C. Nelson, Y. Huang, K. Fujimara & A. Rosenfeld. 1986. Recent developments in quadtree–based geographic information systems. In Proceedings Second International Symposium on Spatial Data Handling. International Geographical Union, Williamsville, N.Y.

SAS Institute. 1982. SAS user's guide: statistics. SAS Institute, Cary, N.C.

Sokal, R. R. & N. L. Oden. 1978. Spatial autocorrelation in biology. 1. Methodology. *Biol. J. Linn. Soc.* 10: 199–228.

Tobler, W. & Z. Chen. 1986. A quadtree for global information storage. *Geog. Anal.* 18: 360–371.

TYDAC Technologies. 1987. Spatial Analysis System. Reference guide. Ottawa, Ontario.

Use of Multi–Dimensional Life Tables for Studying Insect Population Dynamics

Andrew M. Liebhold[1] and Joseph S. Elkinton[2]

ABSTRACT In classical life table analyses, density and mortality are measured for many generations as average values across an area. We present here as an alternative the measurement of density and mortality from a multi–dimensional spatial matrix of sample points. This method is applied to studies of gypsy moth, *Lymantria dispar*, population dynamics. Analysis for key factors and density–dependence may yield substantially different results between the two methods. Identification of a key mortality factor from classical life table data indicates that the mortality is correlated with between generation changes in density. In contrast, identification of a key factor from spatial data indicated that the mortality explains changes in the spatial heterogeneity of density. Conclusions about key factors and density–dependence derived from spatial data often vary among generations. Multidimensional life tables may be used to detect density–dependence that occurs on a finer spatial scale than the classic approach would detect. The multi–dimensional approach is viewed as a technique that complements, rather than replaces, traditional life table data.

The classical approach to understanding the numerical behavior of populations is the collection of life table data at one or more sites over several years (Varley et al. 1973). In this type of study, data is collected on animal density at successive stages within a generation. Simultaneously, data is collected about the impact of specific mortality agents. Together, these data are analyzed to quantify mortality over specific age intervals. K–factor analysis (Varley & Gradwell 1960, 1968) is one of many techniques used to assign the importance of various mortality sources and intervals to total generation mortality. Such data are also used to evaluate which periods and specific sources of mortality are acting in a density–dependent fashion. Positive density dependence is though to contribute to population stability (Varley et al. 1973).

Many field studies have not detected density–dependent processes (Dempster & Pollard 1986; Morrison & Strong 1980; Stiling 1987). This has caused some to conclude that populations densities are determined by density–independent or "density–vague" mortality factors (Strong 1986). Hassell (1985, 1986) has pointed out that previous life table studies may have been inadequate because they failed to recognize variation of prey densities through space. In many situations, pest populations are patchy, in part due to the spatial variation of host plants. Hassell (1985, 1986) observed that natural enemies may be responding in a density–dependent fashion to these localized densities and that by averaging across an area, as is done in most life table studies, one may miss this density dependence.

---

[1]USDA Forest Service, Northeastern Forest Experiment Station, Box 4360, Morgantown, WV 26505.

[2]Department of Entomology, University of Massachusetts, Amherst, MA 01003.

Unfortunately, quantifying spatial heterogeneity of density and mortality often requires intensive sampling. Largely for this reason very few studies have attempted to quantify these relationships. Heads & Lawton (1983), in their study of spatial patterns of mortality of the holly leaf–miner showed that density–dependent patterns of mortality may vary among different spatial scales. Hassell et al. (1987) re–analyzed viburnum whitefly life table data collected by Southwood and Reader (1976) and showed that using the classical averaging analysis, no mortality factors operated in a density–dependent fashion but density–dependence was identifiable among spatial subsets within generations.

Hassell (1985) described a need to collect life–table information from a sampling design that is stratified in space as well as time. A solution we present here as an alternative to the classic series of life tables, is a matrix of life tables, stratified by two (horizontal)dimensions in space within a generation (Fig. 1). We have applied this technique to the study of gypsy moth, *Lymantria dispar* (L.) populaitons. In this study we used techniques that allowed us to detect density dependent mortality on a scale larger than achieved in previous studies of spatial density–dependence. Densities were estimated from a matrix of cells throughout a plot. We also attempted to account for causes of larval mortality occurring in spatial subplots. These matrices of density and mortality data were analyzed for patterns within and among populations.

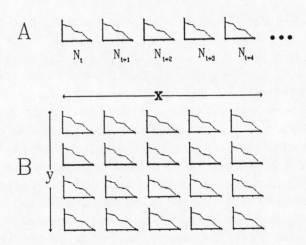

Fig. 1. Spatio–temporal descriptions of sampling schemes for life table studies. A. Classic multi–year design. B. Single year, two–dimensional design.

Methods and Materials

Two 9 ha and one 16 ha study site were located on Otis Airbase, Cape Cod, Massachusetts. Stands were composed of small–diameter *Quercus velutina* Lam., *Q. Alba* L. and *Pinus rigida* Mill.

(Liebhold et al. 1988). At the two 9 ha sites 13 x 13 grids (25 m between each point) were established and a 5 x 17 grid (50 m between each point) was located at the 16 ha stie (Fig. 2). Densities of successive gypsy moth life stages were estimated at each grid point. All densities were expressed as numbers per ha.

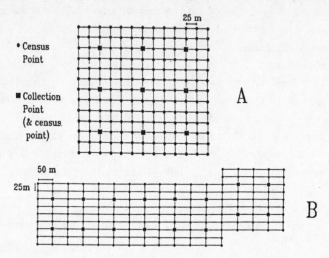

Fig. 2. Location of density estimation census points and larval collection points. A. 9 ha sites (1 & 2). B. 16 ha site (3).

Egg–mass densities were estimated by counting all new egg–masses within 5 m of each point. The density of larvae hatching from each mass was estimated as the product of egg mass density and first instars per egg mass (estimated from 40 field–collected masses [Buonaccorsi & Liebhold 1988]). Fourth instar densities were estimated at each point as the ratio of frass drop and frass yield (rate of frass production per insect) estimates (Liebhold & Eklinton 1988a, 1988b, Buonaccorsi & Liebhold 1988). Frass drop was measured by placing a conical frass trap at each grid point. Frass yield was measured by individually caging a cohort of 40 larvae from the site on host foliage and counting the number of pellets produced per larva over the same period as frass drop measurement. Though L4 was the predominant stage when these samples were taken, other instars occurred at a lower frequency. However, because other samples showed that density decreased little during this period, we feel that these estimates provide an adequate measure of L4 recruitment. Pupal, adult and adult female densities were estimated by locating all pupae in 2 x 5 m quadrats, marking them with an acrylic paint, and recording their positions. These qaudrats were located at each grid point. Pupae were marked once a week during the period that larvae were entering the pupal stage. When all adults had eclosed, the quadrats were revisited and the condition of the pupal remains (i.e. if it was missing, predated, parasitized or had successfully emerged) was recorded. Density estimates from adjoining census points were averaged to form estimates for 1 ha subplots (9 subplots at sites 1 & 2; 16 subplots at site 3). All densities were transformed using $\log_{10}(x+1)$ and k values were calculated (Fig. 3)(Varley & Gradwell 1960,

1968). Since most gypsy moth dispersal occurs during the first instar, k_1 represents changes in density due to mortality and dispersal. Adult mortality is negligable. Thus, k_4 represents a sex–ratio effect, rather than mortality.

Within–plot variation in larval parasitism and disease incidence was measured by making collections of larvae from the center of the 1 ha subplots (Fig. 2). Either weekly or semi–weekly, 50–100 larvae were collected in each subplot. These larvae were placed on artificial diet (Bell et al. 1981) and checked for parasitism and mortality biweekly. The presence of nuclear polyhedrosis virus, NPV, was determined by microscopic examination of cadavers for polyhedral inclusion bodies. Mortality occurring between collections was totaled for each agent and expressed as k (Varley & Gradwell 1960, 1968). No attemp was made to adjust marginal probabilities for simultaneously occurring mortality agents (Royama 1981).

$$L1 \longrightarrow L4 \longrightarrow Pupa \longrightarrow Adult \longrightarrow Egg\text{-}mass$$
$$\quad k_1 \qquad k_2 \qquad k_3 \qquad k_4$$

Fig. 3. Within–generation intervals used in life table analyses.

Results and Discussion

The general pattern of within–generation survival was quite similar among years and sites (Fig. 4) k_2–k_4 generally were greater than k_1. This pattern is similar to what Campbell (1973) found elsewhere in the northeastern United States. During the course of the three years, densities declined at all sites. The major sources of mortality in larvae were NPV, parasitism by *Parasetigena sylvestris*, parasitism by *Cotesia melanoscelus* and mortality due to unascribed causes.

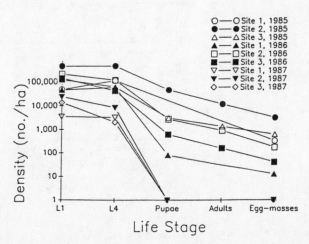

Fig. 4. Densities of successive gypsy moth life stages at each site from 1985–1987.

Considerable within–plot variability in density was frequently observed. For example at site 3 in 1985, one end of the plot had distinctly higher densities through all life stages (Fig. 5). Liebhold et al..(1988) showed that some of the variation in gypsy moth density within a plot can be explained by differences in host foliage density. However, the difference seen in Fig. 5 is clearly not explained only on the basis of host foliage density (Fig. 6). Application of k–factor analysis to the site–averaged measures of density and mortality indicated that k_2, mortality during the late–instar period, was the most highly correlated with generation mortality, K (Table 1). This mortality was, however, inversely density dependent (Table 2).

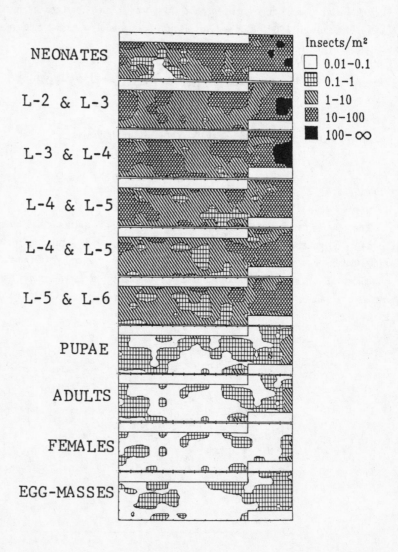

Fig. 5. Spatial distribution of successive gypsy moth life stages at site 3, 1985. Data is interpolated from 86 grid point estimates using a Bessel interpolation.

Site–averaged measures of density and mortality indicated that only unascribed mortality during the late larval period was positively correlated with generation mortality. Parasitism by *Phobocampe disparis* was inversely density dependent and parasitism by *Blepharipa pratensis* was positively density dependent (Table 2). Both agents, however, were minor sources of mortality.

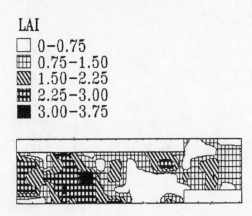

LAI

☐ 0-0.75
▦ 0.75-1.50
▨ 1.50-2.25
▦ 2.25-3.00
■ 3.00-3.75

Fig. 6. Spatial distribution of oak leaf area (expressed as leaf area index [LAI] at site 3. Data from 86 canopy photographs (Liebhold et al. 1988).

Table 1. Results of regression of within–site averaged estimates of mortality (k) to generation mortlaity (K).

Mortality	Slope	Prob. of a greater F	Adjusted R^2
k_1	0.286	0.062	0.330
C. melanoscelus	0.018	0.215	0.145
unascribed	0.007	0.779	0
NPV	−0.004	0.645	0
k_2	1.255	0.007	0.757
B. pratensis	−0.003	0.129	0.277
C. melanoscelus	0.029	0.242	0.112
P. disparis	0.009	0.052	0.475
P. sylvestris	0.055	0.260	0.092
unascribed	0.221	0.022	0.621
NPV	0.008	0.855	0
k_3	0.047	0.729	0
k_4	0.130	0.460	0

Table 2. Regression of within—site averaged mortality (k) to log initial density.

Mortality	Slope	Prob. of a greater F	Adjusted R^2
k_1	−0.063	0.774	0
C. melanoscelus	−0.021	0.198	0.167
unascribed	0.014	0.593	0
NPV	0.012	0.228	0.128
k_2	−1.230	0.005	0.780
B. pratensis	0.003	0.040	0.524
C. melanoscelus	−0.018	0.482	0
P. disparis	−0.011	0.005	0.792
P. sylvestris	−0.041	0.393	0
unascribed	−0.172	0.104	0.330
NPV	0.015	0.724	0
k_3	0.333	0.766	0
k_4	−0.023	0.889	0

Results of k—factor analysis were quite different when averaged within subplots at a site on a given year. As an example, results from site 1, 1985 are shown in Tables 3 & 4. Neither correlation with K nor correlation with density were consistent among years and sites. Similarly, Hassell et al. (1987) found that spatial density dependence was not consistent among years.

Table 3. Results of regression of mortality (k) to generation mortality (K) using estimates from 1 ha subplots at site 1, 1985.

Mortality	Slope	Prob. of a greater F	Adjusted R^2
k_1	0.692	0.001	0.807
C. melanoscelus	−0.018	0.269	0.053
unascribed	−0.007	0.647	0
NPV	0.029	0.299	0.031
k_2	−0.135	0.928	0
C. melanoscelus	−0.278	0.004	0.372
P. sylvestris	−0.009	0.889	0
unascribed	−0.057	0.070	0.140
NPV	0.494	0.028	0.223
k_3	−0.406	0.322	0.017
k_4	−0.179	0.002	0.731

Table 4. Regression of mortality (k) to log initial density using estimates from 1 ha subplots at site 1, 1985.

Mortality	Slope	Prob. of a greater F	Adjusted R²
k_1	0.612	0.001	0.858
C. melanoscelus	−0.188	0.242	0.073
unascribed	0.019	0.900	0
NPV	0.241	0.396	0
k_2	−0.207	0.711	0
C. melanoscelus	−0.906	0.194	0.117
P. sylvestris	−0.502	0.212	0.100
unascribed	−0.123	0.575	0
NPV	3.235	0.014	0.550
k_3	−0.046	0.943	0
k_4	−0.913	0.498	0

Discussion

The reason why analyses of site averages from several years (temporal stratification) were different from those of 1 ha subplots within a site and year (spatial stratification) was that the two sampling schemes addressed different relaitonships. Identification of a key mortality factor from site–averaged data from several years indicates that the mortality is correlated with between generation changes in density. In contrast, identification of key factor from spatial data indicates that the mortality explains changes in the spatial heterogeneity of density. Conclusions about key factors and density–dependence derived from spatial data are unique to the generation under observation. Multi–dimensional life tables may detect density dependence that occurs on a finer spatial scale that the classic approach would not detect. For example, the fine scale, aggregative response of a parasitoid to host densities could only be detected from multi–dimensional data. In contrast, the classical approach to life tables will succeed in detecting the numerical response (in the sense of reproduction rather than dispersal) of mortality agents. Thus, the classical approach has an advantage in that it can be used to detect delayed density–dependent relaitons. Because of the different nature of information derived from the two types of studies, multi–dimensional life tables may complement, but not replace classic life table studies.

Collection of multi–dimensional life tables data at the same sites where long–term life table data is collected poses new possibilities for gaining an understanding of mortality patterns in insect populations. Clearly, sampling strategies other than the two–dimensional grid system used here may be appropriate in other situations. The spatial stratification of the habitat should be designed so as to detect spatial heterogeneity in host densities and mortality. Generation of a matrix of life tables every year, entails a substantial outlay of sampling effort. Presumably spatially stratified data need not be collected over as long of a period as classical life table data is collected.

Acknowledgements

This research was funded inpart by grants 85–CRCR–1–1814 and 87–CRCR–1–2498 from the USDA Competitive Research Grants Program. We are grateful to K.D. Murray for field assistance. We thank Roy Van Driesche and Robert Voss for reviewing this manuscript.

References

Bell, R. A., C. D. Owens, M. Shapiro & J. R. Tardif. 1981. Development of mass rearing technology, pp 599–633. In C. C. Doane & M. L. McManus [eds.], The gypsy Moth: research toward integrated pest management. *U.S. Dep. Agric. For. Serv. Tech. Bull.* 1584.

Buonaccorsi, J. P. & A. M. Liebhold. 1988. Statistical methods for estimating ratios and products in ecological studies. *Environ. Ent.* 17: 572–580.

Campbell, R. W. 1973. Numerical behavior of a gypsy moth population system. *For. Sci.* 19: 162–167.

Dempster, J. P. & E. Pollard. 1986. Spatial heterogeneity, stocasticity and the detection of density dependence in animal populations. *Oikos* 46: 413–416.

Hassell, M. P. 1985. Insect natural enemies as regulating factors. *J. Anim. Ecol.* 54: 323–334.

Hassell, M. P. 1986. Detecting density dependence. *Trends Ecol. Eval.* 1: 90–93.

Hassell, M. P., T. R. E. Southwood & P. M. Reader. 1987. The dynamics of the viburnum whitefly (*Aleurotrachelus jelinekil*): a case study of population regulation. *J. Anim. Ecol.* 56: 283–300.

Heads, P. A. & J. H. Lawton. 1983. Studies on the natural enemy complex of the holly leaf–miner: the effects of scale on the detection of aggregative responses and the implications for biological control. *Oikos* 40: 267–276.

Liebhold, A. M. & J. S. Elkinton. 1988a. Estimating the density of larval gypsy moth, *Lymantria dispar* (*Lepidoptera: Lymantriidae*), using frass drop and frass production measurements: development of sampling techniques. *Environ. Entom.* 17: 381–384.

Liebhold, A. M. & J. S. Elkinton. 1988b. Estimating the density of larval gypsy moth, *Lymantria dispar* (*Lepidoptera: Lymantriidae*), using frass drop and frass production measurments: quantifying sources of variation and sample size. *Environ. Entom.* 17: 385–390.

Liebhold, A. M., J. S. Elkinton, David R. Miller & Y. Wang. 1988. Estimating leaf area index and gypsy moth, *Lymantria dispar* (*Lepidoptera: Lymantriidae*), defoliation using canopy photographs in Cape Cod oak forests. *Environ. Entom.* 17: 560–566.

Morrison, G. & D. R. Strong, Jr. 1980. Spatial variations in host density and the intensity of parasitism: some empirical examples. *Environ. Entom.* 9: 149–152.

Royama, T. 1981. Evaluation of mortality factors in insect life table analysis. *Ecol. Monog.* 51: 495–505.

Southwood, T. R. E. & P. M. Reader. 1976. Population census data and key factor analysis for the Viburnum whitefly, *Aleurotrachelus* jelinekii (Frauen.) on three bushes. *J. Anim. Ecol.* 45: 313–325.

Stiling, P. D. 1987. The frequency of density dependence in insect host–parasitoid systems. *Ecology* 68: 844–856.

Strong, D. R. 1986. Density–vague population change. *Trends Ecol. Evol.* 1: 39–42.

Varley, G. C. & C. R. Gradwell. 1960. Key factors in population studies. *J. Anim. Ecol.* 29: 399–401.

Varley, G. C. & G. R. Gradwell. 1968. Population models for the winter moth. p. 132–142. In T. R. E. Southwood [ed.], Insect abundance. *Symp. R. Entom. Soc. London* No. 4.

Varley, C. G., G. R. Gradwell & M. P. Hassell. 1973. Insect Population Ecology. University of California Press. 212 pp.

Measures of the Dispersion of a Population Based on Ranks

N. I. Lyons[1] and K. Hutcheson[2]

ABSTRACT Gini's (1912) coefficient of concentration is recommended as the basis for a significance test to compare the dispersions of two negative binomial populations. The data are assumed to be quadrat counts, and the jackknife technique is used to estimate standard error. Simulation studies indicate that as few as 20 quadrats may be adequate for some negative binomial populations. Application is made to the comparison of dispersions of the adult southern green stinkbug (*Nezara Viridula*) on three row crops.

An important property of insect populations is the degree of aggregation or clumping of the individuals of a species over the habitat. Commonly used measures of aggregation based on quadrat counts are Morisita's I (1962), Lloyd's (1967) indices of mean crowding and patchiness, and the variance–to–mean ratio. Statistical inference procedures suggested for these indices are based on the large sample estimation of their standard errors. The indices are closely related mathematically and large sample hypotheses tests for comparison between two populations give similar conclusions. For large samples from randomly distributed populations, the mean crowding index is approximately equal to the mean density, and the others are approximately one. The Poisson distribution is taken as a standard, and each index indicates the degree of departure from this standard. Paloheimo & Vukov (1976) note that none of the indices are invariant to quadrat size for a wide variety of alternative underlying distributions, although Reed (1983) suggests transformations to make the variance independent. In fact it may be unreasonable to expect invariance from any statistic based on counts from the same sized quadrats.

The purpose of this paper is to propose as alternatives, two related indices suggested by Gini (1912), the mean absolute difference and the coefficient of concentration. These indices are neither independent of quadrat size, nor do they have a standard value under Poisson sampling. Their primary application will be in comparing the dispersions of two populations, but more importantly they are based upon the empirical distribution function, and as such provide a basis for a more general approach to the statistical analysis of dispersion.

The Lorenz Curve and Gini's Index

Let $X_1, X_2, ..., X_q$ represent the counts of individuals for each of q quadrats, and let $X_{(1)} \leq ... \leq X_{(q)}$ denote the order statistics. The sample can be displayed graphically by plotting and

---

[1]Department of Statistics, University of Georgia, Athens, GA 30602

[2]Institute of Ecology, University of Georgia, Athens, GA 30602.

connecting the points $(i/q, S_i/S_q)$, $i = 0,1,2,...,q$, where $S_0 = 0$ and $S_i = \sum_{j=1}^{i} X_{(j)}$. An example is shown in Fig. 1. The shape of the curve can indicate the dispersion of the individuals over the sample area. At one extreme, a completely uniform sample will result in a straight line passing through $(0,0)$ and $(1,1)$. Sampling from populations with a high degree of clumping should result in a line with sharp downward bend. The curve is known as the Lorenz (1905) curve first used on the study of disparity of distribution of wealth (see, e.g., Marshall & Olkin, 1979, pp.5). The dispersion of a sample can be interpreted as a measure of the variability of, and thus the degree of inequality among, the sample of quadrat counts. It is this interpretation which makes clear the connection between measures of variability, dispersion, and species diversity. In fact Morisita's index I is the equivalent of Simpson's (1949) unbiased estimate of diversity, in which species counts take the place of quadrat counts. Fager (1972), using a variation of this graph, defines the number of moves index of diversity. Solomon (1979), Patil & Taillie (1982), and Lyons & Hutcheson (1988) extend and generalize Fager's result. The number of moves index and its scaled version are directly related to Gini's (1912) mean absolute difference

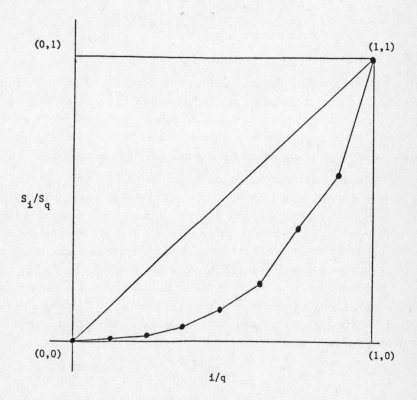

Fig. 1. Lorenz Curve for the sample: 1, 2, 3, 6, 10, 18, 20, 40.

$$g = \frac{1}{q(q-1)} \sum_{i \neq j} \sum |X_i - X_j|$$

$$= \frac{1}{q(q-1)} \sum_{i=1}^{q} (2i-q-1)X_{(i)},$$

and coefficient of concentration $d = g/2\overline{X}$. Gini's coefficient d is equal to twice the area between the graph and the 45° line. The expected value of g is

$$E(g) = 4 \int_{-\infty}^{\infty} x[F(x) - 1/2]dF$$

$$= 4E[xF(x)] - 2E(x),$$

where $F(x)$ is the cumulative distribution function of the underlying probability distribution.

An excellent summary of the results concerning the statistical properties of g is given by David (1968). Hoeffding (1948) proves asymptotic normality of both g and d for any underlying distribution with finite variance. Assuming a normal distribution Nair (1936), Lomnicki (1952), and Barnett, et al., (1967) obtain the first four moments of g in closed form, and fit its distribution by a Pearson type III curve. Downton (1966) notes that $\sqrt{\pi g}$ has high relative efficiency compared to the usual s as an unbiased estimator of σ in normal populations. Ramasubban (1958, 1959) provides series expansions for the mean of g for Poisson, binomial, and negative binomial distributions. Although the moments of d have not been obtained in closed form, Hoeffding gives expressions for the asymptotic mean and variance to order $O(1/q)$ in terms of the moments of the underlying distribution. However, these expressions must be approximated by numerical integration and cannot be used for estimation. In the next section the results of several Monte Carlo simulation studies are presented to examine the adequacy of the normal approximation, the jackknife as an estimator of standard error for g and d, and the power for some two—population hypothesis tests.

A Simulation Study

The first study simulated sampling from each of several negative binomial populations of counts with parameters N and P. Samples of size 20, 50, and 100 were used. The values of g and d, along with their jackknife estimates of standard error were computed for each of 500 replications of the experiment. Table 1 presents the standard deviation, skewness, and kurtosis of the 500 simulated values, along with the average of the 500 jackknife estimates of standard error. The results suggest that 50 quadrats is adequate for both the normal approximation and the jackknife estimator for the index d, however, the jackknife underestimates the standard error of g slightly for q less than 100.

Table 1. Standard deviation, (jackknife estimate), skewness and kurtosis for 500 iterations

(N,P)	q Index	20 g	20 d	50 g	50 d	100 g	100 d
(N,P) (5,1.)	μ 5	.665 (.609) .36 3.3	.057 (.054) .21 3.0	.397 (.393) .21 2.8	.033 (.033) .26 3.0	.292 (.283) .34 3.1	.024 (.024) .14 3.0
(10,1.)	10	.924 (.848) .24 2.8	.041 (.040) .23 2.9	.553 (.523) .03 2.8	.025 (.024) .13 2.9	.402 (.380) .22 2.9	.017 (.017) .19 2.9
(2,.5)	1.	.311 (.279) .33 3.2	.088 (.090) −.20 3.6	.178 (.176) .37 3.5	.055 (.054) .04 3.3	.123 (.127) .26 3.1	.037 (.038) .05 3.1
(10,.5)	5	.525 (.521) .06 2.9	.048 (.049) .16 3.0	.348 (.329) .25 3.2	.033 (.030) .10 3.0	.227 (.235) .04 3.3	.021 (.022) .07 3.1
(20,.5)	10	.748 (.726) .18 2.8	.036 (.036) .17 3.1	.452 (.456) −.04 3.0	.022 (.022) −.14 3.0	.332 (.322) .12 3.0	.016 (.015) −.05 3.0
(1,1.)	1.	.377 (.344) .50 3.4	.092 (.086) −.35 3.1	.233 (.229) .41 3.2	.053 (.052) .11 2.7	.167 (.166) .17 2.9	.037 (.037) −.02 2.9

In the second study the differences between two values of each index computed from independent samples from the same negative binomial population were simulated. The same moments were computed for these differences, with the estimate of the variance of the difference taken to be the sum of the individual jackknife estimates of variance. As expected there was a noticeable improvement in the skewness of the differences over that of the individual indices. The adequacy of the normal approximation appears to be good for both indices even for q = 20. The jackknife estimate is adequate for d with q = 20, however q = 100 is required in most cases for the index g. This suggests the test statistic $z = (d_1-d_2)/(s_{j_1}^2+s_{j_2}^2)^{1/2}$ for comparing two population dispersions.

In the final study the power of the test procedure for comparison of two population dispersions was estimated for q = 50. Jackknife estimators were used in all cases. The results appear in Table 3. All of the estimated significance levels except one (N = 10, P = .5, for the g index) are within 95% limits of α = .05. Clearly both of the Gini indices have significantly higher power in most cases, with the index d outperforming g in this respect in all but one case.

Table 2. Standard deviation, (jackknife estimate), skewness and kurtosis of differences for 500 iterations

q Index	20 g	20 d	50 g	50 d	100 g	100 d
(N,P) (5,1.)	.949 (.892) −.04 3.2	.075 (.077) −.02 2.8	.570 (.561) −.24 3.1	.048 (.048) −.09 3.2	.392 (.405) .29 3.0	.035 (.034) .11 3.2
(10,1.)	1.28 (1.21) −.09 3.3	.056 (.056) −.10 3.8	.765 (.763) .11 3.5	.036 (.035) .11 3.1	.518 (.542) .01 2.8	.024 (.025) .00 2.7
(2,.5)	.415 (.398) −.10 3.4	.119 (.127) .03 2.9	.264 (.260) .01 2.8	.074 (.077) .10 2.8	.181 (.183) .03 3.0	.054 (.054) −.09 2.8
(10,.5)	.724 (.757) −.03 3.2	.066 (.070) −.12 3.0	.492 (.468) −.03 3.2	.043 (.043) .04 3.2	.341 (.334) −.15 3.0	.031 (.030) −.21 3.2
(20,.5)	1.02 (1.02) −.23 3.3	.050 (.050) −.17 3.1	.668 (.645) .04 2.8	.031 (.031) .04 2.8	.452 (.456) −.11 2.9	.022 (.022) −.05 3.0
(1,1.)	.540 (.517) −.13 3.4	.125 (.123) −.19 2.9	.325 (.332) −.16 3.4	.072 (.075) .17 2.9	.256 (.237) .30 3.7	.053 (.053) .00 2.8

An Application

Three crops (mustard, tobacco, and sorghum) were censused at different times during the 1980 growing season for the presence of the southern green stinkbug (*Nezara viridula*). Quadrats of one yd^2 were used. The frequency distributions of adult bugs appear below:

Counts	0	1	2	3	4	5	6	7	8	Total
Mustard	2740	93	31	17	5	4	–	–	–	2890
Tobacco	1054	36	22	10	3	2	2	1	–	1130
Sorghum	680	198	76	27	9	6	2	1	1	1000

Table 3. Estimated significance levels for two–population tests

(N,P) (μ,σ²)	Index	(5,1) (5,10)	(10,1) (10,20)	(2,.5) (1,1.5)	(10,.5) (5,7.5)	(20,.5) (10,15)	(1,1) (1,2)
(5,1) (5,10)	g	5.4	59.4	100.0	12.6	27.2	99.6
	d	4.0	63.8	98.2	16.0	90.2	100.0
	I	3.4	13.4	5.8	15.0	29.8	21.4
	v	4.8	4.8	17.4	16.0	14.0	4.6
(10,1) (10,20)	g	–	4.0	100.0	87.4	15.4	100.0
	d	–	5.6	99.8	29.0	14.4	100.0
	I	–	3.8	10.4	5.8	13.8	47.6
	v	–	4.0	16.8	14.6	14.6	9.8
(2,.5) (1,1.5)	g	–	–	6.0	99.6	100.0	6.8
	d	–	–	3.8	99.6	100.0	12.0
	I	–	–	4.0	10.6	15.6	9.4
	v	–	–	6.0	6.8	4.8	11.4
(10,.5) (5,7.5)	g	–	–	–	7.0*	51.4	98.4
	d	–	–	–	5.4	76.6	100.0
	I	–	–	–	6.4	7.8	36.8
	v	–	–	–	6.4	5.6	11.0
(20,.5) (10,15)	g	–	–	–	–	5.2	100.0
	d	–	–	–	–	4.8	100.0
	I	–	–	–	–	4.0	45.6
	v	–	–	–	–	3.6	8.2
(1,1) (1,2)	g	–	–	–	–	–	5.4
	d	–	–	–	–	–	4.0
	I	–	–	–	–	–	3.4
	v	–	–	–	–	–	4.8

The negative binomial provides an adequate fit to all three distributions. Samples of $q = 30$ and 50 were selected from these frequency distributions and the test for equality of dispersion was performed 500 times. Table 4 indicates the percentage of rejections. The presence of the large proportion of zeroes clearly affected the results by providing a large percentage of samples with no frequencies greater than one. These cases were rejected because the jackknife estimate of standard error is zero.

Table 4. Percentage of rejections for comparison of three row crops.

	Mustard vs. Tobacco		Mustard vs. Sorghum		Tobacco vs. Sorghum	
q	30	50	30	50	30	50
(cases)	(95)	(224)	(145)	(271)	(250)	(396)
g	2.1	5.4	60.0	87.8	43.6	73.7
d	3.2	6.3	82.1	97.8	75.2	97.7
I	0.0	0.0	0.0	2.2	0.8	4.3
v	5.3	12.9	9.7	14.8	14.0	17.9

Acknowledgements

The authors thank Dr. Glenn Ware, Station Statistician, University of Georgia for providing the data. This research was partially supported by Air Force Office of Scientific Research Grant No. AFOSR–85–0161.

References

Barnett, F. C., K. Mullen, & J. G. Saw. 1967. Linear estimates of a population scale parameter. *Biometrika* 54: 551–4.

David, H. A. 1968. Gini's mean difference rediscovered. *Biometrika* 55: 573–5.

Downton, F. 1966. Linear estimates with polynomial coefficients. *Biometrika* 53: 129–41.

Fager, E. W. 1972. Diversity: a sampling study. *Amer. Natural.* 106: 293–310.

Gini, C. 1912. Variabilitae Mutabilita. Studi Economico–Giuridici della R. Universita de Caglri, Anno 3, Part 2, pp. 80.

Hoeffding, W. 1948. A class of statistics with asymptotically normal distribution. *Ann. Math. Statist.* 19: 293–325.

Lloyd, M. 1967. Mean crowding. *J. Anim. Ecol.* 36: 1–30.

Lomnicki, Z. A. 1952. The standard error of Gini's mean difference. *Ann. Math. Statist.* 23: 635–7.

Lorenz, M. O. 1905. Methods of measuring concentration of wealth. *J. Amer. Statist. Assoc.* 9: 209–19.

Lyons, N. I. & K. Hutcheson. 1988. Distributional properties of the number of moves index of diversity. *Biometrics* (to appear).

Marshall, A. W., & I. Olkin. 1979. *Inequalities: The Theory of Majorization.* Academic Press, New York.

Morisita, M. 1962. I_δ index, a measure of dispersion of individuals. *Res. Pop. Ecol.* 4: 1–7.

Nair, U. S. 1936. The standard error of Gini's mean difference. *Biometrika* 28: 428–36.

Paloheimo, J. E., & A. M. Vukov. 1976. On measures of aggregation and indices of contagion. *Math. Biosci.* 30: 69–97.

Patil, G. P., & C. Taillie. 1982. Diversity as a concept and its measurement. *J. Amer. Statist. Assoc.* 77: 548–67.

Ramasubban, T. A. 1958. The mean difference and the mean deviation of some discontinuous distributions. *Biometrika* 45: 549–56.

Ramasubban, T. A. 1959. The generalized mean differences of the binomial and Poisson distributions. *Biometrika* 46: 223–9.

Reed, W. J. 1983. Confidence estimation of ecological aggregation indices based on counts – a robust procedure. *Biometrika* 39: 987–98.

Simpson, E. H. 1949. Measurement of diversity. *Nature* 163: 688.

Solomon, D. L. 1979. A comparative approach to species diversity in Ecological Diversity. In Theory and Practice [eds.] Grassle, J. F., et al. International Cooperative Publishing House. Fairland, Maryland.

A Model of Arthropod Movement Within Agroecosystems

Linda J. (Willson) Young[1] and Jerry H. Young[2]

ABSTRACT The negative binomial distribution often describes the distribution of arthropods within agroecosystems. When mobile arthropods are disrupted, such as may occur during chemical applications or changes in climatic conditions, they tend to move until they are again in a negative binomial distribution. While arthropod movement continues, the distribution remains negative binomial. This paper gives a mathematical model for this behavior. The movement of each arthropod is modeled using a Markov process based on a common transition function which is doubly stochastic and irreducible. The arthropods move independently except that the time until an arthropod moves is exponentially distributed with a mean proportional to the number of arthropods on the plant. Using this type of interaction, the Bose–Einstein distribution (which assigns equal probability to distinguishable spatial patterns) is the equilibrium distribution. If the number of arthropods and plants are each as large as 100, this distribution is closely approximated by the geometric distribution which is the limiting distribution. The geometric is a special case of the negative binomial distribution ($k = 1$) and is often observed when the sampling unit is the minor habitat of the arthropod. Parallels are drawn between the study of particles in a phase space and the distribution of arthropods in agroecosystems.

Movement by biological organisms results in a continually changing environment. Models have been developed to explain various types of movement, including immigration (Young & Willson 1987), dispersion (Skellan 1973, Motro 1982a, b, 1983), and relationships in predators and prey (Leslie & Gower 1960, Murdoch & Oaten 1975, Hamilton & May 1977, Hassell 1978, Hassell & May 1986). The effects of competition (Pianka 1981, Rankin & Singer 1984) and mating behaviour (Dingle 1974, Rankin & Singer 1984), especially as they relate to migration, have also been examined. In this work, we propose a stochastic model of arthropod movement within agroecosystems based on interacting Markov processes. It follows that of Spitzer (1970) who suggested it as a possible model of particle movement. We will focus on the biological interpretation of the model and the resulting equilibrium distribution.

Agroecosystems represent even–aged monocultures which receive uniform water, fertilizer, and cultivation. There are usually few arthropod species within an agroecosystem. In this work, we will consider only those species that are unrestricted in movement and have no apparent physical or biological constraints on behavior. Further, we will restrict our attention to those species that have a well–defined, discrete minor habitat, such as a plant, a leaf, or a terminal. The minor habitat will be the sampling unit. Thus we have an extremely simple ecosystem to study arthropod movement and distribution.

---

[1]Department of Statistics, Oklahmoa State University, Stillwater, OK 74078.
[2]Department of Entomology, Oklahoma State University, Stillwater, OK 74078.

Maxwell–Boltzmann Versus Bose–Einstein Statistics

Examination of arthropod distribution parallels that of the study of the distribution of particles in a phase space. Suppose we have an agroecosystem with C arthropods and N minor habitats which we will designate to be plants for simplicity of discussion. The traditional (or Maxwell–Boltzmann) approach assumes that each of the N^c possible permutations of the arthropods on the plants is equiprobable. While the arthropods may be in any distribution with mean $C/N = \mu$ at a stated point in time, the most probable distribution of the number of arthropods on a plant is the Poisson. In fact, the probability is overwhelming that they will be in the Poisson or a distribution very close to the Poisson. However, the Poisson is rarely found in nature (Taylor 1984). Failure of this model to describe arthropod distribution has been attributed to climatic conditions, geographical factors, and to the biology of the arthropods such as food–gathering traits or mating behavior. Maxwell–Boltzmann statistics have also failed to describe the distribution of particles in a phase space for any known body of matter; in no case are all of the N^c possible arrangements equiprobable. This led Bose and Einstein to suggest that equal weight should be given to distinguishable states (Pathria 1972). Although arthropods within a species are indistinguishable, minor habitats may be distinguished by their position within a field. Therefore, Bose–Einstein statistics would give equal probability to each of the $\begin{bmatrix} N+C-1 \\ C \end{bmatrix}$ distinguishable spatial patterns. In this case, the geometric is the most probable distribution of the number of arthropods on a plant. Just as Bose–Einstein statistics has been shown to apply for photons, nuclei, and atoms containing an even number of elementary particles, so too the geometric has been found when sampling the minor habitat of arthropods in agroecosystems (Willson et al. 1987, Young & Willson 1987). The geometric is a negative binomial distribution with k = 1 and the sum of independent negative binomial distributions is again negative binomial. Thus, we have a possible explanation of the reason that the negative binomial is frequently encountered when sampling arthropods in agroecosystems.

Table 1 presents data collected by sampling the minor habitats of arthropods in agroecosystems. For example, goodness–of–fit tests were used to verify that the negative binomial fits the data. However, a number of negative binomial distributions could adequately describe the data. Thus, emphasis was given to precise estimation of k. Prior to 1981, k was estimated using the method of moments; that is,

$$\hat{k} = \frac{\overline{x}^2}{s^2 - \overline{x}}$$

where $\overline{x}$ is the sample mean and s^2 is the unbiased sample estimate of the variance. After 1981, a multistage method of estimating k was employed (Willson et al. 1984). In it, five observations are collected and the method of moments estimate of k computed. Observations are added to the sample in increments of five. After each increment, the method of moments estimate of k based on all observations in the sample at that point in time is computed. When two successive estimates of k differ by less than 0.05, sampling is discontinued, and the last estimate of k is used as the final estimate. This method produces estimates with smaller biases and variances than either method of moments or maximum likelihood.

Table 1. Arthropod distribution in agroecosystems with emphasis on estimation of the negative binomial parameter k. [a]

Insect	Crop	Sampling unit	No. of fields sampled	Sample size range	Range of estimates of k
Greenbug	seedling oats	single seedling	57	300 – 1600	0.96–1.04
Collops spp.	cotton	single plant	182	106–1/4 acre	1.1–1.3
Convergent lady beetle	cotton	single plant	186	92–1/4 acre	0.7–1.2
Notuxus monodon F.	cotton	single plant	186	106–1/4 acre	0.92–1.6
cotton fleahopper nymphs	cotton	single terminal	28	198–222	0.8–1.1
cotton fleahopper adults	cotton	single terminal	28	240–242	1.1–1.4

[a] This table is reported in Young & Willson (1987).

The strongest support for the Bose–Einstein approach comes from a 3–year study of greenbugs, *Schizaphis graminum* (Rondani), in seedling oats. This is the most homogeneous agroecosystem we sampled. The oats were homozygous and homogenous and thus are genetically identical. The greenbugs reproduce parthenogenetically, resulting in very similar individuals. Fifty–seven samples, ranging in size from 300 to 600, were taken. The estimates of k ranged from 0.96 to 1.04. The estimates of k for *Collops* spp. were all a little more than one, indicating they may not be following Bose–Einstein statistics.

Particle and Arthropod Movement

A further parallel exists between the distributions of particles in a phase space and arthropods in an agroecosystem. In statistical mechanics, it is known that particles will move toward the most probable distribution regardless of their initial distribution. In agroecosystems, mobile arthropods should move toward a geometric distribution unless a limiting factor such as restricted carrying capacity or mating behavior stops them. Thus, a study was undertaken to determine what happens to the spatial distribution of populations of cotton fleahoppers, *Pseudatomoscellis seriatus* (Reuter), when they are disrupted.

A 2 ha field near Chickasha, Oklahoma, was planted to Paymaster 145 cotton cultivar at 18 pounds of seed per acre in 40–inch rows on June 10, 1986. Normal tillage practices for the area were observed. On August 4 and 5, preapplication counts of cotton fleahopper density were made using a sequential estimation procedure which controls the coefficient of variation of the sample mean (Willson & Young 1983). The terminal of the cotton plant is the minor habitat and was taken as the sampling unit. On August 5, random mosaic patterns, covering 20% of the field, were sprayed with 0.2 lb. active ingredient per acre of fenvalerate, a synthetic pyrethroid insecticide. Density counts of cotton fleahoppers were made on August 6, 7, 10, 19, and 20, using the same sequential process.

Based on chi–square goodness–of–fit tests, the negative binomial distribution fit each sample. The method of moments was used to estimate k. On August 4 and 5, estimates of k were close to one (Table 2), the value which would be associated with the geometric distribution and Bose–Einstein statistics. The day after disruption of the population by insecticide application, August 6, the estimate of k rose, but it began to decrease within 48 hours, returning to about one, where it remained. The results are consistent with the hypothesis that mobile arthropods will move toward the most probable Bose–Einstein distribution (the geometric) after disruption of spatial distribution unless some limiting factor prevents such movement. The proposed model of arthropod movement will result in the geometric distribution regardless of the initial distribution. Further, movement toward the geometric is rapid.

Table 2. Random samples of cotton fleahoppers in a cotton field near Chickasha, Oklahoma.

Date	Sample Mean	Estimate of k
Aug. 4, 1986	0.62	1.06
Aug. 5, 1986	0.65	1.03
Aug. 6, 1986	0.46	1.88
Aug. 7, 1986	0.76	1.49
Aug. 10, 1986	1.10	0.94
Aug. 19, 1986	1.02	1.01
Aug. 20, 1986	1.10	0.97

Heuristic Explanation of the Model

Most biological organisms respond to other members of the same species. This interaction is the focal point of the proposed model. Suppose that arthropods interact in the following manner. The time until an arthropod moves is exponentially distributed with a mean proportional to the number of arthropods on the plant. Thus, the time until each of the two arthropods on a plant moves is independently and exponentially distributed with a mean twice that of the mean time until an arthropod who is the sole inhabitant of a plant moves.

The movement of an arthropod from one plant to another occurs according to any Markovian transition function which has three features. First, the probability that an arthropod will move from the minor habitat it is on to another minor habitat in the model (which may be the same minor habitat) is one. Second, the probability that an insect arriving at a given minor habitat has come from another minor habitat in the model is one. These first two requirements mean that the transition function is doubly stochastic. Third, the transition function must be one which permits an arthropod to begin on any plant x, inhabit any given plant y after a finite number of movements, and again return to plant x after a finite number of movements (an irreducible transition function). The probability that an arthropod will eventually move from plant x, to plant y, and back to plant x may be extremely small, or it may require many movements. However, if it is possible, then this condition is satisfied.

Consider an agroecosystem with one species of arthropod. The arthropods may be in any initial distribution. For example, they may be on one end of the field, or they may be distributed evenly throughout the field. Suppose they move according to a common transition function satisfying the three conditions given above. Further, assume that they interact in such a way that the time until an arthropod moves is exponential with a mean proportional to the number on a plant. As the arthropods move, the distribution changes until it becomes Bose–Einstein where every spatial pattern is equiprobable. The spatial arrangement of the arthropods may continue to fluctuate, but the distribution will remain Bose–Einstein. If the number of arthropods and plants are each as large as 100, this distribution can be closely approximated by the geometric which is the limiting distribution. Note that it is the interaction among the members of the species that determines the equilibrium distribution, not the transition function upon which the interaction is superimposed.

Mathematical Statement of the Model

We will now make a more formal mathematical statement of the model. Consider an agroecosystem with C insects and N plants. Suppose the ecosystem is closed to emigration and immigration, and birth and death processes are equal. We will model the movement of each arthropod using a Markov process with common transition function $P_t(x,y)$; that is, $P_t(x,y)$ is the probability a given arthropod moves from plant x to plant y at time t. Further assume P_t is doubly stochastic and irreducible, i.e.,

$$P_t(x,y) \geq 0.$$

$$\sum_x P_t(x,y) = \sum_y P_t(x,y) = 1$$

and (1)

$$P_t^{n_1}(x,y) \cdot P_t^{n_2}(y,x) > 0$$

for all x,y and for suitable $n_1 \geq 0$ and $n_2 \geq 0$ which may depend on x and y. Note that P_t is an N by N matrix.

Arthropod movement is according to independent Markov processes except the time until an arthropod moves is exponentially distributed with a mean proportional to the number of arthropods on the plant. Using this type of interaction, Spitzer (1970) showed that the Bose–Einstein distribution (all distinguishable spatial patterns are equiprobable) is an invariant measure under Markovian translation according to (1). More intuitively, suppose the arthropods are initially placed on the plants according to the Bose–Einstein distribution. Let them move, with the type of interaction described above, as Markov processes with the common transition function P_t, satisfying (1). Then for each $t > 0$, the probability law of the arthropod distribution on the plants will be that of the Bose–Einstein distribution. Further, if the initial distribution of the arthropods is "reasonable" (and not necessarily Bose–Einstein), then by the ergodic theorems, the probability measure converges to that of the Bose–Einstein distribution (Spitzer 1970). Since the time until an arthropod moves is influenced only by other arthropods on the same plant (not those on neighboring plants) and the Bose–Einstein distribution is the invariant measure, this type of interaction has been referred to as zero–range interaction at Bose–Einstein speeds (Waymire 1980).

Simulating the Model

The nature of the exponential distribution simplifies application of this model. First, if we have r independent observations from an exponential distribution with mean rc ($c > 0$), the distribution of the first order statistic is exponential with mean c. Applying this to the model, we have the time until the first arthropod moves from a plant being exponentially distributed with a mean of one time unit regardless of the number of arthropods on that plant. Hence, within an agroecosystem, each inhabited plant is equally likely to have the next arthropod that moves. Therefore, a plant can be chosen at random. If it has at least one arthropod, we select one of them (at random) to move according to the transition function in (1).

Once an arthropod moves, the distribution of the time until the next arthropod moves changes for those remaining on the plant the arthropod left and those on the plant the arthropod selects for habitation. Initially this appears to present some serious modeling problems. However, the exponential distribution has the "memoryless" property, i.e., $P(T > a + b \mid T > a) = P(T > b)$, where T is the time until an arthropod moves and a and b are positive constants. Thus we may reset the time for all arthropods after each movement, ignoring what has happened in the past. The fact that a particular arthropod has been on its plant for 100 time units and another has moved 5 times during that time span has no effect on which one will move next.

The model has been developed for both IBM PC's and the IBM 3081K mainframe computer. The version for the PC is limited to 100 plants (which represent the minor habitats). The number of arthropods is specified by the user. The initial distribution may have the arthropods evenly dispersed in the field, all at one end of the field, or all on a single plant. The transition function may be one for which it is equally likely that an arthropod leaving a plant will go to any plant in the field or one for which an arthropod is more likely to go to a neighboring plant than to other plants in the field. As would be expected, the closer the initial distribution is to Bose–Einstein, the more rapidly the distribution becomes Bose–Einstein. Thus, when the mean number of arthropods per plant is one and they are evenly dispersed, the distribution becomes Bose–Einstein within 500 movements, often passing through a distribution which looks like the Poisson during the process. If the arthropods are on one end of the field, about 1000 movements are required before the equilibrium distribution is obtained. Several thousand moves are needed before the equilibrium distribution occurs when the arthropods are initially on a single plant. The transition function seems to have little impact on the rate at which the equilibrium distribution is obtained. The results from the mainframe which involve a larger number of plants and arthropods have the same trends.

Discussion

The negative binomial often fits biological data well and have been used in some model development (Anscombe 1949, Bliss 1971, Murdoch & Oaten 1975). Researchers have proposed numerous biological models which generate the negative binomial distribution (Boswell & Patil 1970). However, it is difficult to determine if any of the proposed models gave rise to a particular observed distribution. The proposed model is a biologically testable one. The forces resulting in the initial distribution influence the time it takes for the system to reach equilibrium. They do not alter the equilibrium distribution. The movement to the distribution is rapid. This would tend to indicate that the assumptions that the system is closed to immigration and emigration and that the birth and death processes are equal may be relaxed to some extent without affecting the results. For example, in the disruption study giving rise to Table 2, it is evident from the sample mean that the insecticide application first reduced the population, but that it began to build, increasing beyond its original level. The high mobility of the cotton fleahopper relative to the time interval between samples and the change in population size could explain the constant movement toward the geometric which compensated for the presence of immigration or birth processes exceeding emigration or death processes.

Numerous questions remain to be answered. Does this model accurately describe arthropod movement? If so, what effect, if any, do other species have on the time between movements of an arthropod? Would the sex of the other arthropods on the minor habitat influence the time until an arthropod moves? These, and other questions, are now under investigation in both laboratory and field environments.

References Cited

Anscombe, F. J. 1949. The statistical analysis of insect counts based on the negative binomial distribution. *Biometrics* 5: 165–173.

Bliss, C. I. 1971. The aggregation of species within spatial units, pp. 311–336. In G. P. Patil, E. C. Pielou, W. E. Waters [eds.], Statistical Ecology, Vol. 1. The Pennsylvania State University Press, University Park, PA.

Boswell, M. T. & G. P. Patil. 1970. Chance mechanisms generating the negative binomial distribution, pp. 3–22. In G. P. Patil, [ed.], Random Counts in Scientific Work, Vol. 1. The Pennsylvania State University Press, University Park, PA.

Dingle, H. 1974. The experimental analysis of migration and life history strategies in insects, pp. 329–342. In L. Barton–Browne [ed.], Experimental Analysis of Insect Behaviour. Springer–Verlag, Berlin.

Hamilton, W. D. & R. M. May. 1977. Dispersal in stable habitats. *Nature* 269: 578–581.

Hassell, M. P. 1978. The dynamics of arthropod predator–prey systems. Princeton Univ. Press, Princeton, N.J.

Hassell, M. P. & R. M. May. 1986. Generalist and specialist natural enemies in insect predator–prey interactions. *J. Anim. Ecol.* 55: 923–940.

Leslie, P. H. and J. C. Gower. 1960. The properties of a stochastic model for the predator–prey type of interaction between two species. *Biometrika* 47:219–234.

Motro, U. 1982a. Optimal rates of dispersal I. Haploid populations. *Theor. Pop. Biol.* 23: 394–411

Motro, U. 1982b. Optimal rates of dispersal II. Diploid populations. *Theor. Pop. Biol.* 21: 412–429.

Motro, U. 1983. Optimal rates of dispersal III. Parent–Offspring conflict. *Theor. Pop. Biol.* 23: 159–168.

Murdoch, W. W. & A. Oaten. 1975. Predation and population stability, pp. 2–131. In A. MacFadyen [ed.]. *Adv. Ecol. Res.* Vol. 9. Academic Press, London.

Pathria, R. K. 1972. Statistical Mechanics. Pergamon Press, Braunschweig.

Pianka, E. R. 1981. Competition and niche theory, pp. 167–196. In R. M. May [ed.], Theoretical Ecology: Principles and Applications, 2nd Ed. Sinauer Associated, Inc., Sunderland, MA.

Rankin, M. A. & M. C. Singer. 1984. Insect movement: Mechanisms and effects, pp. 185–216. In C. B. Huffaker and R. L. Rabb [eds.], Ecological Entomology. John Wiley and Sons, New York.

Skellam, J. G. 1973. The formulation and interpretation of mathematical models of diffusionary processes in population biology, pp. 63–85. In M. S. Bartlett and R. W. Hiorns [eds.], The Mathematical Theory of the Dynamics of Biological Populations. Academic Press, New York.

Spitzer, F. 1970. Interaction of Markov processes. Advances in Math. 5: 246–290.

Taylor, L. R. 1984. Assessing and interpreting the spatial distribution of insect populations. Ann. Rev. Entomol. 29: 321–357.

Waymire, E. 1980. Zero–range interaction at Bose–Einstein speeds under a positive recurrent single particle law. Ann. Prob. 8: 441–450.

Willson, Linda J., J. L. Folk & J. H. Young. 1984. Multistage estimation compared with fixed–sample–size estimation of the negative binomial parameter k. Biometrics 40: 109–117.

Willson, Linda J. & J. H. Young. 1983. Sequential estimation of insect population densitites with a fixed coefficient of variation. Environ. Entomol. 12: 669–672.

Willson, L. J., J. H. Young & J. L. Folks. 1987. A biological application of Bose–Einstein statistics. Commun. Statist. Theor. Meth. 16: 445–459.

Young, J. H. & L. J. Willson. 1987. The use of Bose–Einstein statistics in population dynamics models of arthropods. Ecol. Model. 36: 89–99.

SECTION IV

GENERAL SAMPLING AND ESTIMATION METHODS

Intervention Analysis in Multivariate Time Series
via the Kalman Filter

David K. Blough[1]

ABSTRACT The Kalman filter has been found useful in fitting ARIMA models. This paper uses the multivariate extension of the state–space approach to fit multivariate time series which include covariates and periodic interventions. An example will be presented in which these techniques are applied to the study of the pink cotton boll worm moth, *Pectinophora gossypiella* (Saunders), (*Lepidoptera: Gelechiidae*). Moth counts were taken at various locations in a large cotton field over time. A special type of multivariate time series will be used, namely a spatial time series. The amount of irrigation water at each location will be included in the model as a covariate and periodic applications of insecticide will be included as interventions. Previously developed techniques for handling missing data, aggregate data, and nonlinear data transformation are also incorporated. Maximum likelihood estimates of the model parameters are obtained by imbedding the filter in a quasi–Newton optimization routine.

In both the physical and social sciences, data are often collected over both space and time. For example, in Entomology investigators might be concerned with the boll worm moth population in a large cotton field during a growing season. Observations would consist of moth counts taken at various locations in the field each day over a number of months.

Of primary interest in such investigations is the identification and estimation of the space–time relationships of the responses under study and the development of a viable mathematical model of the system. Various classes of spatial time series model have been developed by statisticians such as Pfeifer & Deutsch (1980), Stoffer (1986) and Haugh (1984). Essentially, they represent a multivariate generalization of the class of autoregressive–moving average time series models developed by Box & Jenkins (1970), and Hannan (1976).

Recently, the use of Kalman filtering in fitting multivariate time series models has come into practice. Phrasing such models in state–space terms has been developed by Gardner et al., (1980), Harvey (1981), and Shea (1984). In addition to being an efficient way to compute likelihood, the Kalman filter can handle differencing (to achieve stationarity), missing values, and aggregate values. Harvey & Pierse (1984), Ashley & Kohn (1983), Kohn & Ashley (1986), and Jones (1980) establish these techniques. The filter has also been extended by Anderson & Moore (1979) to accomodate nonlinear data transformations. Harvey & Phillips (1979) modified the state–space form of the model to include covariates. For example, in the boll worm investigation, it would be advantageous to include the amount of irrigation water at each field site and time in the model. Water might be thought useful in managing insects in the field and its effects would be of primary importance.

[1]Department of Agricultural Economics, University of Arizona, Tucson, Arizona 85721.

Box & Tiao (1975) developed intervention analysis for univariate time series. An intervention is an event which occurs at one point in time. It affects the response being measured over time, and the result of its effects can be long–term or transitory. With prior knowledge of the nature of the effect of an intervention, Box and Tiao construct models which allow interventions to be taken into account. In the boll worm moth example, an application of insecticide to the cotton field would be an intervention. Its effect would be felt immediately in terms of a drop in the moth population, and then decay, possibly at an exponential rate. The appropriate spatial time series model should include the effects of such interventions.

It is the purpose of this paper to unite Kalman filtering techniques and multivariate time series models with intervention analysis. Section 2 specifies the necessary extension of the state–space formulation of multivariate time series models include intervention effects. The aforementioned boll worm moth investigation is presented in Section 3 as an application of the state–space form intervention analysis.

State Space Formulation of Interventions

The state–space formulation of multivariate time series is well known. See, for example, Harvey (1981), Shea (1984), and Stoffer (1986). Thus, it is clear that spatial time series models can be phrased similarly. Assume that the state space form of the spatial time series model under consideration is

$$\underline{\alpha}_t = T \, \underline{\alpha}_{t-1} + R \, \underline{\varepsilon}_t$$

$$\underline{y}_t = H' \, \underline{\alpha}_t$$

where $\underline{\alpha}_t$ is the state vector and $\underline{y}_t$ is an Nx1 vector. Each component y_{it} of the vector $\underline{y}_t$ is an observation taken at location i and time t, i = 1, 2, ..., N; t = 1, 2, ..., n.

The Kalman filter can accomodate time–varying regression coefficients. In the work of Harvey & Phillips (1981), and Liu & Hanssens (1981), regression models have been developed which allow for the inclusion of randomly varying coefficients, coefficients which vary according to their own ARMA process. But in the spirit of the intervention analysis of Box and Tiao, we seek regression coefficients which vary deterministically; that is, in accord with the assumed nature of the intervention effects. It is possible to include various intervention effects as follows:

Let t* denote the time of an intervention. At time t*+1 and all subsequent times, augment $\underline{\alpha}_t$ to

$$\underline{\alpha}_t^* = \begin{bmatrix} \underline{w} \\ \underline{\alpha}_t \end{bmatrix},$$

where $\underline{\omega}$ is an Nx1 vector representing the initial intervention step; augment H to

$$H^* = \begin{bmatrix} I \\ H \end{bmatrix},$$

where I is the NxN identity matrix; augment R to

$$R^* = \begin{bmatrix} 0 \\ R \end{bmatrix},$$

where 0 is the NxN matrix of zeros. The augmentation of the matrix T depends on the nature of the intervention effect. Some examples (in the terminology of Box and Tiao) follow:

Example (2.1) "Step" response: $T_t^* = \begin{bmatrix} I & 0 \\ 0 & T \end{bmatrix}$, $t = t^*+1, t^*+2, ..., n$.

Example (2.2) "Ramp" response: $T_t^* = \begin{bmatrix} (t-t^*)I & 0 \\ 0 & T \end{bmatrix}$, $t = t^*+1, t^*+2, ..., n$.

Example (2.3) "First order" response: $T_t^* = \begin{bmatrix} \sum\limits_{k=1}^{t-t^*} \phi^{k-1}I & 0 \\ 0 & T \end{bmatrix}$, $t = t^*+1, t^*+2, ..., n; 0 < \phi < 1$.

Example (2.4) "Transitory" response, exponential decay:

$$T_{t^*+1}^* = \begin{bmatrix} I & 0 \\ 0 & T \end{bmatrix}; T_t^* = \begin{bmatrix} \phi I & 0 \\ 0 & T \end{bmatrix}, t = t^*+2, t^*+3, ..., n; 0 < \phi < 1.$$

In all cases I is the NxN identity matrix. Generalizations for more complex intervention effects can be obtained by combining the above examples of generalizing T_t^* as deemed appropriate.

As is proved in Harvey & Phillips (1979), conditional on the ARMA parameters and ϕ, the estimator of $\underline{\omega}$ will be the generalized least squares estimator. By imbedding the Kalman filter in an optimization routine, maximum likelihood estimates of the ARMA parameters and ϕ are obtained. The estimated inverse Hessian matrix provides asymptotic standard errors for these parameters. The final iteration of the Kalman filter will provide standard errors for the elements of $\underline{\omega}$.

Examples

As an example of intervention analysis in spatial time series, an analysis of data supplied by Paul Borth and Roger Huber of the Department of Entomology at the University of Arizona will be presented. Data were collected daily from $N = 3$ locations within a southern Arizona cotton field over $n = 103$ days from 27 May to 26 September, 1986. Observations at location i on day t consist of $y_{it} =$ the number of trapped pink boll worm moths, *Pectinophora gossypiella* (Saunders), (*Lepidoptera: Gelechiidae*), and $z_{it} =$ the amount of irrigation water present. Let $\underline{y}_t$ denote the 3x1 vector of moth counts at the three locations, $\underline{z}_t$ the 3x1 vector of irrigation water amounts at the three locations.

No data were collected on days 13–21, 36–45, 57–60, nor on day 81. On days 53, 58, 65, 80, and 87 insecticide (Cymbush ($\frac{1}{44}$ gallon/acre) + Fundal ($\frac{1}{30}$ gallon/acre)) was applied to the field and hence no moth counts were taken on these days. This resulted in two–day aggregate observations taken on days 54, 66, and 88. Day 82 was a three–day aggregate. Day 72 was also a two–day aggregate.

Models of the class discussed in Pfeifer & Deutsch (1980) were fit with $\underline{z}_t$ representing the covariate and insecticide applications being interventions. Logarithms of moth counts and coded water amounts were used. One difference was needed to achieve stationarity.

Identification of the model was obtained by first considering data prior to the first intervention. The techniques of Pfeifer & Deutsch (1980b) were applied to the ordinary least squares residuals of $\log(\underline{y}_t + \underline{1}) - \log(\underline{y}_{t-1} + \underline{1})$ regressed on $\underline{z}_t$. (Here, $\underline{1}$ is a 3x1 vector of ones and the logarithm of a vector is done componentwise.) Due to missing observations, identification of the model by way of the space–time autocorrelation function and the space–time partial autocorrelation function was difficult, but a STMA(1_2) model did not appear to be inconsistent with the identification results. Thus, the following model was fit:

$$\log(\underline{y}_t + \underline{1}) = \log(\underline{y}_{t-1} + \underline{1}) + \Lambda \underline{z}_t + \Theta \underline{\varepsilon}_{t-1} + \underline{\varepsilon}_t \qquad (2.5)$$

where Λ is a 3x3 matrix of regression coefficients,

$$\Theta = \theta_0 \begin{bmatrix} 1 & 0 & 0 \\ 0 & 1 & 0 \\ 0 & 0 & 1 \end{bmatrix} + \theta_1 \begin{bmatrix} 0 & 1 & 0 \\ \frac{1}{2} & 0 & \frac{1}{2} \\ 0 & 1 & 0 \end{bmatrix} + \theta_2 \begin{bmatrix} 0 & 0 & 1 \\ 0 & 0 & 0 \\ 1 & 0 & 0 \end{bmatrix}$$

is the spatial weight matrix of order 2, and $\underline{\varepsilon}_t$ is a 3x1 "noise" vector; i.e.,

$$E(\underline{\varepsilon}_t) = \underline{0}; \; E(\underline{\varepsilon}_{t+j} \cdot \underline{\varepsilon}_t) = \begin{cases} 0 & \text{if } j \neq 0 \\ \sigma^2 Q & \text{if } j = 0 \end{cases}.$$

Interventions are implicitly included in this model by way of the state–space formulation of the model and its extension as presented in section 2.

All variables were corrected for their mean. By imbedding the Kalman filter in a quasi–Newton optimization routine (see, for example, Gill & Murray (1972), Bard (1974), and Kennedy & Gentle (1980)), maximum likelihood estimates for the model parameters were obtained. In a preliminary fit of model (2.5), two of the five insecticide interventions were found to significantly affect response (confidence intervals for the remaining three were found to include zero), and hence only these two (at times $t = 53$ and $t = 65$) were included in the final model. They were deemed to have effects corresponding to example (2.4). The corresponding intervention steps $\underline{\omega}_1$ and $\underline{\omega}_2$, at times $t = 53$ and $t = 65$ were estimated as well as their two respective rates of decay ϕ_1 and ϕ_2 using the techniques of section 2.

The results follow (approximate standard errors are in parentheses):

$$\hat{\Lambda} = 10^{-3} \begin{bmatrix} 5.0 & 3.5 & -15.9 \\ (11.2) & (15.3) & (13.1) \\ -13.2 & 29.7 & -18.0 \\ (11.3) & (15.2) & (13.2) \\ -13.8 & 3.7 & 11.3 \\ (14.6) & (20.1) & (17.2) \end{bmatrix},$$

$$\hat{\Theta} = \begin{bmatrix} -0.739 & 0.161 & -0.032 \\ (0.004) & (0.003) & (0.002) \\ 0.084 & -0.739 & 0.084 \\ (0.002) & (0.004) & (0.002) \\ -0.032 & 0.161 & -0.739 \\ (0.002) & (0.003) & (0.004) \end{bmatrix},$$

$$\hat{\underline{\omega}}_1 = \begin{bmatrix} -3.46 \\ (0.52) \\ -3.42 \\ (0.56) \\ -3.35 \\ (0.68) \end{bmatrix}, \qquad \hat{\underline{\omega}}_2 = \begin{bmatrix} -3.58 \\ (0.48) \\ -3.91 \\ (0.67) \\ -4.42 \\ (0.67) \end{bmatrix},$$

$$\hat{\phi}_1 = \begin{matrix} 0.949 \\ (0.027) \end{matrix}, \qquad \hat{\phi}_2 = \begin{matrix} 0.784 \\ (0.077) \end{matrix},$$

$$\hat{\sigma}^2 \hat{Q} = \begin{bmatrix} 0.459 & 0.329 & 0.353 \\ & 0.465 & 0.496 \\ (\text{sym}) & & 0.792 \end{bmatrix},$$

$$\hat{\sigma} = 0.057721.$$

An investigation of the space–time autocorrelation function and the space–time partial autocorrelation function of the residuals of this model showed no obvious deficiency in the STMAX(1_2) model.

The model indicates that irrigation water has no significant effect on the moth population (all confidence intervals for the regression coefficients contain 0). For the Entomologist, control of moths with irrigation does not look promising.

Figures $1 - 3$ provide plots of the amount of irrigation water in the field over time at each of the three locations. Figures $4 - 6$ show the observed and predicted moth counts (transformed) over time at each of the three locations. In Figures $4 - 6$, the "*" 's on the time axis represent the time of the two insecticide interventions.

It should be noted that the predicted values in Figures $4 - 6$ were obtained by using the Kalman filter to obtain "one step ahead" forecasts. Fixed point smoothing as presented by Harvey & Pierse (1984) was used to estimate missing values. It was assumed that the application of insecticide would effectively kill the entire resident moth population in the field. Hence, the intervention steps were initialized at $-\underline{y}_t{}^*$. The initial variances were taken to be 10^{12}, and the initial covariances were taken to be zero. This is all in accord with the computational suggestions given in Harvey & Phillips (1979).

At time $t = n$, the final state vector contains a component of the form $\hat{\phi}_1^{n-t_1^*-1}\hat{\underline{\omega}}_1$ for the first intervention, and a component of the form $\hat{\phi}_2^{n-t_2^*-1}\hat{\underline{\omega}}_2$ for the second intervention. In addition, the corresponding final covariance matrix P_t contains the variances and covariances of the components of $\underline{\alpha}_t$. Hence, $\underline{\omega}_1$ and $\underline{\omega}_2$ were obtained by dividing the corresponding final state vector components by $\hat{\phi}_1^{n-t_1^*-1}$ and $\hat{\phi}_2^{n-t_2^*-1}$, respectively. Standard errors were obtained from P_t by dividing the appropriate elements (their square roots) similarly. For any type of intervention, parameter estimates can be obtained in a similar manner.

Conclusion

The state–space formulation of STARMAX models (and more generally, multivariate ARMAX models) provides an efficient way to compute likelihood in the presence of missing values and aggregate values. Inasmuch as the Kalman filter can accommodate time–varying parameters, it has been shown that interventions of the type discussed by Box & Tiao (1975) can also be incorporated. This is accomplished by appropriately augmenting the component vectors and matrices of the stage–space model from the time immediately after an intervention and onward. It should be noted that in the aforementioned example, further analysis of the data is certainly warrented. The effect of irrigation water might be lagged and so inclusion of terms such as z_{t-1} and z_{t-2} is an area for future investigation.

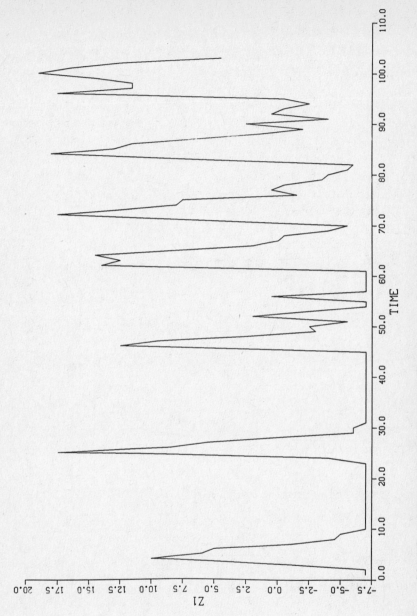

FIGURE 1:CODED IRRIGATION WATER AMOUNT FOR REGION 1.

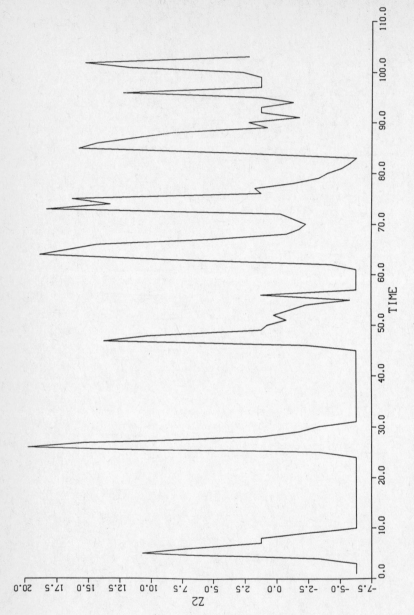

FIGURE 2:CODED IRRIGATION WATER AMOUNT FOR REGION 2.

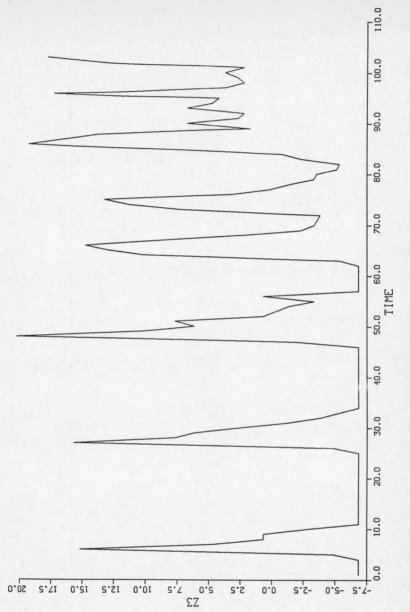

FIGURE 3:CODED IRRIGATION WATER AMOUNT FOR REGION 3.

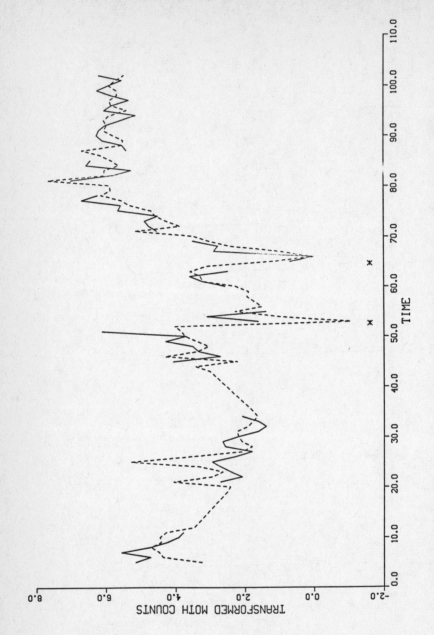

FIGURE 4: OBSERVED (SOLID LINE) AND PREDICTED (DASHED LINE) MOTH COUNTS FOR REGION 1.

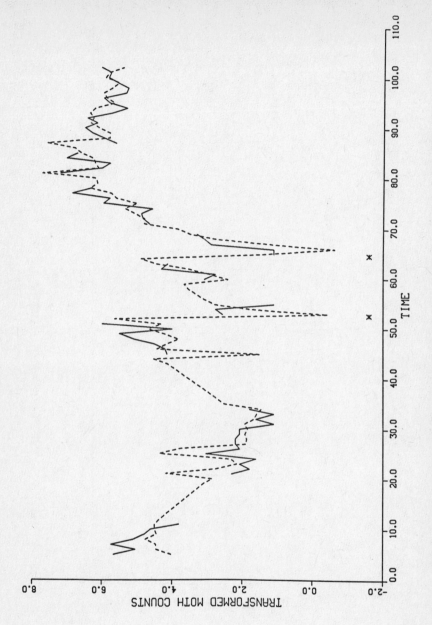

FIGURE 5: OBSERVED (SOLID LINE) AND PREDICTED (DASHED LINE) MOTH COUNTS FOR REGION 2.

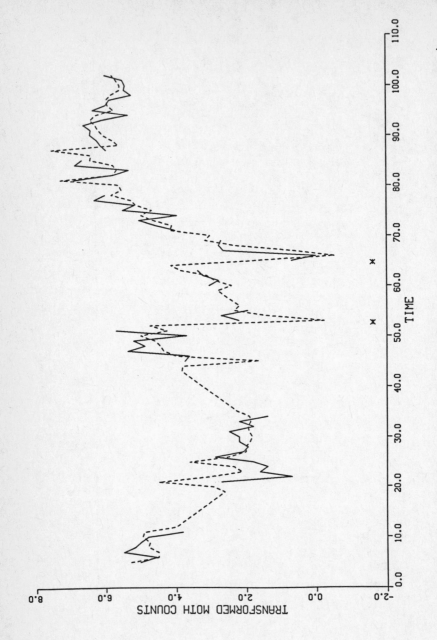

FIGURE 6: OBSERVED (SOLID LINE) AND PREDICTED (DASHED LINE) MOTH COUNTS FOR REGION 3.

References Cited

Anderson, B. D. O. & J. B. Moore. 1979. Optimal Filtering. Englewood Cliffs: Prentice—Hall.

Ansley, C. F. & R. Kohn. 1983. Exact likelihood of a vector autoregressive—moving average process with missing or aggregate data. *Biometrika* 70: 275–278.

Bard, Y. 1974. Nonlinear Parameter Estimation. New York: Academic Press.

Box, G. E. P. & G. M. Jenkins. 1970. Time Series Analysis, Forecasting, Control. San Francisco: Holden—Day.

Box, G. E. P. & G. C. Tiao. 1975. Intervention analysis with applications to economic and environmental problems. *J. Amer. Statist. Assoc.* 70: 70–79.

Gardner, G., A. C. Harvey & G. D. A. Phillips. 1980. An algorithm for exact maximum likelihood estimation of autoregressive—moving average models by means of kalman filtering. *Appl. Statist.* 29: 311–322.

Gill, P. E. & W. Murray. 1972. Quasi—Newton methods for unconstrained optimization. *J. Inst. Math. & Applic.* 9: 91–108.

Hannan, E. J. 1976. The identification and parameterization of ARMAX and state—space forms. *Econometrics* 44: 713–723.

Harvey, A. C. 1981. The kalman filter and its applications in econometrics and time series analysis." In Methods of Operations Research, 44, [ed.] R. Henn, et al., Cambridge: Oelgeschlager, Gunn and Hain.

Harvey, A. C. & G. D. A. Phillips. 1979. Maximum likelihood estimation of regression models with autoregressive—moving average disturbances. *Biometrika* 66: 49–58.

Harvey, A. C. & G. D. A. Phillips. 1981. The estimation of regression models with time—varying parameters. In M. Deistler, E. Furst, and G. Schwodiauer, [eds.], Games, Economic Dynamics, and Time Series Analysis, Physica—Verlag, Wein, pp. 306–321.

Harvey, A. C. & R. G. Pierse. 1984. Estimating missing observations in economic time series. *J. Amer. Statist. Assoc.* 79: 125–131.

Haugh, L. D. 1984. An introductory overview of some recent approaches to modeling spatial time series. In Time Series Analysis: Theory and Practice, [ed.] O. D. Anderson, New York: North—Holland.

Jones, R. H. 1980. Maximum likelihood fitting of ARMA models to time series with missing observations. *Technometrics* 22: 389–395.

Kennedy, W. J., Jr. & J. E. Gentle. 1980. Statistical Computing. New York: Marcel Dekker.

Kohn, R. & C. F. Ansley. 1986. Estimation, prediction, and interpolation for ARMA models with missing data. *J. Amer. Statist. Assoc.* 81: 751–761.

Liu, L. M. & D. M. Hanssens. 1981. A bayesian approach to time—varying cross—sectional regression models. *J. Econom.* 15: 341–356.

Pfeifer, P. E. & S. J. Deutsch. 1980. A three—stage iterative procedure for space—time modeling. *Technometrics* 22: 35–47.

Pfeifer, P. E. & S. J. Deutsch. 1980b. Identification and interpretation of first—order space—time ARMA models. *Technometrics* 22: 397—408.

Shea, B. L. 1984. Maximum likelihood estimation of multivariate ARMA processes via the kalman filter." In Time Series Analysis: Theory and Practice, [ed.] O. D. Anderson, New York: North—Holland.

Stoffer, D. S. 1986. Estimation and identification of space—time ARMAX models in the presence of missing data. *J. Amer. Statist. Assoc.* 81: 762—772.

Appendix: Invertibility of the STMA(1_2) Process

In the example in the text, a STMA(1_2) process was deemed appropriate for the error terms in model (2.5). It is necessary that the parameter estimates yield a model which is invertible. This occurs when the roots of the determinantal equation

$$0 = \mid m \, I - \Theta \mid$$

are less than 1 in absolute value. For the STMA(1_2) model,

$$p(m) = \mid m \, I - \Theta \mid = a_3 m^3 + a_2 m^2 + a_1 m + a_0 \, , \quad \text{where}$$

$a_3 = 1$, $a_2 = 3\theta_0$, $a_1 = 3\theta_0^2 - \theta_1^2 - \theta_2^2$, $a_0 = \theta_0^3 + \theta_1^2 \theta_2 - \theta_0 \theta_1^2 - \theta_0 \theta_2^2$, I is the 3x3 identity matrix, and

$$\Theta = \theta_0 \begin{bmatrix} 1 & 0 & 0 \\ 0 & 1 & 0 \\ 0 & 0 & 1 \end{bmatrix} + \theta_1 \begin{bmatrix} 0 & 1 & 0 \\ \frac{1}{2} & 0 & \frac{1}{2} \\ 0 & 1 & 0 \end{bmatrix} + \theta_2 \begin{bmatrix} 0 & 0 & 1 \\ 0 & 0 & 0 \\ 1 & 0 & 0 \end{bmatrix} .$$

by algebraic methods, it is readily seen that the roots of $p(m)$ are

$$m_1 = \theta_2 - \theta_0$$

$$m_2 = \tfrac{1}{2}(\sqrt{4\theta_1^2 + \theta_2^2} - \theta_2 - 2\theta_0)$$

$$m_3 = \tfrac{1}{2}(\sqrt{4\theta_1^2 + \theta_2^2} + \theta_2 + 2\theta_0) \, .$$

In the example then, the final parameter estimates were

$$\hat{\theta}_0 = -0.739$$
$$\hat{\theta}_1 = 0.161$$
$$\hat{\theta}_2 = -0.032$$

and so

$$m_1 = 0.707$$
$$m_2 = 0.917$$
$$m_3 = 0.593$$

all less than 1 in absolute value. Hence the STMA(1_2) process fit is invertible.

Estimating the Size of Gypsy Moth Populations Using Ratios

John P. Buonaccorsi[1] and Andrew M. Liebhold[2]

ABSTRACT: Estimates of gypsy moth populations can be obtained using a ratio of mean frass drop from a forest canopy to mean frass production for individually caged larvae. Appropriate statistical methods for point estimation and confidence intervals have been developed. Those methods included the use of two resampling techniques, the jackknife and the bootstrap. Exact theoretical comparisons of the proposed methods are essentially impossible. This paper evaluates the techniques via computer simulations for a limited number of situations.

Recently, Liebhold & Elkinton (1988a,b) made use of frass counts in order to estimate larval gypsy moth densities. The rate of frass drop from the forest canopy was measured in a large number of funnels. At the same time frass production (yield) rates were observed for individually caged larvae feeding under similar circumstances. An estimate of the population density per unit area is then proportional to the ratio of the mean frass drop per trap to the mean frass yield per larva. Since the resulting estimate involves the ratio of two random variables, some care must be exercised when making statements about the properties of the density estimate and in obtaining estimated standard errors and confidence intervals.

There is a rich history of the use of ratios in a variety of statistical problems; see Miller (1986) and Buonaccorsi (1985) for a general discussion and references. A detailed discussion of the problem to be treated here is given by Buonaccorsi & Liebhold (1988). Since that paper contains extensive detail, our introductory comments will be somewhat limited and we will not reproduce all of the formulas. The objective of this paper is to begin the process of evaluating and comparing, via computer simulations, the procedures described there.

The general setting is the following. There are two independent samples. The X sample consisting of n values is assumed to come from a population distribution with mean u_x and variance v_x. The sample mean and variance are denoted by $\bar{x}$ and s_x^2 respectively. The Y sample of size m is from a population with u_y and variance v_y with sample mean $\bar{y}$ and sample variance s_y^2. The objective is to estimate the ratio $D = u_x/u_y$.

For the density estimation problem, an X observation arises from a funnel while each Y observation results from a larva. In each case, the response is number of pellets. The ratio D equals the population density expressed as number of larvae per $0.14m^2$ (the size of the funnel). Conversion to a density expressed on a particular per unit basis involves just multiplication by a constant.

---

[1]Department of Mathematics and Statistics, University of Massachusetts, Amherst, Massachusetts 01003.

[2]Department of Entomology, University of Massachusetts, Amherst, Massachusetts 01003.

Point Estimation

The most obvious estimate is $R = \bar{x}/\bar{y}$ but this is known to produce a biased estimate of D. There are a number of ways to try to estimate and remove the bias. We consider two, jackknifing and bootstrapping. For the jackknife, observations are eliminated one at a time from the sample. Each time an observation is eliminated, the ratio is formed from the remaining observations and a pseudo–value is then constructed. A weighted combination of the pseudo values yields a jackknife estimator R_J which eliminates (in a certain sense) first order bias.

For the bootstrap, one treats the two sets of sample data as if they were populations and examines the behavior of the ratio obtained when we repeatedly sample from these two "populations". This is done by taking B bootstrap samples, where each bootstrap sample consists of taking a sample of size n with replacement from the X data and an independent sample of size m, also with replacement, from the Y data. For the ith bootstrap sample, denote the ratio of the resulting means by R_{bi}. The bootstrap estimate of bias is $R^* - R$ where R^* is the mean of the R_{bi} values. A naive estimator which attempts to adjust for bias is $R_B = R - (R^* - R)$.

Standard Errors and Confidence Intervals

Of course a point estimate by itself is not very informative and one would like some measure of the precision of the estimate (usually provided via a standard error) and/or a confidence interval for the item of interest. Large sample theory suggests

$$SE = [(s_x^2/n) + (R^2 s_x^2/m)]^{1/2}/|\bar{y}|$$

as a standard error for R and

$$R \pm z \cdot SE \tag{1}$$

as an approximate confidence interval for D where z denotes the appropriate standard normal table value. This is often referred to as the delta method. In our simulations we consider only 95% confidence intervals for which $z = 1.96$.

There is a variation of a result due to Fieller (1954) which suggests another approximate confidence region based on the joint approximate normality of $\bar{x}$ and $\bar{y}$. Computational details are given in Buonaccorsi & Liebhold (1988). We refer to this as the Fieller method. We note that in theory this method has the unappealing property of sometimes yielding a confidence region which is infinite in length. This pathological behavior can be a problem when the coefficient of variation for Y is large. For the settings considered in our simulations, we always obtained finite intervals. In other situations, however, this will need attention.

Using the pseudo–values described in the preceding section, one can form a jackknife standard error which we denote SE_J and using the bootstrap replicates one can form a bootstrap estimate of standard error which we denote SE_B. Both of these quantities were originally developed as estimators for the standard deviation of R but they can also be used as estimators for the standard deviation of R_J or R_B. There are a number of options for combining the different point estimates with the different standard errors.

The above confidence intervals rely on the approximate normality of the point estimator or in the case of Fieller's method the joint normality of $\bar{x}$ and $\bar{y}$. One of the features of the bootstrap method is that the empirical bootstrap distribution (EBD) based on the B bootstrap values provides an estimate of the distribution of R. A number of ways have been proposed for using the EBD for nonparametric interval estimation; see Efron & Tibshirani (1986). We restrict our attention to the use of the percentile method.

Simulation

The full inferential setting is determined by a description of the two population distributions which are being sampled and the two sample sizes involved. Reasonable population distributions were selected by examination of the field data collected by Liebhold & Elkinton (1988b).

For the larva yield distribution, the data indicated that any model should have a relatively large mass associated with the zero value and the nonzero values were somewhat bell shaped. We started by generating values according to a normal distribution with mean 17 and variance 97. Generated values were rounded to the nearest integer and any negative values were rounded to zero. This procedure results in a population distribution concentrated on the integers 0, 1, . . . as portrayed in Fig. 1C. The resulting mean and variance are $u_y = 17.1686$ and $v_y = 90.012$. This distribution was used throughout for the Y values.

For the funnel drop values (the X values) we utilized a truncated exponential distribution. That is, variates were generated according to an exponential distribution with parameter c but limited to the range of zero to an upper value Q. As before, the resulting values were rounded off to the nearest integer. Two distributions were used. The first with c = 40 and Q = 100 yielded the distribution in Fig. 1B with mean $u_x = 31.1675$ and variance $v_x = 630.139$. Combined with the Y distribution above the resulting density is D = 1.8154. The second X distribution arose from c = 2 and Q = 50 producing the distribution in Fig. 1A with mean $u_x = 1.979$, variance $v_x = 4.1638$ and resulting density D = 0.1153.

A total of four settings were considered by using two sets of sample sizes (n=m=10 and n=m=20) in conjunction with each of the two X distributions. For each setting 1000 replicates of the experiment were simulated. For each replicate each of the three point estimates were computed along with seven confidence interval procedures. Five of the intervals arose from a simple combination of a point estimator with a standard error similar to what appears in equation

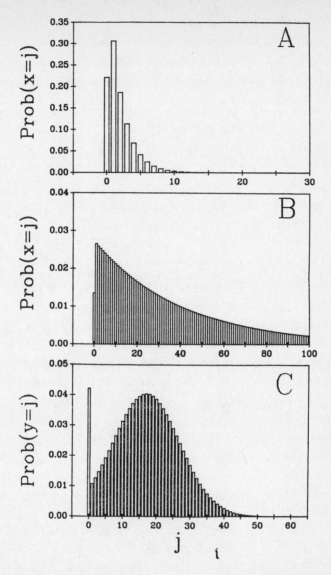

Fig. 1. Display of population distributions used. A is the frass drop distribution with $u_x = 1.969$, B is the frass drop distribution with $u_x = 31.1675$ and C is the single frass yield distribution used.

(1). We used R in conjunction with SE, SE_J, and SE_B; R_J with SE_J and R_B with SE_B. The other two procedures were the Fieller method and the Bootstrap percentile method described earlier. For the bootstrap method, on each replicate we took B = 500 bootstrap samples. Since we consider 95% confidence intervals, the bootstrap percentile method uses the 12th and 488th largest bootstrap values to form the interval.

Results

Table 1 contains descriptive statistics for each of the three point estimators. The relative bias column expresses the estimated bias as a percent of D with the sign giving the direction of the bias. The column labeled MSE gives the estimated mean squared error which equals the estimated bias squared plus the standard deviation squared. On average R overestimates D while both R_J and R_B underestimate but often with smaller relative bias than R. With respect to MSE, the differences are minor although R_J usually has the smallest value. Further information about the sampling distribution of the point estimators is given by the extreme values and selected percentiles. If one examines the median (50th percentile), in most cases the estimators have a median which is smaller than the true density. Hence the percentage of time that one obtains an underestimate is usually greater than 50. In this regard R seems preferable. The distribution of both R_J and R_B is shifted further to the left than that of R. From a practical viewpoint, there is little that distinguishes among the three estimators for the settings considered here. We remind the reader also that the values in Table 1 are estimates and are themself subject to sampling error.

An analysis of the behavior of the confidence intervals is a bit more complicated. Typically different intervals are contrasted by estimating the confidence level (the probability that the interval contains the true value) and giving the average interval length. While this information is useful, it does not tell the whole story. Our analysis consists of estimating P(W) = probability that the interval contains W for a number of different choices of W, including the true value D. Fig. 2 and 3 display plots of the percentage of time the interval contained W for a number of different values of W (obtained by taking varying percentages of D). The point on the x–axis underneath the vertical line corresponds to the true density. For easy comparisons we have produced in Table 2 numerical quantities for selected values of W along with the average interval lengths.

Throughout our target confidence level was 0.95. Ideally, a good interval procedure covers the true value the desired 95% of the time and has small coverage probabilities for other values. Examination of the results lead to the following conclusions.

1. None of the procedures do very well in attaining the desired 95% confidence level. Consistently, the estimated confidence levels are below 0.95. For the large samples sizes, Fieller's method might be acceptable.

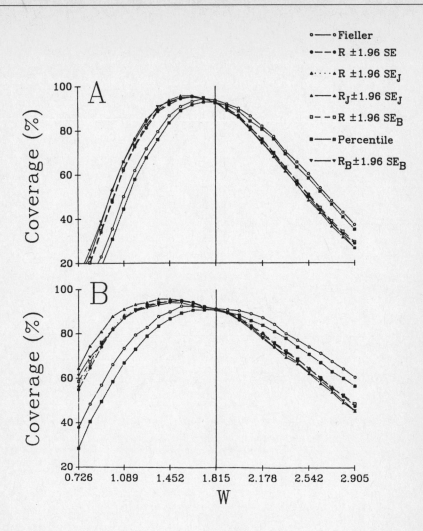

Fig. 2. Observed coverage rates (% of time the interval contained the value W) in 1000 replicates when D = 1.8154.

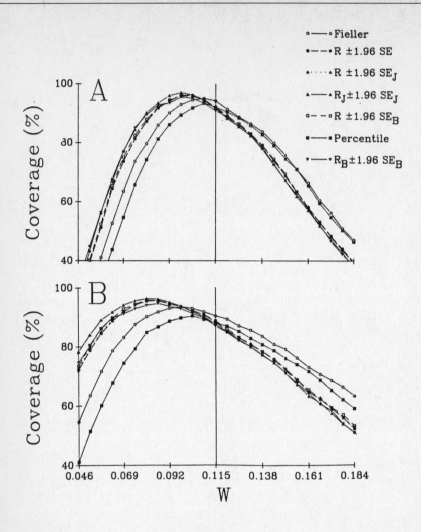

Fig. 3. Observed coverage rates (% of time the interval contained the value W) in 1000 replicates when D = .1153.

Table 1. Descriptive statistics for three point estimators based on 1000 replicates. Rel. bias denotes relative bias. MSE denotes estimated mean squared error as described in text. n is common sample size.

D	n	Est.	Mean	Rel Bias(%)	SD	MSE	Min	25	50	75	Max
									Percentiles		
1.8154	10	R	1.85	2.13	0.6034	0.3665	0.61	1.40	1.79	2.21	4.93
		R_J	1.78	−1.51	0.5692	0.3243	0.59	1.36	1.73	2.12	4.77
		R_B	1.80	−1.03	0.5997	0.3598	0.57	1.36	1.71	2.15	5.08
	20	R	1.82	0.25	0.4074	0.1667	0.88	1.53	1.79	2.08	3.54
		R_J	1.79	−1.40	0.3970	0.1579	0.86	1.51	1.76	2.04	3.47
		R_B	1.79	−1.23	0.4010	0.1610	0.86	1.51	1.75	2.05	3.43
0.1153	10	R	0.119	2.95	0.0455	0.0021	0.03	0.087	0.111	0.145	0.351
		R_J	0.114	−0.78	0.0429	0.0018	0.029	0.083	0.107	0.139	0.321
		R_B	0.115	−0.69	0.0449	0.0220	0.029	0.082	0.107	0.139	0.354
	20	R	0.118	2.16	0.0304	0.0009	0.046	0.096	0.115	0.137	0.237
		R_J	0.116	0.45	0.0297	0.0009	0.045	0.094	0.113	0.134	0.235
		R_B	0.116	0.45	0.0301	0.0009	0.043	0.093	0.113	0.134	0.230

2. In terms of confidence level, the Fieller method is superior having achieved the highest coverage of the true value in all four cases.

3. In a number of cases the confidence intervals are biased which means that there is a higher probability of covering some false values than of covering the true value. This is true in all four cases for the methods based on a point estimator and a standard error. It is also a problem, though to a lesser degree, for the Fieller and percentile method when D is small. This bias problem is clearly connected to the asymmetry observed in the methods.

4. The Fieller and percentile method fare worse than the other five intervals for false values larger than D (in a sense of having higher coverage rates) but the situation is clearly reversed for values smaller than D.

At the larger density the Bootstrap has an estimated confidence level close to that of Fieller and it has considerably smaller coverage of false values. Hence, it would be favored. At the smaller density though the inability of the bootstrap method to get close to the desired confidence level recommends against its use. Of course, one needs to be concerned with the inability of all of the procedures to achieve the desired confidence level and the natural question is how to improve the situation. For the bootstrap technique the bias corrected percentile methods described by Efron & Tibshirani (1986) should be tried. Computationally these are more intensive. This is not much of a problem when analyzing a single data set but it can pose some problems in large scale simulations. All of the other methods rely on a normal approximation of one sort or another. A natural way to try and correct the situation is with the use of an approximation based on a t distribution which will serve to widen all of the intervals. The choice of a degrees of freedom here is somewhat complicated by the fact that there are two samples and our interest is in a nonlinear function. There are ways to obtain an approximate degrees of freedom based on the work of Satterthwaite (1946). This technique should certainly be investigated in future simulations. We note that this modification will in no way help in correcting the asymmetry and bias of the five simple methods. Thus, it seems the attention should be focused on the Fieller method.

Discussion

It is important to emphasize that our simulations have considered a rather limited number of settings. This is due to space and time limitations along with the desire to present fairly detailed analyses for those settings that were examined. In particular, we wanted to emphasize the need for a more comprehensive assessment of confidence interval behavior than what is traditionally found. In terms of point estimation the three techniques are not different enough for the settings considered to conclusively favor any one. Of course, for alternate settings this could change drastically. For confidence intervals, the Fieller method and bootstrap methods based on the empirical bootstrap distribution seem most promising. However, as described in the preceding section, modifications are needed. Of course, as the sample size increases the odd behavior of the other five interval techniques will disappear but it remains to be seen at what point this happens. At this point, we recommend that these methods not be used.

Only cases with equal sample sizes have been examined. Buonaccorsi & Liebhold (1988) developed optimal allocation of resources under cost considerations. Even with equal costs for each type of observation, equal numbers of samples is not necessarily the best thing to do. Future studies should also examine more closely the effects of the allocation used.

In our tables we have summarized the information about the X and Y distribution used only through the density D. We caution the reader not to be mislead by this. In general, the behavior of all of the methods will depend heavily on the exact form of the X and Y distribution present. For example, there are an infinite number of distributions that have the same means, u_x and u_y, that we used. A change in the population variances could result in different conclusions regarding the individual and comparative behavior of the methods. Since $\bar{y}$ occurs in the denominator of R, cases where v_y is large relative to u_y (i.e., a large coefficient of variation) should be given special attention.

Table 2. Summary of confidence interval performance. Except for last column, entries are the number of times out of 1000 replicates that the given interval contained the value at top of column. Values under column labeled D are estimated confidence levels. Len gives average interval length. n is common sample size.

D	n	Method	0.4D	0.6D	0.8D	D	1.2D	1.4D	1.6D	Len
1.815	10	R–SE	549	872	947	912	804	647	477	2.26
		R–SE$_J$	566	885	952	912	807	649	479	2.29
		R$_J$–SE$_J$	644	911	957	912	787	625	455	2.29
		R–SE$_B$	586	873	945	910	794	645	488	2.35
		R$_B$–SE$_B$	613	871	940	906	780	629	457	2.35
		Fieller	379	734	899	914	875	744	609	2.53
		Boot	285	669	868	910	841	711	569	2.32
	20	R–SE	109	621	918	925	757	503	289	1.56
		R–SE$_J$	112	622	921	925	760	507	290	1.60
		R$_J$–SE$_J$	133	661	938	926	746	483	268	1.60
		R–SE$_B$	114	628	927	925	759	510	295	1.58
		R$_B$–SE$_B$	136	660	931	928	740	493	267	1.58
		Fieller	65	502	868	937	818	583	373	1.64
		Boot	50	446	836	927	804	583	352	1.58
0.1153	10	R–SE	722	919	945	884	783	646	522	0.166
		R–SE$_J$	727	922	946	884	781	649	527	0.168
		R$_J$–SE$_J$	783	942	950	876	772	633	510	0.168
		R–SE$_B$	735	929	945	887	783	654	532	0.173
		R$_B$–SE$_B$	747	912	936	870	770	642	509	0.173
		Fieller	546	831	930	904	834	731	633	0.186
		Boot	412	745	886	879	806	717	590	0.170
	20	R–SE	257	733	941	923	796	590	400	0.036
		R–SE$_J$	257	736	942	923	796	593	403	0.036
		R$_J$–SE$_J$	293	770	960	920	790	579	384	0.036
		R–SE$_B$	270	750	948	918	796	589	402	0.038
		R$_B$–SE$_B$	327	772	936	913	785	585	389	0.038
		Fieller	187	630	901	942	843	672	478	0.038
		Boot	107	541	859	919	833	660	474	0.037

The problem we have considered is a special case of a problem where the main interest is in a nonlinear function of the population parameters. Simulations similar to ours have been carried out recently for a number of problems encountered in ecological studies. These include the estimation of niche overlap (Mueller & Altenberg 1985), uncertainty in population growth rates (Meyer et al. 1986) and estimation of the Gini index of inequality (Dixon et al. 1987).

As noted at the beginning, ratios occur in a variety of statistical problems. Simulations concerned with ratio estimation in regression problems are given by Hinkley (1969), Weber & Welsh (1983) and Buonaccorsi & Iyer (1984). The first two papers are concerned mainly with a comparison of jackknife estimators and the "usual" estimator. The latter paper compares Fieller's method and the delta method for obtaining confidence intervals and reaches the same conclusions given in this paper. Cochran (1977) summarizes some studies concerned with ratio estimation in finite population sampling problems.

Acknowledgement

The authors are grateful to Joe Elkinton and Hari Iyer for reviewing this paper and for their helpful comments.

References Cited

Buonaccorsi, J. P. 1985. Ratios of linear combinations in the general linear model. *Commun. Statist.—Theor. Meth.* 14: 635–650.

Buonaccorsi, J. P. & H. K. Iyer. 1984. A comparison of confidence regions and designs in estimation of a ratio. *Commun. Statist.—Simul. Computa.* 13: 723–741.

Buonaccorsi, J. P. & A. M. Liebhold. 1988. Statistical methods for estimating ratios and products in Ecological studies. *Environ. Entomol.* 17: 572–580.

Cochran, W. G. 1977. Sampling techniques, Third Edition. Wiley & Sons, New York.

Dixon, P. M., J. Weiner, T. Mitchell—Olds & R. Woodley. Bootstraping the Gini coefficient of inequality. *Ecology* 68: 1548–1551.

Efron, B. & R. Tibshirani. 1986. Bootstrap methods for standard errors, confidence intervals, and other measures of statistical accuracy. *Statistical Science* 1: 54–77.

Fieller, E. C. 1954. Some problems in interval estimation. *J.R. Statist. Soc. Ser. B.* 16: 175–185.

Hinkley, D. V. 1977. Jackknifing in unbalanced situations. *Technometrics* 19: 285–292.

Liebhold, A. M. & J. S. Elkinton. 1988a. Techniques for estimating the density of late—instar gypsy moth, (*Lepidoptera: Lymantriidae*), populations using frass drop and frass production measurements. *Environ. Entomol.* 17: 381–384.

Liebhold, A. M. & J. S. Elkinton. 1987b. Estimating the density of larval gypsy moth (*Lepidoptera: Lymantriidae*), using frass drop and frass production measurements: sources of variation and sample size. *Environ. Entomol.* 17: 385–390.

Meyer, J. S., C. G. Ingersoll, L. L. McDonald & M. S. Boyce. 1986. Estimating uncertainty in population growth rates: jackknife vs. bootstrap techniques. *Ecology* 67: 1156–1166.

Miller, R. G., Jr. 1986. Beyond ANOVA, basics of applied statistics. John Wiley, New York.

Mueller, L. D. & L. Altenberg. 1985. Statistical inference on measures of niche overlap. *Ecology* 66: 1204–1210.

Satterthwaite, F. E. 1946. An approximate distribution of estimates of variance components. *Biometrics* 2: 110–114.

Weber, N. C. & A. H. Welsh. 1983. Jackknifing the general linear model. *Aust. J. Statist.* 25: 425–436.

Numerical Survival Rate Estimation for Capture–Recapture Models Using SAS PROC NLIN

Kenneth P. Burnham[1]

ABSTRACT   This paper discusses and illustrates recent developments in the modeling and analysis of capture–recapture data for open populations. Results are presented using a unified theory of cohort survival processes. The statistical models for survival rate estimation are represented as products of conditionally independent multinomials; this is achieved by always conditioning on the known number of releases initiating the cohorts. This approach facilitates extension of the Jolly–Seber model to control–treatment studies, or other contexts where comparison of population survival rates is of most interest. Numerical computation of maximum likelihood survival and capture rate estimators, under any model, is easily achieved using iteratively–reweighted nonlinear least squares in SAS PROC NLIN. In his classical paper, Jolly (1965) illustrated his methods using summary statistics on female black–kneed capsids (*Blepharidopterus angulatus*). Here I use the entire data set on males and females to illustrate survival rate estimation and flexible modeling of capture–recapture data.

In the analysis of capture–recapture data, successful parameter estimation depends critically on using the correct model. However, we do not know the correct model; consequently, much of the effort in a "good" analysis of capture–recapture data should be directed at model selection. By "model" I primarily mean the structure of the expected recapture rates, given the releases, expressed as functions of survival and capture probabilities (ϕ and p, respectively). Also relevant, but not considered here, is the structure of the expected captures of unmarked animals as functions of ϕ, p and abundance parameters (N, or B).

The Jolly–Seber model provides a good starting point for considering open population models (Jolly 1965, Seber 1965). The capture–recapture literature since 1965 is dominated by attempts to formulate alternative models, both more general than, and special cases of, the Jolly–Seber model, as well as explicit systems of models with associated formal model selection procedures (see, e.g., Robson 1969, Pollock 1975, Arnason & Baniuk 1980, Buckland 1980, Cormack 1981, Pollock 1981a, Sandland & Kirkwood 1981, Jolly 1982, Brownie & Robson 1983, Crosbie & Manly 1985, Brownie et al. 1986, Burnham et al. 1987 and Clobert et al. 1987). Moreover, it is clear that what is needed, and being sought, is a convenient, unified theory for capture–recapture which allows easy model formulation and analysis in the context of some standard statistical theory, such as the log–linear approach of Cormack (1981). Recently, Brownie & Robson (1983) pioneered the basis of another possible unified approach to modeling and analysis of marked–animal cohort survival processes (see also Brownie et al. 1985: 170–175; also note that Cormack 1964 is relevant here); they showed that capture–recapture models can be

---

[1]USDA–Agricultural Research Service, and Department of Statistics, Box 8203, North Carolina State University, Raleigh, NC 27695.

formulated using multinomial and binomial models for first recaptures following the known releases. Standard maximum likelihood analysis is then used. Burnham et al. (1987) elaborate on this approach for the analysis of survival rates (in capture–recapture experiments.)

Burnham et al. (1987) was in page proofs one year ago. Since then I have realized how to implement numerical maximum likelihood estimation of ϕ and p in these models using iteratively reweighted nonlinear least squares. In particular, PROC NLIN in SAS (SAS 1985) can be used as a basis for easy, flexible estimation of the ϕ and p parameters in capture–recapture models. The objectives of this paper are to (1) outline the underlying unified theory, and (2) illustrate flexible, numerical analysis of capture–recapture data (re ϕ and p) with the full black–kneed capsids data of C. R. Muir (Muir 1957). Summary statistics of these female capsids data were used as an example by Jolly (1965), and in fact it was this study by Muir which got George Jolly involved in capture–recapture (Jolly, personal communication).

Some General Capture–Recapture Theory

The notation of Burnham et al. (1987) is used here. Let k be the number of capture–release occasions, indexed as occasion i = 1 to k. At each occasion there are releases; let h(i) stand for a capture history at release time (occasion) i. When the value of i is clear from the context, this notation is reduced to just "h." One format for the complete data set is the complete capture history array: a listing of every observed h(k) (i.e., capture history at the end of the study) and the corresponding number of animals, X_h, with final capture history h. When there are different groups of animals the notation is extended by a group index, v = 1, ..., V. For example with males and females we have {h, X_{mh}, X_{fh}} (see, e.g., Table 1). A negative X_{vh} denotes that the animals with that history were lost on the last capture indicated in the string of 1's (captured) and 0's constituting "h." The complete capture history array is a convenient way to summarize

Table 1. Example capture histories for the female (X_{fh}) and male (X_{mh}) black–kneed capsids data of Muir (1957); the entire data set, as complete capture histories, has 252 distinct histories.

Capture History, h	X_{fh}	X_{mh}
0000000000010	36	14
0000000001100	4	1
0000000101000	7	1
0000001000000	74	58
0000100110000	2	0
0000111001100	0	-1
0100110000000	0	2
1000000000000	30	97
1001000000000	3	5

the data. Models and analysis methods can be based on this data representation, for example the loglinear method of Cormack (1981) and the multinomial modeling of Crosbie & Manly (1985). There are, however, strong advantages to basing modeling and analysis on the data representation outlined below.

Let u_i be the number of unmarked animals captured at time i. Any of these animals released at time i have history $h(1) = 1$ (for $i = 1$), or $h(2) = 01$ (for $i = 2$), or $h(3) = 001$ (for $i = 3$), and so forth. Let R_{ih} be the number of animals released at time i that have capture history h ($\equiv h(i)$). Let m_{ijh} be the number of animals, of the R_{ih}, captured again at time j and not captured at times $i + 1$ to $j - 1$. Let r_{ih} be the sum $m_{i,i+1,h} + \cdots + m_{ikh}$ = the total number of animals, of the R_{ih}, that are ever recaptured again. Conversely, $R_{ih} - r_{ih}$ animals of this release subcohort are never captured again. The m_{ijh} represent mutually exclusive events; a reasonable model to assume is that these m_{ijh}, $j = i + 1$, ..., k, are multinomial random variables given the known releases R_{ih}. The data from a capture–recapture study with only one group (category) of animals can be presented, without loss of information, as the statistics.

$$u_i, i = 1, ..., k ,$$
$$R_{ih}, m_{ijh}, j = i + 1, ..., k, h(i) \in H(i), i = 1, ..., k - 1 .$$

Here H(i) is just a symbol for the set of all observed capture histories in release cohort i. Each h(i) defines a subcohort. If there are 2 or more groups we add the index $v = 1$, ..., V, e.g., u_{vi}, R_{vih}, m_{vijh}. Numerous summary statistics can be defined from these basic variables, such as cohort totals R_i and m_{ij} obtained by summing over all subcohorts at time i, and the total recaptures on each occasion: $m_2 = m_{12}$, $m_3 = m_{13} + m_{23}$, and in general $m_j = m_{1j} + \cdots + m_{j-1,j}$, $j = 3$, ..., k.

The probability distribution of a capture–recapture data set takes the conceptual form $Pr\{data\} = Pr\{first\ captures\}*Pr\{releases | captures\}*Pr\{recaptures | releases\}$. An explicit form for this $Pr\{data\}$ can be given (Burnham 1987. A unified approach to animal release–resampling studies of survival processes and population estimation. North Carolina State University, Institute of Statistics Mimeograph Series 1698, Raleigh, NC). It suffices here to say that $Pr\{first\ captures\} = Pr\{u_i, i = 1, ..., k\}$ and this term is the only one involving the population abundance and recruitment parameters (N and B in Jolly 1965). Unless we make very strong assumptions about the recruitment process, there is no information in this first term about the survival and capture parameters ϕ and p. The second term, $Pr\{releases | captures\}$, is equivalent to the distribution of losses on capture given the captures (if there are no such losses this second term vanishes). Except under extraordinary circumstances, all the information about ϕ and p is in the term, $Pr\{recaptures | releases\}$. For purposes of this paper, this term will now be expressed as 3 new terms, which will be formally labelled as terms 1, 2 and 3:

$$\Pr\{\text{recaptures}\,|\,\text{releases}\} = \Pr\{JS\_MSS\}*\Pr\{\text{cohorts}\,|\,JS\_MSS\}*$$
$$\Pr\{\text{subcohorts}\,|\,\text{cohorts}\} \tag{1}$$
$$= (\text{term 1})*(\text{term 2})*(\text{term 3}) \ .$$

Here, JS_MSS denotes the component of the minimal sufficient statistic, under the Jolly–Seber model, which comes from the recaptures given the releases. Cohorts refers to the m_{ij} data (given the R_i); subcohorts refers to the m_{ijh}.

Jolly–Seber Model

The Jolly–Seber model assumes that the survival probability between capture times i and $i+1$ ($i=1, ..., k-1$), and the capture probability at time i are the same for all animals in the population at risk of capture. These parameters are ϕ_i and p_i respectively. For a full discussion of the assumptions in capture–recapture see Jolly (1965) or Pollock (1981b). The structure of the model, in terms of the m_{ij} given the R_i is

$$E(m_{i,i+1}) = R_i(\phi_i p_{i+1}) \qquad\qquad ,i=1, ..., k-1$$
$$E(m_{i,j}) \ = R_i(\phi_i q_{i+1})\cdots(\phi_{j-2}q_{j-1})(\phi_{j-1}p_j) \quad ,i=1, ..., k-1, j=i+1, ..., k,$$

where $q_i = 1-p_i$. There are $2k-3$ estimable parameters: ϕ_i, $i=1, ..., k-2$, p_i, $i=2, ..., k-1$, and the product $\phi_{k-1}p_k$.

Given the assumption that the parameters depend only upon time and that the recaptures (m) given releases (R) have independent multinomial distributions, the minimal sufficient statistic (JS_MSS) is the sums r_i, $i=1, ..., k-1$ and m_j, $j=2, ..., k-1$. This JS_MSS has dimension $2k-3$. For proofs and more details see Brownie & Robson (1983).

The distribution of JS_MSS can be expressed as the product of $2k-3$ conditionally independent binomials. Before that distribution can be given we need some more notation. Define the totals T_i, $i=2, ..., k-1$ as $T_2 = r_1$ and $T_j = T_{j-1}-m_{j-1} + r_{j-1}$, $j=3, ..., k-1$. Define two functions of the basic parameters: $\lambda_i = E(r_i\,|\,R_i)/R_i$, $i=1, ..., k-1$, and $\tau_j = E(m_j\,|\,T_j)/T_j$, $j=2, ..., k-1$. As a function of the parameters, λ_i is best expressed recursively:

$$\lambda_i = \phi_i(p_{i+1} + q_{i+1}\lambda_{i+1}) \qquad ,i=1, ..., k-1 \ ,$$

using the convention that $\lambda_k = 0$. Note that $\lambda_{k-1} = \phi_{k-1}p_k$. It is easy to compute these λ_i using backwards recursion. Finally,

$$\tau_j = \frac{p_j}{(p_j + q_j\lambda_j)} \qquad ,j=2, ..., k-1 \ .$$

Let $y \mid n \sim \mathrm{bin}(n, \theta)$ denote that given n, y is a binomial random variable with sample size n and parameter θ. Then

$$
\begin{aligned}
r_i \mid R_i &\sim \mathrm{bin}(R_i, \lambda_i) &&, i = 1, ..., k{-}1 \;, \\
m_j \mid T_j &\sim \mathrm{bin}(T_j, \tau_j) &&, j = 2, ..., k{-}1 \;.
\end{aligned}
\tag{2}
$$

This defines the distribution of JS_MSS, hence term 1 in formula (1). Under the assumption that the data fit the Jolly–Seber model, terms 2 and 3 in (1) are products of multiple hypergeometric distributions and these terms provide the basis of a fully–efficient goodness of fit test to the Jolly–Seber model (Pollock et al. 1985, Burnham et al. 1987). For inference about ϕ and p, assuming the Jolly–Seber model fits, all that is needed are these summary statistics and their distribution in (2) above. Note that any sub–model that is a special case of Jolly–Seber can be dealt with by starting from (2). Such sub–models are, e.g., the p_i and/or the ϕ_i constant (Jolly 1982, Brownie et al. 1986) or modeling the survival rates as functions of any time–varying variable (Clobert et al. 1987).

An immediate generalization is to allow V data sets, each one fitting the Jolly–Seber model with parameters ϕ_{vi} and p_{vi}, $v = 1, ..., V$. In particular, we might have data on adult males and females, or control and treatment population groups. Burnham et al. (1987) give an extensive treatment of this situation: testing goodness of fit, testing equality of the parameters over groups and parameter estimation for a class of models having closed–form solutions. Also, Burnham et al. (1987) discuss using program SURVIV (White 1983) to fit special cases of the Jolly–Seber model using numerical methods. Properly implemented, the SAS PROC NLIN approach below, and use of program SURVIV will give the same results. The advantages of this PROC NLIN approach is that SAS is a readily available package on a wide variety of machines and it is fairly easy to use, especially for anyone who has already used SAS.

Analysis Using SAS

Given the distribution in (2) we can in principle impose constraints on the parameters (then estimators are no longer closed–form) and compute the maximum likelihood estimates (MLE's). The problem is how to do this easily. Using PROC NLIN (nonlinear least squares) in SAS (1985) provides one solution to this computing problem.

For many statistical distributions, including binomials, the MLE's can be found using iteratively reweighted nonlinear least squares (see e.g., Jennrich & Moore 1975, Green 1984). The binomial case will be outlined here. Let $y_i \sim \mathrm{bin}(n_i, \pi_i(\underline{\theta}))$, independently for $i = 1, ..., g$. Let $\underline{\theta}$ be an f–dimensional parameter vector with $f \le g$, and let the π_i be known, well–behaved functions. Define weights w_i by

$$
w_i \frac{n_i}{\pi_i(\underline{\theta})(1 - \pi_i(\underline{\theta}))} \qquad , i = 1, ..., g;
$$

note that w_i is the reciprocal of the variance of the ratio y_i/n_i.

The MLE can be found as the value of $\underline{\theta}$ which minimizes the weighted residual sum of squares,

$$\sum_{i=1}^{g} w_i((y_i/n_i) - \pi_i(\underline{\theta}))^2 ,$$

in terms of $\underline{\theta}$, provided that the weights, as functions of the parameters, are recomputed at each step in the numerical optimization based on the current value of $\underline{\theta}$. PROC NLIN in SAS has the capability to do this iterative reweighting. PROC NLIN "thinks" it is solving the regression problem wherein the model is

$$y_i/n_i = \pi_i(\underline{\theta}) + \sigma\epsilon_i \quad , i = 1, ..., g,$$

with the variance of ϵ_i being $1/w_i$ and σ (or σ^2) being an unknown parameter to be estimated from the residual mean square. Subject to the Jolly–Seber assumptions, theoretically $\sigma^2 = 1$. The SIGSQ = 1 option is used to force $\sigma^2 = 1$ rather than estimate it. This results in PROC NLIN computing theoretical variances and covariances which are identical to the appropriate estimated likelihood theory variances.

Version 5 of SAS, and version 6.03 for personal computers, is even more powerful. In these versions we can declare an alternative loss function to be minimized (denoted in PROC NLIN as _LOSS_). Thus the negative of the log–likelihood function can be used directly as the minimization criterion, and SAS then prints out this (negative) log–likelihood value. It is still necessary to declare the "regression" version of the model and to use the SIGSQ = 1 option to force SAS to set $\sigma^2 = 1$.

Another more powerful advantage of PROC NLIN is that bounds can be put on the parameters. In particular, we can bound a rate parameter to be between 0 and 1. This is useful even for the fully–parameterized Jolly–Seber model because the closed–form MLEs of the ϕ_i can exceed 1, in which case there are benefits to constraining them (Buckland 1980).

Examples Using the Black–kneed Capsids Data

The Capsids Data.

In Aug. through Sept. 1955, Mr. C. R. Muir conducted a capture–recapture study of adult black–kneed capsids in an apple orchard at the East Malling Agricultural Research Station (Muir 1957). This insect was important as a predator of orchard insect pests; this capture–recapture study was just one part of a research program in the orchard. There were k = 13 capture occasions. It often took 2 days to complete an "occasion"; treating the midpoint of 2 days as the time instant of the occasion, capture occasions averaged 3.5 days apart. The times between occasions varied from 3 to 4 days. Capsids were individually marked and their sex determined. To analyze the data Mr. Muir asked the station statistician, Mr. G. M. Jolly, for help. At that time Mr. Jolly had only a passing acquaintance with capture–recapture. Out of this request for help, came the now classic 1965 paper of Jolly which uses summary statistics from the female capsids data as an example.

In the course of work on Burnham et al. (1987) I asked Mr. Jolly for the full capsids data. He was able to secure for me a copy of the original field records. The data were decoded and entered to a data file as capture histories. Table 1 shows a few example records. For just those records, Table 2 shows the data reformatted as subcohort releases and recaptures. The full data set consists of 252 distinct capture histories at occasion 13.

Using program RELEASE, which implements theory in Burnham et al. (1987), I tested the fit of the Jolly–Seber model. Chi–square goodness of fit tests are computed based on terms 2 and 3 in formula (1); these are TESTS 2 and 3 in RELEASE. The Jolly–Seber model is not rejected by either test for either sex (the overall chi–square goodness of fit test statistic is 130.2 on 130 df). Burnham et al. (1987) also give a test of the equality of the male and female survival and capture rates (TEST 1; it is based on term 1 in (1)). That test produces a chi–square test statistic of 259.3 on 23 df (P = 0). It is clear that the male and female capsids had different parameters, but separately they fit the time–specific Jolly–Seber model.

Table 3 gives the usual summary statistics and the statistics needed for the distribution of JS_MSS. The data for the females differ some from what Jolly (1965) used. Upon seeing this, I carefully checked the raw data and the data entry. There were no errors. The discrepancy cannot be resolved. However, the parameter estimates under the Jolly–Seber model here and in Jolly (1965) are very similar.

Table 2. The full m–array representation of female captures from the nine example capture histories data in Table 1; capture history is conditional on release occasion, i; R_{fih} denotes number released; m_{fijh} denotes (first) recaptures, at occasion j, after release at occasion i.

Capture History, h(i)	i	R_{fih}	j=2	3	4	5	6	7	8	9	10	11	12	13
1	1	33	0	0	3	0	0	0	0	0	0	0	0	0
1001	4	3	.	.	.	0	0	0	0	0	0	0	0	0
00001	5	2	.	.	.	.	0	0	2	0	0	0	0	0
0000001	7	74	.	.	.	.	.	.	0	0	0	0	0	0
00000001	8	7	.	.	.	.	.	.	.	0	7	0	0	0
00001001	8	2	.	.	.	.	.	.	.	2	0	0	0	0
000010011	9	2	.	.	.	.	.	.	.	.	0	0	0	0
0000000001	10	4	.	.	.	.	.	.	.	.	.	4	0	0
0000000101	10	7	.	.	.	.	.	.	.	.	.	0	0	0
00000000011	11	4	.	.	.	.	.	.	.	.	.	.	0	0
000000000001	12	36	.	.	.	.	.	.	.	.	.	.	.	0

The Jolly–Seber model in PROC NLIN.

Fig. 1 shows the necessary code to implement the Jolly–Seber model numerically in PROC NLIN of SAS using the JS_MSS (as per the representation of formulae (2)) as input. The numerical solution converged properly. Fig. 2 shows the key output: the sum of loss line (2368.882) is the negative log–likelihood function at its unconstrained minimum; parameter estimates, and their standard errors, agree with results using explicit formulae for these MLE's. Note that p_{13} is set to 1, so S12 is really S12*P13 $(= \phi_{12}p_{13})$.

Table 3. Summary statistics as an (reduced) m–array for female and male black–kneed capsids from the study of Muir (1957); under the Jolly–Seber model (Jolly 1965, Seber 1965) the r_i and m_j are minimal sufficient statistics, i = 2, ..., 13, j = 2, ..., 12.

i	R_{fi}	m_{fij} j=2	3	4	5	6	7	8	9	10	11	12	13	r_{fi}
1	54	10	3	5	2	2	1	0	0	0	1	0	0	24
2	143		36	18	8	4	7	4	2	0	2	1	1	83
3	166			33	14	8	5	0	4	1	3	3	0	71
4	202				30	20	10	3	2	2	1	1	2	71
5	185					32	22	9	7	2	0	1	3	76
6	196						52	17	10	4	4	2	3	92
7	230							42	27	17	8	6	2	102
8	164								49	21	11	4	10	95
9	160									33	16	11	9	69
10	122										28	15	12	55
11	117											26	18	44
12	118												35	35
m_{fj}		10	39	56	54	66	97	75	101	80	74	70	95	
T_{fj}		24	97	129	144	166	192	197	217	185	160	130	95	

i	R_{mi}	m_{mij} j=2	3	4	5	6	7	8	9	10	11	12	13	r_{mi}
1	134	19	10	6	1	0	0	0	0	1	0	0	0	37
2	156		24	8	9	3	1	0	1	1	0	1	0	48
3	173			24	2	5	4	0	0	0	0	0	0	35
4	190				23	6	1	4	0	0	1	0	0	35
5	171					15	7	2	3	2	0	0	0	29
6	140						17	5	2	0	0	0	1	25
7	93							4	1	3	1	0	0	9
8	84								6	2	1	0	0	9
9	48									3	2	1	0	6
10	24										3	1	1	5
11	24											0	0	0
12	17												0	0
m_{mj}		19	34	38	35	29	30	15	13	12	8	3	2	
T_{mj}		37	66	67	64	58	54	33	27	20	13	5	2	

Starting values (lines 350–380 in Fig. 1) must be supplied. The BOUNDS statement was inactivated in this run; a BOUNDS statement is optional. Statements 440–490 compute the current values of λ and τ. Lines 550–560 set up the negative log–likelihood function. When weights are used to get a weighted residual sum of squares for the model, SAS then computes a weighted loss function, $\Sigma(\_LOSS\_*\_WEIGHT\_)$; summation is over the observations in the data set. Thus, it is necessary in line 560 to define $\_LOSS\_$ to SAS as the negative log–likelihood divided by the weight. Also, it is advisable to add code to guard against the case of $r_i = 0$ or $m_j = 0$ which would otherwise result in an infinite weight and a zero argument to a log function. However, in Fig. 1, I want to show the bare minimum code required.

```
00010   DATA;
00050      INPUT K INDEX$ I SIZE COUNT;
00060      RATIO=COUNT/SIZE;
00070   CARDS;
00080   13 R  1  54 24
 • • •    • • •  • • •        (data for ri given Ri , i=2, ..., 11 not shown)
00190   13 R 12 118 35
00200   13 T  2  24 10
 • • •    • • •  • • •        (data for mj given Tj , j=3, ..., 11 not shown)
00300   13 T 12 130 70
00320   PROC NLIN METHOD=DUD SIGSQ=1.;
00330   ARRAY S{12} S1-S12;
00340   ARRAY P{13} P1-P13;
00350   PARAMETERS S1=.8  S2=.8  S3=.8   S4 =.8  S5 =.8  S6 =.8
00360              S7=.8  S8=.8  S9=.8   S10=.8  S11=.8  S12=.8
00370              P2=.25 P3=.25 P4=.25  P5 =.25 P6 =.25 P7 =.25
00380              P8=.25 P9=.25 P10=.25 P11=.25 P12=.25;
00390   * BOUNDS   0<S1<1, 0<S2<1, 0<S3 <1, 0<S4 <1, 0<S5 <1, 0<S6 <1,
00400              0<S7<1, 0<S8<1, 0<S9 <1, 0<S10<1, 0<S11<1, 0<S12<1,
00410              0<P2<1, 0<P3<1, 0<P4 <1, 0<P5 <1, 0<P6 <1, 0<P7 <1,
00420              0<P8<1, 0<P9<1, 0<P10<1, 0<P11<1, 0<P12<1;
00430
00440   P1=0; P13=1.0;  LAMBDA=0.0;
00450   DO J=K-1 TO I BY -1;
00460      LAMBDA=S{J}*(P{J+1}+((1-P{J+1})*LAMBDA));
00470   END;
00480   IF INDEX='R' THEN ERATIO=LAMBDA;
00490      ELSE ERATIO=P{I}/(P{I}+((1.-P{I})*LAMBDA));
00500
00510   MODEL RATIO = ERATIO;
00520
00530   _WEIGHT_=SIZE/(ERATIO*(1.-ERATIO));
00540
00550   TEMP=(COUNT*LOG(ERATIO))+((SIZE-COUNT)*LOG(1-ERATIO));
00560   _LOSS_=-TEMP/_WEIGHT_;
```

Fig. 1. The minimum SAS code needed to compute the maximum likelihood estimates of ϕ and p for the Jolly–Seber model; the input data used (not all lines of it are shown) are the minimal sufficient statistics for the female black–kneed capsids (Muir 1957, Jolly 1965) under the Jolly–Seber model; line numbers and blank lines were added to aid in presentation and discussion of the code.

A critical point not so far mentioned is that there is in PROC NLIN a solution method that does not require analytical partial derivatives: method DUD (Ralston & Jennrich 1978). If analytical partial derivatives (of the model structure) were needed with respect to the parameters, this analysis method would not be practical. This is because it can be very difficult, and time consuming, to get the correct derivatives and to code them. With this approach, using SAS one can try different models in the course of a few minutes.

The unconstrained MLE of ϕ_2 is 1.045 (se = 0.111). Re—running the job with the bounds statement active (just remove the leading "*" on line 390, Fig. 1) gives, as expected, S2 = 1.0. Also affected are S1, S3, P2 and P3, but only slightly, e.g. S3 becomes 0.895, compared to 0.874 without the constraint.

NON—LINEAR LEAST SQUARES SUMMARY STATISTICS,   DEPENDENT VARIABLE: RATIO

SOURCE	DF	WEIGHTED SS	WEIGHTED MS
REGRESSION	23	2841.4297359	123.5404233
RESIDUAL	0	0.0000000	0.0000000
UNCORRECTED TOTAL	23	2841.4297359	
(CORRECTED TOTAL)	22	63.4524645	
SUM OF LOSS		2368.8824582	(= - log-likelihood)

PARAMETER	ESTIMATE	ASYMPTOTIC STD. ERROR
S1	0.631860776	0.10628284876
S2	1.044788956	0.11099508876
S3	0.874287849	0.10806684315
S4	0.666549961	0.08178264958
S5	0.690566732	0.07490871725
S6	0.760837669	0.07271115787
S7	0.64295461	0.05809526318
S8	0.987653795	0.09616624802
S9	0.729416467	0.08765730567
S10	0.852843232	0.11839846444
S11	0.787677281	0.13115340975
S12	0.296610169	0.04204845639
P2	0.293079096	0.08705161056
P3	0.223360490	0.03915847541
P4	0.212370473	0.03374631217
P5	0.197745013	0.03067532532
P6	0.236522281	0.03175587944
P7	0.311680947	0.03479260099
P8	0.262595364	0.03120009381
P9	0.272983664	0.03250080282
P10	0.255665304	0.03405035730
P11	0.244481160	0.03662964003
P12	0.257082895	0.04316148561

NOTE: STANDARD ERRORS COMPUTED USING SIGSQ = 1

Fig. 2. Key output from the SAS code in Fig. 1; the iterative procedure to find the estimates converged; the estimates and their standard errors are identical to results produced with the closed—form formulae for the maximum likelihood results under the Jolly—Seber model.

Restricted Models for the Female Capsids.

To fit a model with a constant capture probability requires only a few changes to the code in Fig. 1. Replace P2 through P12 in the PARAMETERS statement with the single PCON, for the constant capture probability, p. Similarly change the bounds statement, if active bounds are desired, to reference only PCON as $0 < PCON < 1$. Next, add lines 434–436 as shown in Fig. 3

```
 •••
 00350  PARAMETERS S1=.8  S2=.8  S3=.8   S4 =.8  S5 =.8  S6 =.8
 00360             S7=.8  S8=.8  S9=.8   S10=.8  S11=.8  S12=.8
>00370             PCON=.2;
>00380
 00390  * BOUNDS   0<S1<1, 0<S2<1, 0<S3 <1, 0<S4 <1, 0<S5 <1, 0<S6 <1,
 00400             0<S7<1, 0<S8<1, 0<S9 <1, 0<S10<1, 0<S11<1, 0<S12<1,
>00410             0<PCON<1;
>00420
 00430
>00434  DO J=2 TO 13;
>00435      P{J}=PCON;
>00436  END;
 00440  P1=0;           LAMBDA=0.0;
 •••
```

SOURCE	DF	WEIGHTED SS	WEIGHTED MS
REGRESSION	13	2828.1104845	217.5469603
RESIDUAL	10	8.7620265	0.8762027
UNCORRECTED TOTAL	23	2836.8725110	
(CORRECTED TOTAL)	22	63.2371756	
SUM OF LOSS		2373.2690762	(= - log-likelihood)

PARAMETER	ESTIMATE	ASYMPTOTIC STD. ERROR
S1	0.650706501	0.10328666951
S2	0.996329491	0.08522365006
S3	0.830993926	0.07626432419
S4	0.637195561	0.05892305338
S5	0.718919377	0.06085674848
S6	0.847713571	0.06603788121
S7	0.610036053	0.04828335586
S8	1.018955533	0.07688920410
S9	0.709766800	0.06446824831
S10	0.834852186	0.08137611661
S11	0.809092992	0.08753249145
S12	1.167308533	0.13152248360
PCON	0.250936690	0.01086588653

Fig. 3. Altered and added code (lines indicated by ">") to change from the Jolly–Seber model code in Fig. 1 to a model with constant capture rate PCON; also shown is a key part of the SAS output from running this model (parameter estimates were not bounded).

to set each P{I} = PCON (in this SAS code, PI and P{I} are the same variable). Finally, remove the P13 = 1.0 in line 440. Under the constant p_i model, ϕ_{k-1} and p_k are not confounded so the identifiability constraint P13 = 1.0 is not needed. The treatment of P1 remains arbitrary here although logically PCON would be assumed to also estimate p_1. (P1 shows up in the SAS code only because I have an array for the p_i which is indexed from 1 to 13).

In the analysis of variance table in Fig. 3 the weighted residual SS is asymptotically distributed as a central chi–square (10 df) if the true model is Jolly–Seber with constant p_i. Thus, a quick test of this hypothesis is provided by noting that 8.8 is not at all significantly large for a chi–square on 10 df. In my opinion a better test is the likelihood ratio test; here that test statistic is 2(2373.27 − 2368.88) = 8.78 (10 df). The reduced model for female capsids fits quite well compared to the full Jolly–Seber model.

Let t_i be the time interval from occasion i to i + 1. To get a model with time–constant survival rate set $\phi_i = \phi^{t_i}$. Here ϕ will be a daily survival rate. Fig. 4 shows the changes made to the code in Fig. 1 to get a fully–reduced model, i.e., constant p and ϕ. The t_i for the capsids study are implicit in Fig. 4. From Fig. 4 this constant p and ϕ model does not fit the female capsids data very well compared to the saturated Jolly–Seber model (e.g., a weighted residual SS of 61.9 on 21 df, whereas the 1% critical level for a central chi–square on 21 df is 38.9). The

```
>00350   PARAMETERS SDAY=.8 PCON=.2;
>00390   * BOUNDS    0<SDAY<1, 0<PCON<1;
>00431   S1=SDAY**3.5;  S2=SDAY**3.0;   S3=SDAY**4.0;   S4=SDAY**3.0;
>00432   S5=SDAY**4.0;  S6=SDAY**3.0;   S7=SDAY**3.5;   S8=SDAY**3.5;
>00433   S9=SDAY**3.5;  S10=SDAY**3.0;  S11=SDAY**4.0;  S12=SDAY**3.0;
>00434   DO J=2 TO 13;
>00435       P{J}=PCON;
>00436   END;
>00440   P1=0;            LAMBDA=0.0;
```

SOURCE	DF	WEIGHTED SS	WEIGHTED MS
REGRESSION	2	2779.5777020	1389.7888510
RESIDUAL	21	61.8774339	2.9465445
UNCORRECTED TOTAL	23	2841.4551359	
(CORRECTED TOTAL)	22	65.6963314	
SUM OF LOSS		2399.5911136	(= - log-likelihood)

PARAMETER	ESTIMATE	ASYMPTOTIC STD. ERROR
SDAY	0.9313858542	0.00349320369
PCON	0.2574917154	0.01109920572

Fig. 4. To get a fully reduced "Jolly–Seber model," i.e., constant p (PCON) and constant daily survival rate (SDAY), replace lines 350–440 in Fig. 1 with the SAS code shown here; also shown here are key parts of the SAS output for this fully reduced model.

likelihood ratio test of this 2 parameter model vs. the 13 parameter model of Fig. 3 is 2(2399.6 −2373.9) = 51.4 (11 df). The model with a constant p and varying daily survival rate is acceptable; a constant p and ϕ model does not fit the data.

The Male Capsids Data.

From Table 3 we see that the male capsids data are sparse for cohorts 10–12. The lack of recaptures for cohorts 11 and 12 ($r_{11} = r_{12} = 0.0$) renders p_{11}, p_{12} and ϕ_{11} nonestimable (division by zero occurs in an attempt to estimate them). However, there are still 23 estimable parameters in the saturated Jolly–Seber model of the male capsids JS_MSS; we may take them to be ϕ_1, ..., ϕ_{10}, λ_{11}, λ_{12}, p_2, ..., p_{10}, τ_{11} and τ_{12}. The SAS code to numerically fit this model is in Fig. 5. Note the parameterization as S1, ..., S10, LAM11, LAM12, P2, ..., P10, TAU11 and TAU12. The negative log–likelihood and estimated parameters are in Fig. 6. Sub–models can be obtained by adding the necessary code and altering the PARAMETERS and BOUNDS statements in Fig. 5. Alternatively, the SAS code in Fig. 1 can be used as the starting point for suitable constrained models because, e.g., if p is constant then ϕ_1 through ϕ_{12} are all mathematically estimable. For this latter model applied to the male data I get CPON = 0.158 (se = 0.0168) and the log–likelihood = −898.552. The likelihood ratio test of the p–constant model vs the p–varying (saturated) model gives a chi–square = 15.7 (10 df, P = 0.108).

If there is a reason to suspect a trend in capture rates p_2 to p_9 a logit–linear model can be used. To model the log–odds of the p_i as linear on occasion (this is reasonable only if occasions are equally spaced in time) replace P2 to P9 in the PARAMETERS statement by A = −1 B = 0. Then add to Fig. 5 the lines

```
00451  DO J=2 TO 9;
00452    TEMP=A+(B*(J–2));
00453    P{J}=1/(1+EXP(–TEMP));
00454  END;
```

Convergence occurs easily with this model; log–likelihood = −981.311, 16 parameters. This further illustrates the ease of modeling using PROC NLIN.

Joint Analysis of Female and Male Data.

The flexibility and power of this analysis method really becomes evident when exploring parsimonious models for 2, or more, related data sets. One such model will be given by way of illustration for the male and female capsids data jointly. From separate analyses, constant capture rates over time (within sex) seem tenable. It is also clear that the survival rates should be allowed to vary by sex and time. However perhaps female and male survival rates are proportional. The simplest such model is to have $\phi_{mi} = \phi_* \phi_{fi}$. Fig. 7 shows SAS code to fit this model to the male and female data (wherin θ is denoted SRMDF for "Survival Rate of Male Divided by Female"). The model is set—up by lines 620 through 720 in Fig. 7. Fig. 8 shows output from this model fitting.

```
00010    DATA;
00050       INPUT K INDEX$ I SIZE COUNT;
00060       RATIO=COUNT/SIZE;
00080    CARDS;
00090    13 R   1    134    37
...      . . . .   .
00310    13 T 12      5    3
00330    PROC NLIN METHOD=DUD SIGSQ=1.;
00340    ARRAY S{12} S1-S12;
00350    ARRAY P{13} P1-P13;
00360    PARAMETERS  S1=.5   S2=.5   S3=.5   S4=.5   S5=.5   S6=.5   S7=.5
00370                S8=.5   S9=.5  S10=.5  LAM11=.0  LAM12=.0
00380                P2=.2   P3=.2   P4=.2   P5=.2   P6=.2   P7=.2
00390                P8=.2   P9=.2  P10=.2  TAU11=.6  TAU12=.6;
00400    BOUNDS      0<S1<1,  0<S2<1,  0<S3<1,  0<S4<1,  0<S5 <1,
00410                0<S6<1,  0<S7<1,  0<S8<1,  0<S9<1,  0<S10<1,
00420                0<P2<1,  0<P3<1,  0<P4<1,  0<P5<1,  0<P6<1,  0<P7<1,
00430                0<P8<1,  0<P9<1,  0<P10<1,  0<TAU11<1,  0<TAU12<1;
00440
00450    P1=0;  P11=1.0;  P12=1.0;  P13=1.0;  S11=0.0;  S12=0.0;
00460    LAMBDA=0.0;
00470    DO J=K-1 TO I BY -1;
00480       LAMBDA=S{J}*(P{J+1}+((1-P{J+1})*LAMBDA));
00490    END;
00500    IF INDEX='R' THEN DO;
00510       ERATIO=LAMBDA;
00520       IF I=11 THEN ERATIO=LAM11;
00530       IF I=12 THEN ERATIO=LAM12;
00540       END;
00550    ELSE DO;
00560       ERATIO=P{I}/(P{I}+((1-P{I})*LAMBDA));
00570       IF I=11 THEN ERATIO=TAU11;
00580       IF I=12 THEN ERATIO=TAU12;
00590       END;
00600
00610    MODEL RATIO = ERATIO;
00620
00630    IF ERATIO>0 AND ERATIO<1 THEN
00640        _WEIGHT_=SIZE/(ERATIO*(1.-ERATIO));
00650        ELSE _WEIGHT_=1;
00660
00670    IF ERATIO>0 THEN AA=LOG(ERATIO);     ELSE AA=0;
00680    IF ERATIO<1 THEN BB=LOG(1.-ERATIO); ELSE BB=0;
00690    TEMP=(COUNT*AA)+((SIZE-COUNT)*BB);
00700    _LOSS_=-TEMP/_WEIGHT_;
```

Fig. 5.   SAS code used to get the maximum likelihood estimates of ϕ and p for the Jolly—Seber model with the male capsids data; the input data are the minimal sufficient statistics (JS_MSS).

The fit of this joint model is marginal. From the Jolly–Seber models separate by sex, we get the log–likelihood of –3259.57 (= –2368.88 – 890.69) for fitting a saturated joint model. From Fig. 7 the log–likelihood for the "proportionality" model is –3283.79. The likelihood ratio statistic is 48.44 on 31 df (P =0.024).

Discussion

Complete flexibility is now possible in fitting models to capture–recapture data. A theory which makes this possible is the modeling of the data using conditional (on release) multinomial distributions, thereby getting a simple, explicit and standard type of likelihood for the survival and capture parameters. The practical, comprehensive analysis of capture–recapture data is made

SOURCE	DF	WEIGHTED SS	WEIGHTED MS
REGRESSION	23	816.38777618	35.49512070
RESIDUAL	0	0.00000000	0.00000000
UNCORRECTED TOTAL	23	816.38777618	
(CORRECTED TOTAL)	22	218.42040425	
SUM OF LOSS		890.69397659	(= – log-likelihood)

PARAMETER	ESTIMATE	ASYMPTOTIC STD. ERROR
S1	0.5783582091	0.10908135019
S2	0.8959040961	0.17193372200
S3	0.5901130189	0.12704913011
S4	0.5929276329	0.13521889098
S5	0.5596491265	0.13785027199
S6	0.9193121581	0.32861992568
S7	0.5366569128	0.24061628362
S8	0.4960317468	0.24170144061
S9	0.3150000003	0.16217537204
S10	0.2083333333	0.08289816935
LAM11	0.0000000000	0.00000000111
LAM12	0.0000000000	0.00000000733
P2	0.2451612902	0.06481033300
P3	0.1769253642	0.04209747325
P4	0.1944444443	0.04544179148
P5	0.1699029118	0.04272007267
P6	0.1515151490	0.04103969646
P7	0.1079136717	0.04032621559
P8	0.0819672042	0.03550586683
P9	0.1039999996	0.05055056786
P10	0.2380952376	0.10988046879
TAU11	0.6153846154	0.13493200297
TAU12	0.6000000000	0.21908902300

Fig. 6. Key SAS output from fitting the saturated Jolly–Seber model to the male capsids data using the SAS code in Fig. 5.

possible by the sophisticated computing power now available, both in terms of hardware and software. In particular, excellent derivative–free optimization algorithms are the key to the software power, especially when packaged in a product like SAS. Implementing MLE via iteratively reweighted nonlinear least squares is just a device for using SAS to get the MLEs; it

```
00010  DATA;
00050     INPUT SEX$ K INDEX$ I SIZE COUNT;
00060     RATIO=COUNT/SIZE;
00070  CARDS;
00080  F 13 R 1    54  24
  •••    •  •  •  •      •
00520  M 13 T 12    5   3
00530  PROC NLIN METHOD=DUD SIGSQ=1.;
00540  ARRAY S{12} S1-S12;
00550  ARRAY P{13} P1-P13;
00560  PARAMETERS S1=.8 S2=.8 S3=.8  S4=.8  S5=.8  S6=.8
00570             S7=.8 S8=.8 S9=.8 S10=.8 S11=.8 S12=.8
00580             PMALE=.15 PFEMALE=.25 SRMDF=.75;
00590  BOUNDS     0<S1, 0<S2, 0<S3, 0<S4, 0<S5, 0<S6, 0<S7, 0<S8, 0<S9,
00600             0<S10, 0<S11, 0<S12, 0<PMALE, 0<PFEMALE, 0<SRMDF;
00610
00620  IF SEX='M' THEN PC=PMALE;   ELSE PC=PFEMALE;
00630  DO J=1 TO 13;
00640     P{J}=PC;
00650  END;
00660  IF SEX='M' THEN FAC=SRMDF;   ELSE FAC=1.;
00670  LAMBDA=0.0;
00680  DO J=K-1 TO I BY -1;
00690     LAMBDA=(S{J}*FAC)*(P{J+1}+((1-P{J+1})*LAMBDA));
00700  END;
00710  IF INDEX='R' THEN ERATIO=LAMBDA;
00720     ELSE ERATIO=P{I}/(P{I}+((1.-P{I})*LAMBDA));
00730
00740  MODEL RATIO = ERATIO;
00750
00760  IF ERATIO>0 AND ERATIO<1 THEN
00770     _WEIGHT_=SIZE/(ERATIO*(1.-ERATIO));
00780     ELSE _WEIGHT_=1.;
00790
00800  IF ERATIO>0 THEN AA=LOG(ERATIO);      ELSE AA=0.0;
00810  IF ERATIO<1 THEN BB=LOG(1.-ERATIO);  ELSE BB=0.0;
00820  TEMP=(COUNT*AA)+((SIZE-COUNT)*BB);
00830  _LOSS_=-TEMP/_WEIGHT_;
00840
00850  OUTPUT OUT=TWO P=ERATIO R=RESIDS;
00860
00870  DATA; SET TWO;      /* COMPUTE SQUARED, NORMALIZED RESIDUALS */
00880     CHISQ=RATIO - ERATIO;
00890     CHISQ=CHISQ*CHISQ*SIZE/(ERATIO*(1-ERATIO));
00900  PROC PRINT;
```

Fig. 7. SAS code used to get the maximum likelihood estimates of ϕ and p under the proportionality model for the joint analysis of the female and male capsids data; input data are the separate minimal sufficient statistics (JS_MSS) for female and males.

would be more convenient to have an equally flexible package directly computing likelihood inference.

The examples only touch the surface of what is possible. As suggested by Clobert et al. (1987) we could reparameterize on logits. For example, define, to SAS, the "survival" parameters as, say, LS and then set $S\{I\} = 1/(1 + EXP(-LS\{I\}))$ before computing ERATIO. Such logits (LS) may be more normally distributed than the p_i and $\hat{\phi}_i$. Good confidence intervals remain a concern in capture–recapture (Manly 1984). If the direct logit transform does not solve the problem (I do not think it will), compute likelihood intervals. With current software it takes some effort to get the likelihood intervals, but it can be done and seems to be one of the few justifiable methods when parameters are bounded and the MLE is on a boundary.

The theory and methods here extend easily to capture–recapture models more general than Jolly–Seber and to other cohort survival processes such as bird banding models (Brownie et al. 1985). A first extension of Jolly–Seber is to allow a one–period ahead effect, on either p and/or ϕ, after a release. The model of Sandland & Kirkwood (1981) is such a generalization wherein after a known release at time i, the capture probability is $p_{a,i+1}$ at time i + 1, and after that (give no capture at time i + 1) it is $p_{b,j}$, j > i + 1. The Sandland & Kirkwood model can be implemented in SAS. Another generalization is to allow age–specific effects if there are identifiable age classes. Maximum likelihood estimation with SAS can be done for the models of Pollock 1981a. The control–treatment release–recapture models in Burnham et al. (1987) can be implemented in SAS. In these examples, the advantage of using the approach given here would be the ability to explore more restricted models, which do not have closed–form MLEs.

An integrated model and analysis that includes ϕ, p and population size, N (or recruitment, B) is possible. The theory exists because the distribution of the u_i is known. However, that distribution of initial captures involves binomials where the sample size is itself a parameter. So far it has not been possible to put that part of the problem into PROC NLIN. This entails no loss of statistical efficiency unless one wants to place restrictions (models) on the recruitment process, such as in Crosbie & Manly (1985).

The method outlined here is the last step in a thorough analysis of a capture–recapture data set. First one should use programs like POPAN (Arnason & Baniuk 1980), RELEASE (Burnham et al. 1987), and JOLLY or JOLLYAGE (Brownie et al. 1986, Pollock et al. in prep) to do extensive goodness of fit testing and comparisons between the standard models fit by these programs. Once the search for a parsimonius and meaningful model has focused on a class of models (such as to the time–specific Jolly–Seber model), then intensive analysis of customized models can begin. Then PROC NLIN in SAS, as illustrated here, can be very useful.

SOURCE	DF	WEIGHTED SS	WEIGHTED MS
REGRESSION	15	3630.4138029	242.0275869
RESIDUAL	31	40.8056670	1.3163118
UNCORRECTED TOTAL	46	3671.2194699	
(CORRECTED TOTAL)	45	502.4760777	
SUM OF LOSS		3283.7880001	(= - log-likelihood)

PARAMETER	ESTIMATE	ASYMPTOTIC STD. ERROR
S1	0.765409878	0.08313774439
S2	1.030986631	0.07375562610
S3	0.832016498	0.06484330243
S4	0.664863549	0.05216473620
S5	0.712054069	0.05354383095
S6	0.856763266	0.06125679050
S7	0.605919798	0.04548736195
S8	0.971788533	0.07106438045
S9	0.711756066	0.06160583043
S10	0.852055508	0.08010304218
S11	0.782686338	0.08296160196
S12	1.132987942	0.12626904575
PMALE	0.166778450	0.01567718261
PFEMALE	0.249997448	0.01084282574
SRMDF	0.747500908	0.02896779703

Fig. 8. Key SAS output from fitting the proportionality model to the female and male capsids data; unconstrained maximum likelihood estimates are shown for S1–S12, the survival rates of the female capsids, SRMDF, the ratio of male to female survival rates (assumed here to be constant over time), and the constant capture probabilities PMALE and PFEMALE.

Acknowledgements

I owe a debt of gratitude to Mr. R. C. Muir (deceased) since it was his study that produced the black–kneed capsids data. Those data were made available to me through the efforts of Mr. G. M. Jolly, who kindly requested them for me from the archives of East Malling Research Station, England, and Dr. J. Flegg who actually retrieved the data and sent them to Mr. Jolly. Thanks also to Mr. M. G. Solomon of East Malling for permission to use the data. The data were made available as original field forms; I thank Ms. B. Knopf for coding the data into usable computerized form. Review comments from Drs. D. R. Anderson, C. Brownie, J. D. Nichols, C. J. Schwarz and G. C. White were helpful in revising this paper; their comments and suggestions are appreciated.

References cited

Arnason, A. N., & L. Baniuk. 1980. A computer system for mark–recapture analysis of open populations. *J. Wildl. Mgt.* 44: 325–332

Brownie, C., D. R. Anderson, K. P. Burnham, & D. S. Robson. 1985. Statistical inference from band recovery data – a handbook, 2nd edition. U.S. Fish and Wildlife Service Resource Publication 156.

Brownie, C., J. E. Hines, & J. D. Nichols. 1986. Constant parameter capture–recapture models. *Biometrics* 42: 561–574.

Brownie, C., & D. S. Robson. 1983. Estimation of time–specific survival rates from tag–resighting samples: a generalization of the Jolly–Seber model. *Biometrics* 39: 437–453.

Buckland, S. T. 1980. A modified analysis of the Jolly–Seber capture–recapture model. *Biometrics* 36: 419–435.

Burnham, K. P., D. R. Anderson, G. C. White, C. Brownie, & K. H. Pollock. 1987. Design and analysis methods for fish survival experiments based on capture–recapture. Monograph 5, American Fisheries Society, Bethesda, Maryland.

Clobert, J., J. D. Lebreton, & D. Allaine. 1987. A general approach to survival rate estimation by recaptures or resightings of marked birds. *Ardea* 75: 113–142.

Cormack, R. M. 1964. Estimates of survival from sighting of marked animals. *Biometrika* 51: 429–438.

Cormack, R. M. 1981. Loglinear models for capture–recapture experiments on open populations. Pages 217–235. In R. W. Hiorns and D. Cooke [eds.]. The mathematical theory of the dynamics of biological populations. Academic Press, London.

Crosbie, S. F., & B. F. J. Manly. 1985. A new approach for parsimonious modeling of capture–recapture studies. *Biometrics* 41: 385–398.

Green, P. J. 1984. Iteratively reweighted least squares maximum likelihood estimation and some robust and resistant alternatives. *J. Royal Statist. Soc.*, Series B 46: 149–192.

Jennrich, R. I., & R. H. Moore. 1975. Maximum likelihood estimation by means of nonlinear least squares. Pages 57–65 in American Statistical Association 1975 Proceedings of the Statistical Computing Section. American Statistical Association, Washington, D. C.

Jolly, G. M. 1965. Explicit estimates from capture–recapture data with both death and immigration – stochastic models. *Biometrika* 52: 225–247.

Jolly, G. M. 1982. Mark–recapture models with parameters constant in time. *Biometrics* 38: 301–321.

Manly, B. F. J. 1984. Obtaining confidence limits on parameters of the Jolly–Seber model for capture–recapture data. *Biometrics* 40: 749–758.

Muir, R. C. 1957. On the application of the capture–recapture method to an orchard population of *Blepharidopterus angulatatus* (Fall.) (*Hemiptera–Heteroptera, Miridae*). East Malling Annual Report 1957: 140–147.

Pollock, K. H. 1975. A K–sample tag–recapture model allowing for unequal survival and catchability. *Biometrika* 62: 577–583.

Pollock, K. H.   1981a.   Capture–recapture models allowing for age–dependent survival and capture rates. *Biometrics* 37: 521–529.

Pollock, K. H.   1981b.   Capture–recapture models: a review of current methods, assumptions, and experimental design. Studies in Avian Biology 6: 426–435.

Pollock, K. H., J. E. Hines, & J. D. Nichols.   1985.   Goodness–of–fit tests for open capture–recapture models. *Biometrics* 41: 399–410.

Pollock, K. H., J. D. Nichols, C. Brownie, & J. E. Hines.   (in prep).   Statistical inference for capture–recapture experiments. Wildlife Monographs, The Wildlife Society, Washington, D. C.

Ralston, M. L., & R. I. Jennrich.   1978.   Dud, a derivative–free algorithm for nonlinear least squares. *Technometrics* 20: 7–14.

Robson, D. S.   1969.   Mark–recapture methods of population estimation. Pages 129–140 in N. L. Johnson and H. Smith, Jr. [eds.].   New developments in survey sampling.   Wiley Interscience, New York.

SAS Institute Inc. 1985. SAS user's guide: statistics, version 5 edition. SAS Institute Inc., Cary, NC 27511.

Sandland, R. L., & P. Kirkwood.   1981.   Estimation of survival in marked populations with possibly dependent sighting probabilities. *Biometrika* 68: 531–541.

Seber, G. A. F.   1965.   A note on the multiple–recapture census. *Biometrika* 52: 249–259.

White, G. C.   1983.   Numerical estimation of survival rates from band–recovery and biotelemetry data. *J. Wildlife Mgt.* 47: 716–728.

Sampling Forest Canopy Arthropods
Available to Birds as Prey

Robert J. Cooper[1]

ABSTRACT A method of sampling canopy arthropods using direct observation is presented whereby the abundance of different types of arthropods (flying, foliage dwelling, etc.) can be directly compared because the same sampling unit (numbers/leaf area) is used. Total leaf surface area in patches of directly observable foliage was determined by using leaf length/surface area equations. Estimates of relative abundance of different arthropod taxa were used to compare with percent composition of those taxa in diets of canopy foraging bird species. A total of 6,802 arthropods was identified in 1404 samples containing 197,964 leaves. Examples of arthropod abundance estimation and assessment of use versus availability of arthropod prey are provided from one month (May, 1986) of the study.

Entomologists and ornithologists frequently have different objectives when conducting field studies that involve arthropod sampling. Although many entomological field studies involve insect communities, many more involve a single species or family of insects (often pest species). Common sampling methods include sticky traps, Malaise traps, pitfall traps, sweep netting, pole pruning, shake cloth sampling, and use of suction traps (Southwood 1978, Cooper & Whitmore, in press). Most of these methods are effective at sampling some but not all types of arthropods. Pole pruning, for example, is an effective way to sample canopy dwelling caterpillars, but is unacceptable for sampling more active insects. A variety of traps exist that sample flying insects effectively, but are unsuitable for more sedentary arthropods like caterpillars. If the study involves a single target organism, however, this is not usually a problem.

Ornithologists may also use the above methods if the objective is to sample a single type of arthropod. More frequently, however, ornithologists are interested in the numbers and types of arthropods available to birds as prey. Many studies are also designed to simultaneously examine use of those prey through diet or foraging behavior analysis (e.g., Tinberegn 1960, Royama 1970, Bryant 1973, Davies 1977, Busby & Sealy 1979, Biermann & Sealy 1982). This presents a sampling problem because different types of prey that individual birds eat (e.g. flying, foliage dwelling, etc.) must be sampled in such a way that their abundances can be directly compared. Most of the above methods are effective at sampling only a portion of the total arthropod fauna available to birds. In addition, the sampling unit used with trapping methods often has no unit of area, weight, or volume associated with it. Comparison of one or more methods involving different sampling units is therefore difficult.

This problem is especially pronounced in forest ecosystems, where a variety of substrates and canopy layers exist and are occupied by different kinds of arthropods. The purpose of this study is

---

[1]Forest Arthropod Research Team, Division of Forestry, Morgantown, WV 26506–6125.

to present sampling and estimation procedures that allow the researcher to sample the different kinds of arthropods available to canopy foraging, foliage gleaning birds of an eastern deciduous forest in such a way that their abundances are directly comparable. This is accomplished by an intermediate step of sampling units of substrate (leaves) in which case the population sampled is the population of leaves. The numbers and types of arthropods present on these units are then counted on the selected units and can be compared directly. These arthropods are then the sample from the population of arthropods. Counting is accomplished by means of direct observation in the forest canopy.

Methods

This study was conducted on Sleepy Creek Public Hunting and Fishing Area (P.H.F.A.) near Berkeley Springs (Morgan Co.), West Virginia. The area is characterized by 40 yr–old oak–hickory forest with sparse understory. Three 60–ha square plots were established in which birds were collected for diet analysis from 13 of May – 22 of July, 1986. Arthropod sampling was performed simultaneously (0600–1200) each day on the same plots.

The locations and relative abundances of different taxa and sizes of arthropods available to forest birds were estimated by direct observation of arthropods in the forest canopy. Randomly selected stout (>15 in dbh) trees were ascended using tree climbing gear. Patches of directly observable foliage (usually branches with 100–200 leaves within arm's reach) were examined for arthropods with great care not to flush or dislodge any individuals. Eventually this involved examining each leaf in the patch individually, including refugia created by leaf rolls, dead leaves, or 2 leaves sandwiched or tied together. This procedure is described in greater detail by Cooper (1988). When patch examination was completed, the number of leaves in the patch was recorded. Leaf lengths were measured for a randomly selected subsample of leaves from the patch. Other pertinent information, such as exact arthropod location (i.e., leaf tops or undersides, branches, etc.), patch height, and tree species, was also recorded. All data were recorded on a portable tape recorder. Arthropods were identified as precisely as possible taxonomically. Level of identification varied between taxa. Generally, arthropods were identified to at least the level of order. An extensive collection of voucher specimens was established that was used for identifying unknown specimens.

Most patches were surveyed at heights of 3–12 m. Thus, the lower canopy was surveyed but not the upper canopy where most foliage gleaning species obtain food. The assumption is therefore made that the distribution of arthropods (at higher taxonomic levels, i.e. orders) is similar in the lower and upper canopy. This assumption is supported by the observations of other reseachers (Holmes et al. 1979, Mason & Loye 1981).

Data were later transcribed from tapes and entered on computer. Arthropod densities were expressed as number of individuals/leaf surface area. Surface areas were determined by first deriving leaf length/surface area equations for each tree species by measuring leaf lengths and surface areas with a leaf area meter for a random sample of approximately 100 leaves. Next, total leaf surface area for a particular patch was determined by entering the average leaf length for the patch sample into the regression equation for that tree species and multiplying the resulting single leaf surface area estimate by the number of leaves in the sample.

Relative abundances of different arthropod taxa were estimated by first averaging the density of different arthropod types for all patches surveyed for a particular tree species. Average densities were then multiplied by the basal area (cross sectional area of the tree bole at breast height) of that tree species in the study area. Here, basal area is assumed to be directly related to crown volume, which is the actual measurement of interest but was unavailable for these tree species. Relative densities of arthropod taxa for the entire study area were then determined by dividing the abundance of a particular arthropod taxon by the total abundance of all arthropods, expressed as a percent. Thus, the sampling procedure actually resembles stratified sampling where the estimates of abundance of each arthropod taxon are weighted by the basal area of each tree species (taxon) measured.

Results

A total of 6,802 arthropods was identified in 1,404 sample patches containing 197,964 leaves. Data from 13–31 of May, 1986, are presented as an example. Samples predominated from chestnut oak (*Quercus prinus*), white oak (*Q. alba*), red oaks (northern red oak, *Q. rubra*, scarlet oak, *Q. coccinea*, and black oak, *Q. velutina*), black gum (*Nyssa sylvatica*), and red maple (*Acer rubrum*) (Table 1).

Table 1. Sample sizes, total leaf area sampled, and basal area of common tree species sampled from 13–31 of May, 1986 on Sleepy Creek P.H.F.A.

Tree species	Number of samples	Total leaf area (m²)	Basal area (cm²/ha)
Black Gum	22	21.7	227
Chestnut Oak	33	28.7	1980
Hickory	4	8.9	907
Red Maple	24	28.6	424
Red Oaks	37	54.3	2880
Sassafras	8	13.6	389
White Oak	28	24.4	1007

Other species sampled were hickories (*Carya spp.*) and sassafras (*Sassafras albidum*). Surface area of substrate sampled was often larger for the oak species because of the larger leaves characteristic of those species. Oaks were also dominant in terms of basal area. Thus arthropod taxa that were common on oaks had higher relative densities (percents) compared with taxa common on less dominant tree species. For example, gypsy moth (*Lymantria dispar*) larvae were common on oak species in May 1986 (Table 2). Other Lepidoptera larvae were relatively uncommon on oaks, but one geometrid species, *Itame pustularia*, was common on red maples. The relatively large oak basal area and low red maple basal area, however, caused gypsy moth larvae to comprise 54% and other larvae to comprise only 9% of the total arthropod fauna (Table 3).

Table 2. Densities (number/m^2 of leaf area) of arthropods sampled from 13–31 of May, 1986 on Sleepy Creek P.H.F.A. Tree species codes are as follows: BGUM = black gum; COAK = chestnut oak; HICK = hickory; RMAP = red maple; ROAK = red oaks; SASS = sassafras; WOAK = white oak.

	Tree species						
Taxon	BGUM	COAK	HICK	RMAP	ROAK	SASS	WOAK
Arachnida	34.7	7.2	24.4	1.3	10.0	0.0	37.7
Orthoptera	3.6	3.1	0.0	6.5	0.0	4.6	0.0
Hemiptera	11.3	2.8	0.0	15.3	5.2	25.5	30.6
Homoptera	0.0	43.3	0.0	41.3	13.4	0.0	13.7
Coleoptera	30.4	38.4	0.0	45.2	19.0	42.1	22.1
Lepidoptera adults	189.4	21.2	9.8	18.2	20.1	48.7	178.7
L. dispar larvae	42.8	479.2	34.2	101.7	455.3	186.3	673.2
Other larvae	22.9	59.1	31.8	330.2	48.8	39.4	79.1
Diptera	32.3	18.8	105.0	72.8	44.2	53.5	25.9
Formicidae	106.9	93.7	61.1	106.1	95.4	18.2	54.7
Other Hymenoptera	19.2	29.1	53.7	24.1	12.7	20.4	29.0
Other	2.5	2.6	0.0	0.0	1.7	0.0	4.4

Table 3. Relative abundances (%) of arthropod categories sampled from 13–31 of May, 1986 on Sleepy Creek P.H.F.A.

Prey taxon	Percent of total
L. dispar larvae	54
Hymenoptera	15
Lepidoptera larvae (other)	9
Diptera	6
Lepidoptera adults	6
Coleoptera	3
Homoptera	3
Arachnida	2
Hemiptera	1
Lepidoptera pupae	1

Sample sizes were adequate to estimate densities within 20% of the true density with 95% confidence for most arthropod taxa. Required sample sizes were mostly in the range n=15–30.

Exceptions were densities of arthropods that tended to be clumped, such as gypsy moth larvae, which required sample sizes of several hundred for the oak species. Sample sizes obtained in the less commonly sampled tree species were probably inadequate.

Relative densities of arthropod taxa were then compared with the percent of the total diet comprised by those taxa. The tufted titmouse (*Parus bicolor*), for example, ate Coleoptera, Arachnida, Homoptera, Lepidoptera pupae, and Lepidoptera larvae in proportions greater than abundance of those taxa, and ate gypsy moth larvae and Hymenoptera in proportions less than the abundance of those taxa (Fig. 1).

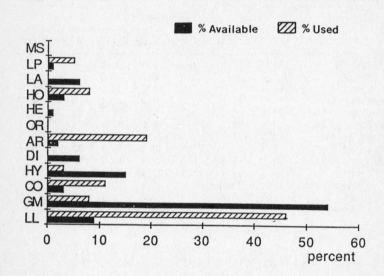

Fig. 1. Availability and use of arthropod prey categories by tufted titmice from 13–31 of May, 1986 on Sleepy Creek P.H.F.A. Prey category codes are as follows: AR = Arachnida; CO = Coleoptera; DI = Diptera; GM = Gypsy moth larvae; HE = Hemiptera; HO = Homoptera; HY = Hymenoptera; LA = Lepidoptera Adults; LL = Lepidoptera larvae; LP = Lepidoptera pupae; MS = miscellaneous prey; OR = Orthoptera.

Availability and use of arthropods were also examined another way. Arthropod sampling results were used to detect trends in abundance of dietarily important prey taxa. For example, Lepidoptera larvae (except gypsy moth larvae) were observed to decrease steadily in abundance over the 10 weeks of the study (Fig. 2). This decline was also observed using pole prune sampling (Martinat et al. 1988). The percentage of Lepidoptera larvae in diets of several bird species, such as the red–eyed vireo (*vireo divaceus*), was also observed to decline over this period (Fig. 3). Thus there is evidence that red–eyed vireos are responding dietarily to abundances of this important prey taxon.

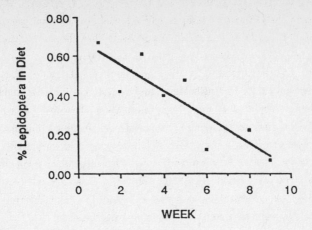

Fig. 2. Abundances of larval Lepidoptera (excluding gypsy moth larvae) from 13 May – 22 July, 1986 on Sleepy Creek P.H.F.A.

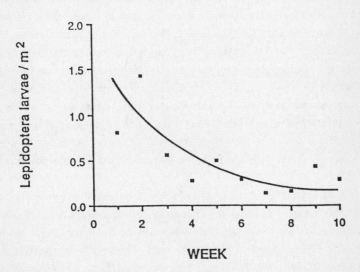

Fig. 3. Percent larval Lepidoptera in red–eyed vireo diets, 13 May – 22 July, 1986 on Sleepy Creek P.H.F.A.

Discussion

The method described in this paper permits the comparison of abundance of different types of arthropods that have different types of behaviors or that occupy different locations in the forest canopy. This method is an improvement over commonly used trapping techniques or methods such as pole pruning that effectively sample only a portion of the total canopy arthropod fauna. Another advantage of this method for ornithologists is that, because arthropods are observed directly, data are collected concerning the locations, substrates, and tree species occupied as well as foraging and escape behaviors used by different groups of arthropods. This information is not likely to be otherwise available to ornithologists and cannot usually be obtained through any other sampling technique. Yet this is important information that can contribute greatly to an understanding of avian diets and foraging behavior. For example, the information collected using this method was used to establish prey categories in a study of dietary similarity among canopy foraging birds (Cooper 1988). Prey categories were thus dependent upon insect ecology rather than somewhat arbitrary taxonomic groupings (orders). Cooper et al. (in press) showed that categorization based on ecological characteristics provided a much clearer picture of dietary patterns among species and over time than taxonomic categorization.

Despite the advantages of this method, there are some obvious shortcomings. First, arthropods occupying some substrates were more likely to be observed than those in other substrates. Flying insects, for example, had to land on or otherwise occupy the patch under examination to be counted and were therefore probably under represented in this study. Second, even among arthropods occupying the same substrate, such as leaves, there are some arthropods that are more likely to be observed than others. Ants, for example, are dark in color and highly active and are therefore easy to detect. Many larval Lepidoptera are cryptically colored and relatively sedentary, making them harder to detect. The technique is therefore biased against arthropods that are detected with lower probability. Verification counts obtained by clipping patches of foliage already sampled with this method indicate that approximately 25% of larger (>10mm) caterpillars were missed using direct observation. Virtually 100% of more active prey taxa (e.g. ants, Homoptera) were accounted for. This bias may be calibrated using verification counts (i.e. a more accurate sampling procedure) to estimate a selection function (McDonald & Manly, in 1988).

Also, as described here, the technique is highly labor intensive, physically demanding, and costly. More effort is expended in the field but no laboratory expense is involved since specimens are "collected" on tape. For this reason, entomological expertise is required to identify arthropods by sight. Identification is frequently only to the level of order. Towers provide a less strenuous alternative to tree climbing and have been used by other researchers (Greenberg & Gradwohl 1980), although these are also expensive and restrictive in terms of sample location.

An alternative method to direct observation that we have experimented with involves pesticide knockdown (Cooper & Whitmore, in press). Using a fogging machine and a pyrethroid pesticide, which is highly effective against many arthropod taxa, the forest canopy can be fogged in a systematic fashion. Dead insects fall to the ground and are collected on sampling cards or in jars at the bottom of funnels made of canvas or plastic. The percent composition of each arthropod taxon can then be computed and compared with the percent each taxon comprises in bird diets. The drawbacks of this method are that arthropod densities (number/leaf area) are difficult to compute and arthropods are not observed until they are collected, so an understanding of their location and behavior must be obtained some other way. Foggers are also expensive. A major advantage is that larger numbers of arthropods are collected (observed) than with direct observation in less time.

We have thus explored several ways to estimate the relative abundance of different arthropod taxa in forest canopies with the ultimate objective of comparing those abundances with percentages of the same taxa in bird diets. These methods each have drawbacks, but represent an improvement over existing methods for examining use versus availability of arthropod prey by birds. Improvements such as estimation of selection functions should make these methods still more effective in bringing together these 2 types of data sets.

Acknowledgments

This research was funded in part by USDA Forest Service Cooperative Agreements 23–972, 23–043, and 23–144, and a Sigma Xi grant–in–aid of research. I thank L. McDonald and K. Smith for reviewing an earlier version of the manuscript. D. Fosbroke, S. Fosbroke, B. Sample, and R. Whitmore reviewed this manuscript. K. Dodge and P. Martinat provided helpful comments. P. Martinat provided entomological expertise. Published with the approval of the Director, West Virginia University Agriculture and Forestry Experiment Station as Scientific Paper No. 2107. This is contribution No. 5 of the Forest Arthropod Research Team.

References Cited

Biermann, G. C., & S. G. Sealy. 1982. Parental feeding of nestling Yellow Warblers in relation to brood size and prey availability. *Auk* 99: 332–341.

Bryant, D. M. 1973. The factors influencing the selection of food by the House Martin (*Delichon urbica*). *J. Anim. Ecol.* 42: 539–564.

Busby, D. G., & S. G. Sealy. 1979. Feeding ecology of a population of nesting Yellow Warblers. *Can. J. Zool.* 57: 1670–1681.

Cooper, R. J. 1988. Dietary relationships among insectivorous birds of an eastern deciduous forest. PhD. dissertation, West Virginia University, Morgantown.

Cooper, R. J., P. J. Martinat, & R. C. Whitmore. In press. Dietary similarity among insectivorous birds: influence of taxonomic versus ecological categorization of prey. In M. L. Morrison, C. J. Ralph, and J. Verner [eds.], Food Exploitation by Terrestrial Birds. Studies in Avian Biology.

Cooper, R. J., & R. C. Whitmore. In press. Arthropod sampling methods in ornithology. In M. L. Morrison, C. J. Ralph, and J. Verner [eds.], Food Exploitation by Terrestrial Birds. Studies in Avian Biology.

Davies, N. B. 1977. Prey selection and the search strategy of the Spotted Flycatcher (*Muscicapa striata*): a field study on optimal foraging. *Anim. Behav.* 25: 1016–1033.

Greenberg, R., & J. A. Gradwohl. 1980. Leaf surface specializations of birds and arthropods in a Panamanian forest. *Oecologia* 46: 115–124.

Holmes, R. T., J. C. Schultz, & P. Nothnagle. 1979. Bird predation on forest insects: an exclosure experiment. *Science* 206: 462–463.

Martinat, P. J., C. C. Coffman, K. Dodge, R. J. Cooper, & R. C. Whitmore. 1988. Effect of Dimilin 25–W on the canopy arthropod community in a central Appalachian forest. *J. Econ. Entomol.* 81: 261–267.

Mason, C. E., & J. E. Loye. 1981. Treehoppers (Homoptera: *Membracidae*) collected at multiple levels in a deciduous woodlot in Delaware. *Entomol. News* 92: 64–68.

McDonald, L. L., & B. F. J. Manly. 1988. Calibration of biased sampling procedures. Proc. Symposium on Estimation and Analysis of Insect Populations. Springer–Verlag, New York. In this volume.

Royama, T. 1970. Factors governing the hunting behavior and selection of food by the great tit (*Parus major*). *J. Anim. Ecol.* 39: 619–660.

Southwood, T. R. E. 1978. Ecological methods, with particular reference to the study of insect populations. Chapman and Hall, London.

Tinbergen, L. 1960. The natural control of insects in pinewoods. I. Factors influencing the intensity of predation by songbirds. *Arch. Neerl. Zool.* 13: 265–343.

Design Based Sampling as a Technique for Estimating Arthropod Populations in Cotton over Large Land Masses

Joe Ellington[1], Morris Southward[2], Karl Kiser[1]

and Gaye Ferguson Faubion[1]

ABSTRACT The Mesilla Valley, New Mexico, cotton arthropod ecosystem is chosen to exemplify the complexity found in large interacting cropping ecosystems. Here arthropod density is greatly modified by migration from neighboring plant ecosystems; the complex group of arthropods exists in three trophic levels; the beneficial complex may be composed of host–specific, host–nonspecific and hyperparasitoid species; and clumping may be related to a variety of cultural and behavioral factors within and between fields. It may not be possible to predict arthropod density in this ecosystem without consistent periodic updates. In order to set economic thresholds, there is a need for an economical means to determine between field variation and density estimates.

Sweepnet, D–vac, clamshell and Insectavac samplers were utilized to evaluate the most suitable arthropod sampling method. A portable computer–based vision system has been developed and used to count and classify insects based on size, shape and color at the field site to an overall 80% level of accuracy, and a model–based sampling method has been proposed to evaluate questions or conjectures regarding the structure of sampling design and biological relationships.

The density of phytophagous insect pests may be difficult to predict in some agricultural ecosystems because numerical density and species diversity are constantly modified in time and space by migration; the intrinsic rate of increase; and diverse mortality factors including beneficial arthropods, diseases, pesticide applications, abiotic factors and genetic feedback mechanisms. Arthropod sampling may be particularly difficult in complex agricultural ecosystems because the spatial patterns of insects may vary in time and space within and between fields. It may be possible to predict density only within a narrow time span and then only with consistent periodic sampling updates. Here we discuss within– and between–field sampling techniques for just such a system – cotton production in the Mesilla Valley, New Mexico.

An Exemplary Case

The Mesilla Valley study area is a large, cultivated land mass of about 100,000 acres surrounded by extensive desert. This is essentially a closed system except for bollworm (*Heliothis zea*) which may be capable of long–range immigration from Mexico. This production area is made

---

[1]Department of Entomology, Plant Pathology and Weed Science, New Mexico State University, Las Cruces, New Mexico 88003.

[2]Department of Experimental Statistics, New Mexico State University, Las Cruces, New Mexico 88003.

up of relatively small fields mostly of cotton, pecans, alfalfa, lettuce, onions, chile and winter grains. Cotton is a perennial grown as an annual. Migration of thrips to cotton occurs from winter grain and spring onions. Alfalfa is a perennial which acts as a natural insectary for *Lygus* spp., thrips and a host of beneficial and innocuous arthropods which migrate to cotton after each cutting during the spring and summer. Pecans may increase populations of lacewing, principally *Chrysopa carnea*, and ladybeetle, principally *Hippodomia convergens*, in cotton. Lettuce might increase populations of cabbage looper, *Trichoplusia ni*, and bollworm; however, this crop is usually intensively sprayed with insecticides. Chile does not apparently contribute insects to cotton.

Arthropod Complexity and Distribution

Basically, integrated pest management has not worked in cotton ecosystems when key pests require weekly pesticide applications for bollworm, pink bollworm (*Pectinophora gossypiella*) or boll weevil (*Anthonomus grandis*) and *Lygus* spp. Pink bollworm, *Lygus* spp. and bollworm are secondary pests, and boll weevil has not been introduced to the Mesilla Valley; however, pink bollworm, *Lygus* spp. and bollworm can become primary pests if mismanaged. Insecticide applications are often made unnecessarily for preventative maintenance. Thus the Mesilla Valley is a prime area to implement an area—wide integrated pest management program in cotton. Such a program should begin by evaluating the major mortality factors already operating in the ecosystem (parasitoids and predators). However, cotton ecosystems are very complex. In Arkansas, about 600 species of predators representing 45 families of insects, 9 families of spiders and 4 families of mites may be associated with cotton (Whitcomb & Bell 1964). In the San Joaquin Valley of California, 300 to 350 arthropod species may be found in cotton (Van den Bosch & Hagen 1966). A complex arthropod group is also found in the Mesilla Valley, composed of three trophic levels — primary consumers, parasitoids and predators, and hyperparasitoids. Some 15–30 genera may be present in sufficiently high numbers to warrant evaluation. Included in this group are four superfamilies of minute, parasitic Hymenoptera, all of which are biologically relatively unknown. Sampling this diverse array of arthropods can be particularly difficult because various life forms occupy different positions in the crop environment and most insect populations tend to be in some way contiguously distributed (Southwood 1978). The spatial patterns of arthropod populations are not fixed. They may vary in time, in response to a variety of weather, host plant, soil, predator, parasitoid and behavioral factors. An observed dispersion pattern may depend on the size of the clumps, distance between clumps, the spatial patterns of clumps and the spatial patterns of individuals in clumps (Elliott 1979). In addition, the detection of a contiguous or patchy distribution may depend on the scale of the pattern relative to the size of the sampling unit. If a quadrat size is steadily increased, the apparent dispersion of a contiguous population may be random, contiguous and finally regular (Elliott 1979). Thus, a wide variety of factors interact to make the adoption of a single sampling strategy difficult.

Arthropod Sampling

No single technique to sample all of the arthropods in cotton fields has been devised. In the Southwest cotton is usually sampled with a 38–cm sweepnet or D–Vac. In an effort to improve sampling technique, a mechanical, high–vacuum, 3,200 CFM machine [the Insectavac (IV)] was built and calibrated (relative/absolute) for 24 groups of cotton insects (Ellington et al. 1984 a,b) and compared to 38–cm sweepnet samples. Relative samples were taken by vacuuming 30.5–m quadrats replicated 13 to 40 times in a randomized complete block design in 3 fields. Eighty–three replicates were taken throughout the sampling period. Absolute samples were obtained by closing a 61–cm–long clam shell device connected to the IV over the cotton plants previously sampled with the IV. Eight or 12 clam shell closures were used per quadrat. The absolute samples were corrected for 30–m row lengths. Fifty sweeps with a 38–cm net were taken across the tops of two cotton rows adjacent to the IV quadrats. The IV sampled more insects per quadrat and gave a more accurate estimate of the population density and variation than the sweepnet. Overall, the IV collected 47% of the insect groups evaluated. The range was from 14–64%.

Optimum Quadrat Size

The optimization of sampling plans for multiple species involves a compromise between the level of precision of the estimate and the cost of obtaining the information for a particular management decision. The value of density estimates depends on many factors, including the cost of sampling and cash value of the crop (size of the area to be sampled, insect damage, cost of treatments, interest rates, etc.).

Statistical analysis should be used to set the necessary level of precision for density estimation. For some statistical methods to be applied to a set of data, the frequency distribution must be approximated by the normal or Gaussian function, the variance should be independent of the mean and the components of the variance should be additive (Snedecor & Cochran 1972). However, for some biological data, the normal distribution is not a suitable model (Southwood 1978) but nonparametric procedures can be used. The jackknife is a nonparametric method which makes no assumption that data conform to a bell–shaped curve (Cox & Hinkley 1974, Efron 1982). In the case of our sampling we were concerned that, because of the clumping aspect of many of the insects, we were not achieving the best (unbiased) estimates of the variance of the mean. Consequently, we decided to try the jackknife procedure.

Ellington et al. (1988) collected fifteen adult and immature groups from three cotton fields 12 times over a 3–year period with an IV insect sampling machine. Quadrats were 30.5 or 61.1 m long and arranged end to end to a maximum of 305.0 m long. There were from 30 to 60 quadrat replicates in a randomized complete block design on each sampling date. A total of 8590 vacuum quadrat samples were taken in the 3 fields throughout the study. In general, the smaller the quadrat, the fewer manhours required to achieve a given confidence limit. The variance obtained from the jackknife was the same as that obtained from an analysis of variance. It was concluded from this analysis that the sampling procedure is random and the resulting estimates unbiased and minimal and therefore nothing was gained by using the jackknife procedure. There was no systematic difference in the estimate of the mean for any arthropod group with quadrat size. Advantages of small quadrats include greater number of quadrats sampled for a fixed amount of labor, resulting in more degrees of freedom and better coverage of habitats within a field, and better representation of average insect densities. On average, six 30.5–m quadrats (less than one manhour) were required to estimate with 80% confidence, confidence intervals of width ± 10% of the true mean for 7 adult insect groups evaluated in this study. Eight, mostly immature, forms required more than six 30.5–m quadrats (more than one manhour) to set the same confidence limits.

Automatic Counting and Classification of Arthropods

This collection requires the counting of some 1500 insects from 15–30 insect groups (about 3 manhours) to evaluate the density of the major arthropod groups from each field. To make population counts useful, density estimates should be made at the field site. In order to facilitate the counting and evaluation process, a portable computer–based vision system has been constructed which can count and classify insects based on size, shape and color at the field site. Composite input images obtained using a Sony DXC–M3 color video camera are split into RGB by an Electrohome NTSC decoder. The intensity levels of pixels within the insect contours are histogrammed for each of the red, green and blue images. The histograms provide color patterns for the insects. Lighting and magnification are calibrated to provide standardization. Contours of insects in the image are located by an edge detection algorithm which thresholds on the range of background intensity levels, conjoining the values in each of the red, green and blue digitized images. The edge detection algorithm produces a set of chain code files. The chain code lengths provide parameter values. Each chain code file is used to produce a set of Fourier coefficients which form a description of the shape of the insect. From the Fourier series other descriptors such as area are derived. Appendages are removed because they add random noise to the image. The entities of the database are represented as vectors of numeric values and include a Fourier coefficient vector efficient search method for the template matching of an unknown insect to those represented in the database. We can currently identify the insects of interest to an 80% level of accuracy using only size and shape descriptors. The objective is to estimate the density of the major arthropod complex in approximately 45 min at the field site.

Host Mortality from the Beneficial Complex

The role of the beneficial complex in suppressing phytophagous populations in cotton ecosystems is presently inadequately defined. Parasitoid effects can be evaluated by counting adults to predict pest mortality. Ehler & Van den Bosch (1974) and Fye (1979) suggest most common predators in cotton fields are opportunists that switch between prey species depending on availability. Switching in predators which attack several prey species can dampen the rate of change in prey populations (Murdock 1969).

Predation rates have been quantified for specific predators in cropping ecosystems in the laboratory (Fye 1979; Laurence & Watson 1979 and others); in large, outdoor cages (Hutchison & Pitre 1983; Lingren et al. 1968 and others); and in the field using radioactive phosphorus (McDaniel & Sterling 1979, 1982; Moore et al. 1974). Serological techniques have been used to qualitatively and quantitatively analyze insect predation in different cropping systems (Kiritani & Demster 1973; Fichter & Stephen 1979; Ragsdale et al. 1981; Leathwick & Winterbourn 1984).

The value of predators is currently being determined in cotton ecosystems by quantitatively assessing the gut contents of predators using electrophoretic techniques. Starved predators, predators fed a known prey and the prey were analyzed together on the same electrophoretic gel. Esterase and malate dehydrogenase (MDH) enzymes were analyzed initially. MDH was considered more suitable for detecting predator–prey interactions because it has a high level of catalytic activity, forming a very intensely staining band. This enzyme also has a tolerance for acidic pH level (pH 6 to 7) and is not destroyed in predator mesenteron as rapidly as more labile enzymes (Ross et al. 1982). Banding patterns are unique to each species. For those species analyzed to date, there are fewer bands per species when compared with esterase banding patterns.

Migration

Repeated absolute IV samples from cotton and alfalfa (Fig. 1) suggest mobile adult *Lygus* and adult predator forms exist in much higher populations than nonmobile immature forms where the abbreviations are defined by:

BWA	=	Bollworm adult, *Heliothis* spp.
BWL	=	Bollworm larvae, *Heliothis* spp.
CUCA	=	Cucumber beetle adult, *Diabrotica* spp.
LYGA	=	Lygus bug adult, *Lygus* spp.
LYGN	=	Lygus bug nymph, *Lygus* spp.
STINK	=	Stink bugs, *Pentatomidae*
TACH	=	Tachinidae
BEBA	=	Big–eyed bug adult, *Geocoris* spp.
BEBN	=	Big–eyed bug nymph, *Geocoris* spp.
COLLA	=	Collops beetle adult, *Collops vittatus*
LADY	=	Ladybeetle adult, *Hippodamia* spp.
LWA	=	Lacewing adult, *Chrysopa* spp.
LWL	=	Lacewing larva, *Chrysopa* spp.
MPBA	=	Minute pirate bug adult *Orius*, spp.
MPBN	=	Minute pirate bug nymph *Orius*, spp.
NABA	=	Nabid adult, *Nabis* spp.
NABN	=	Nabid nymph, *Nabis* spp.
SYRA	=	Syrphid fly adult, Syrphidae
ANAPA	=	Anaphes
BETHA	=	Bethyloidea
CHALA	=	Chalcidoidea

CYNA = Cynapoidea
ICHNA = *Schneumonoides*
PROCA = Proctotrupoidea

An expanding resident insect population should have more immature forms than adult forms.

The average adult/nymph ratio of several insect genera from two years of sampling in cotton was 8.2 and in alfalfa, 2.7 (Table 1). A much higher ratio of adults to nymphs occurred in cotton, suggesting much of the insect population is migratory, probably from alfalfa because alfalfa is the only perennial with a complex arthropod population grown extensively in the area. The alfalfa ratios, although much smaller, also indicated on the average more adults than immatures. Cutting alfalfa every 4 to 5 weeks kills the immature forms and forces the adult forms to other fields. The adult/nymph ratio in cotton was higher on every sampling date while the adult/nymph ratio in alfalfa varied on sampling dates. This appears to be a sampling artifact dependent on sampling date. The immature mortality from cutting and adult migration between fields on cutting dates probably accounts for the larger adult populations in alfalfa.

Table 1. Average Adult/Nymph Ratio of Insects in 30–cm x 30–m rows, Cotton and Alfalfa Plots, Mesilla Valley, New Mexico, 1984 and 1985.

	LYG	BW	BEB	LW	MPB	NAB	Avg.
Cotton	4.2	2.1	9.2	8.3	4.6	21.1	8.2
Alfalfa	1.6	1.5	5.3	3.1	2.8	1.7	2.7

Insects were collected by taking 240 absolute Insectavac samples from cotton in 1982 and 216 absolute Insectavac samples from alfalfa over a 2 year period. Designated abbreviations: LYG = lygus, *Lygus* spp.; BW = bullworm, *Heliothis zea*; BEB = big eyed bug, *Geocoris* spp.; LW = lacewing, *Chrysopa carnea*; MPB = minute pirate bug, *Orius* spp.; NAB = nabid, *Nabis* spp.

Complex Interactions Affecting Arthropod Density Estimates

In the Mesilla Valley cotton ecosystem, arthropod density is greatly modified by migration from neighboring plant ecosystems, the complex group of arthropods exists in three trophic levels, the beneficial complex may be composed of host–specific, host–nonspecific and hyperparasitoid species and clumping may be related to a variety of cultural and behavioral factors within and between fields. It may not be possible to predict arthropod density in this ecosystem without consistent periodic updates. In order to set economic thresholds, there is a need for an economical means to determine between field variation and density estimates.

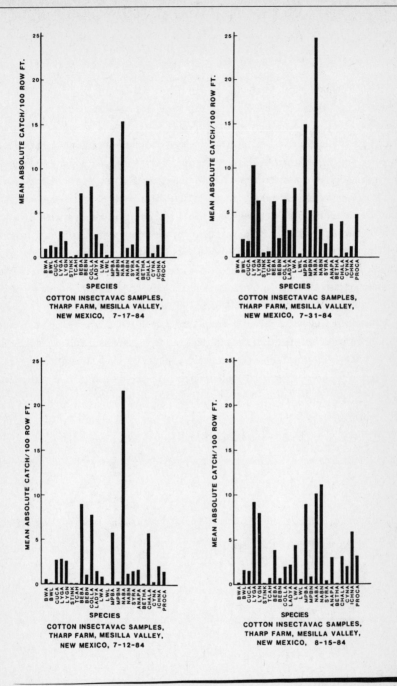

Fig. 1. Density of insects collected from 30–cm x 30–m rows absolute (corrected from relative) Insectavac samples from cotton.

Design—Based Sampling over Large Land Masses

Prediction of insect population density over large land masses requires information on the geographical distribution of the population, the magnitude of migration, and information on the size and age composition of the population. This information is obtained by some sort of sampling program.

From a statistical viewpoint, estimating insect populations consists of obtaining a part of a population (a sample), making estimates of insect density using data from the sample, and then inferring the condition of the entire population from the estimates of the density derived from the sample. Because the process relies on sample data, there will be uncertainty in the estimates: The random variability which causes the uncertainty and is associated with data is introduced either by the methods used to collect the data or by assumption of the stochastic properties of the population being sampled. Traditional design—based sampling techniques, such as those given by Cochran (1977), deal with randomization procedures applied to a fixed finite population. Most of These techniques are aimed at estimating means, totals, and variances of these. An alternative sampling technique is the model—based approach (Cassel et al. 1977), in which the population unit is associated with a random variable for which a stochastic structure is specified.

Design—based sampling relies on the randomization used in selecting elements of a finite population to induce a probability structure into a sample. This is evident when one looks at the expected value of the sample mean and sample variance resulting from unrestricted random sampling. A mean, $\bar{y}$, and variance, s^2, which are based on sample data are random variables and are point estimates of the population parameters $\bar{Y}$ and S^2, respectively. An expectation of a random variable is the weighted average of the random variable where the weighting factor is the probability of occurrence. In this case the expectation of $\bar{y}$ is $\bar{Y}$. In unrestricted random sampling, the probability of choosing a random sample of size n from a population of size N is given by the term $1/\binom{N}{n}$, where $\binom{N}{n}$ is the combinatoric symbol for the number of ways that n items can be chosen from a group of N items; thus $1/\binom{N}{n}$ is the probability of $\bar{y}$ from any given sample and the expectations of the mean and variance are

$$E(\bar{y}_n) = \Sigma \bar{y}_n / \binom{N}{n} = \bar{Y}$$

$$E(s^2) = \Sigma \left[\frac{y_i - \bar{y}_n}{n-1} \right]^2 / \binom{N}{n} = S^2 \ .$$

In more complex sampling designs which involve restricted random sampling, the probability of selection is utilized in an analogous manner, but the probability would be more complicated.

The relationship between levels of the hierarchical structures in design—based sampling determines the levels of the random variability in the data. If the sampling design is unrestricted, there is only one source of random variability, i.e., variations among sampling units. However, if the sampling design has been restricted in several ways, there will be a source of variability for each restriction. In design—based sampling, the inference from the sample to the population is confined to the finite population that was sampled. Such inferences are said to have internal validity when there are no conflicting interpretations for the observed results. Such is the case when the sampled population and the target population are the same (Koch & Gillings 1983).

Inference to larger populations or to populations that were not part of the sampled population in design—based sampling depends on the scientific principle of reproducible results. For example, if the average of a certain species of insect was nearly the same for a number of randomly chosen samples by randomly chosen samplers in a given area with similar sampling methods, one could infer from the reproducibility of the average that the results would apply to other samplers using the same sampling methods in the same area.

In model—based sampling the probability structure is formally assumed via the assumption that the data are equivalent to a specified probability distribution, say the Gaussian normal. In practical applications this is equivalent to assuming that the data can be modeled by a mathematical statement. The simplest form is

$$Y = \text{mean} + \text{unexplained random variability}$$

where the unexplained random variability is assumed to have normal distribution with mean of zero and variance sigma squared. Formal structure can be specified for both the mean and the random part of the model. Since the assumption of normality leads to the assumption that the target population is infinitely large, the inference from sample data is to a much larger population than that possible from a design—based sampling scheme. Continuing with the above example of samples from a randomly chosen set of samplers, the inference would be to an infinitely large set of samples from samplers using the same sampling methods in the same area. An appeal to the scientific principle of reproducibility would not be needed.

In addition to having a different base of inference, model—based sampling allows the structure of the random variability to be examined and studied (Hocking 1985). Examine two sampling schemes, one design—based and the other model—based. To keep the example manageable, assume that the population consists of 5 fields and 5 samplers. Further, assume that for the design—based sampling scheme we will choose 8 sampler × field combinations from the possible 25. Also, we will use unrestricted random sampling. There are 1,081,575 possible samples of size 8 that could be drawn. This example deals with only one of these samples (Table 2). It would be possible to make some comparisons between fields for sampler E and some comparisons among samplers collecting on field 1, but other comparisons would be very limited. Most importantly, the inferences would be to these 5 samplers and these 5 fields. Random variability would be based on the differences among the samples of these 8 observations. Extensions of these results to other samplers and other fields would be difficult.

Table 2. One possible sample based on randomization

Sampler	Field 1	2	3	4	5
A					
B	II				II
C					
D	II		II		
E	II		II	II	II

Method 2 = II

Now examine the same problem but use model–based sampling and the same amount of sampling effort. Assume a factorial structure among the samplers and the fields, and assume that the samplers and fields are random factors in a linear model of the form

$$Y_{ijk} = M + S_i + F_j + S \times F_{ij} + E_{ijk},$$

where i refers to sampler, j refers to field, k refers to replicates, M refers to the average sample, S refers to sampler effect, F refers to field effect, and S × F refers to the interaction of samplers and fields. We now have 2 samplers and 2 fields, samplers and fields randomly chosen from the set of 5 samplers and 5 fields. Each sampler × field combination would be repeated twice. Because of the structure we have built into the sampling scheme, there are covariances as well as variances that have to be accounted for. The variance–covariance structure is:

Covariance $(y_{ijk}, y_{i'j'k'}) =$

$V_E + V_{SF} + V_F + V_S$,   i=i', j=j', k=k'   variance of an observation,

$V_{SF} + V_F + V_S$,   i=i', j=j', k≠k'   covariance between observations from same field and sampler,

V_F,   i≠i', j=j', k≠k'   covariance between observations in same field

V_S,   i=i', j≠j', k≠k'   covariance between observations from same sampler.

The factorial structure enables us to examine and estimate the variance–covariance component of variability that is introduced into our data by the interaction of sampler and field, by sampler alone, and by field alone. Various combinations of samplers and field can be evaluated by forming linear combinations of these terms. The appropriate standard error needed for confidence intervals or tests of these comparisons can be obtained readily from the variance–covariance matrix, again using the linear combination of the observed data to select the appropriate elements from the variance–covariance matrix.

Analysis of data obtained from a design–based sampling scheme is illustrated by the following example using hypothetical data. Consider the factorial arrangement of fields and samplers which would produce a 4–cell table (density of insects are shown for each cell):

	Field 1	Field 2
Sampler 1	25	9
	24	7
Sampler 2	28	16
	24	15 .

These data produce the following AOV table:

Source	df	SS	MS	E(MS)
Total	7	434.0		
Sampler	1	40.5	40.50	$\sigma^2 + r\sigma_{SF}^2 + F\sigma_S^2$
Field	1	364.5	364.50	$\sigma^2 + r\sigma_{SF}^2 + rs\sigma_F^2$
SxF	1	18.0	18.00	$\sigma^2 + r\sigma_{SF}^2$
Error	4	11.0	2.750	σ^2 .

The analysis of variance method of estimating variance components is used here (Lentner & Bishop 1986). This method equates the mean square to the expected mean square and solves algebraically for the desired variance component. In this example,

estimate of inherent error is the mean square error, MSE $= 2.75$,
estimate of the interaction component is (MSSxF − MSE)/2 $= 7.625$,
estimate of the field component is (MSF − MSSxF)/4 $= 86.625$,
estimate of the sampler component is (MSS − MSSxF)/4 $= 5.625$.

This example was constructed to illustrate a situation where the field variability is considerably higher than either sampler or interaction variability. In this case the variance component for fields is an order of magnitude greater than that for sampler or interaction. Had this been real data, an investigator would know that field variation was a highly important factor in a sampling design and in future studies would need to account for field variation in the design of any sampling study.

Design—based and model—based sampling are both valid forms of sampling, but they address different questions and provide inference to different populations. Design—based sampling requires much weaker assumptions, namely random sampling, than model—based sampling. However, simple design—based sampling schemes are often not general enough to answer specific questions of scientific interest with ease when limitations of field operations are considered. In situations where there is agreement the stochastic distribution assumed, model—based sampling often will allow examination of questions directed towards the structure of the sampling process or biological relationships among the insects of interest.

References Cited

Cassel, Claes—Magnus, Carl—Erik Sarndal & Jan Hakan Wretman. 1977. Foundations of inference in survey sampling. John Wiley and Sons, New York.

Cochran, William G. 1977. Sampling Techniques. John Wiley and Sons, New York.

Cox, D.R. & D.U. Hinkley. 1974. Theoretical Statistics. Chapman and Hall.

Efron, B. 1982. The jackknife, the bootstrap and other resampling plans. Soc. for Industrial and Applied Math.

Ellington, J., K. Kiser, M. Cardenas, J. Duttle & Y. Lopez. 1984a. The Insectavac: A high—clearance, high—volume arthropod vacuuming platform for agricultural ecosystems. *Environ. Entomol.* 13: 259–265.

Ellington J., K. Kiser, G. Ferguson & M. Cardenas. 1984b. A comparison of sweep—net, absolute and Insectavac sampling methods in cotton ecosystems. *J. Econ. Entomol.* 77: 599–605.

Ellington, J., K. Kiser, M. Southward & G. Ferguson. 1988. In prep. An evaluation of sampling strategies with a high—vacuum insect sampling machine in complex cotton ecosystems.

Elliott, J. M. 1979. Some methods for the statistical analysis of samples of benthic invertebrates. Freshwater Biological Assoc., Scientific Publ. No. 25.

Ehler, L. E. & R. van Den Bosch. 1974. An analysis of the natural biological control of *Trichoplusia ni* (Lepidoptera: Noctuidae) on cotton in California. *Can. Entomol.* 106: 1067–1073.

Fichter, B. L. & W. P. Stephen. 1979. Selection and use of host—specific antigens. *Misc. Pub. Entomol. Soc. Amer.* 11: 25–33.

Fichter, B. L. & W. P. Stephen. 1981. Time related decay in prey antigens ingested by the predator *Podisus maculiventris* (Hemiptera: Pentatomidae) as detected by ELISA. *Oecologia* 51: 404–407.

Fye, R. E. 1979. Cotton insect populations. USDA Tech. Bull. 1952.

Hocking R. R. 1985. The analysis of linear models. Brooks/Cole Publishing Co. Monterey, California.

Hutchison, W. D. & H. N. Pitre. 1983. Predation of *Heliothis virescens* (Lepidoptera: Noctuidae) eggs by *Geocoris punctipes* (Hemiptera: Lygaeidae) adults on cotton. *Environ. Entomol.* 12: 1652–1656.

Kiritani, K. & J. P. Dempster. 1973. Different approaches to the quantitative evaluation of natural enemies. *J. Appl. Ecol.* 10: 323–330.

Koch G. G. & D. B. Gillings. 1983. Inference, design based vs. model based. In Encyclopedia of Statistical Sciences, Vol. 4, Kotz and Johnson [eds.], John Wiley and Sons, New York.

Laurence, R. K. & T. F. Watson. 1979. Predator–prey relationshiop of *Geocoris punctipes* and *Heliothis virescens*. *Environ. Entomol.* 8: 245–248.

Leathwick, D. M. & M. J. Winterbourn. 1984. Arthropod predation on aphids in a lucerne crop. *New Zealand Entomol.* 8: 75–80.

Lentner, M. & T. Bishop. 1986. Experimental design and analysis. Valley Book Company, Blacksburg, Virginia.

Lingren, P. D., R. L. Ridgway & S. L. Jones. 1968. Consumption by several common arthropod predators of eggs and larvae of two *Heliothis* species that attack cotton. *Ann. Entomol. Soc. Amer.* 61: 613–618.

McDaniel, S. G. & W. L. Sterling. 1979. Predator determination and efficiency in *Heliothis virescens* eggs in cotton using [23]P[2]. *Environ. Entomol.* 8: 1083–1087.

McDaniel, S. G. & W. L. Sterling. 1982. Predation of *Heliothis virescens* (F.) eggs on cotton in east Texas. *Environ. Entomol.* 11: 60–66.

Moore, S. T., M. F. Schuster & F. A. Harris. 1974. Radiosotope technique for estimaitng lady beetle consumption of tobacco budworm eggs and larvae. *J. Econ. Entomol.* 67: 703–705.

Murdock, W. W. 1969. Switching in general predators experiments on predator specificity and stability of prey populations. *Ecol. Monogr.* 39: 334–335.

Ragsdale, D. W., A. D. Larson & L. D. Newsom. 1981. Quantitative assessment of the predators and *Nezara viridula* eggs and nymphs within a soybean agroecosystem using an ELISA. *Environ. Entomol.* 10: 402–405.

Ross, H. R., C. S. Ross & R. P. Ross. 1982. A Textbook of Entomology, 4th ed. John Wiley and Sons, New York.

Snedecor, G. W. & W. G. Cochran. 1972. Statistical Methods. Iowa State University Press, Ames.

Southwood, T. R. E. 1978. Ecological methods, with particular reference to the study of insect populations. John Wiley and Sons, New York.

Van den Bosch, R. & K. S. Hagen. 1966. Predaceous and parasitic arthropods in California cotton fields. Calif. Agric. Exp. Sta. Bull. 820.

Whitcomb, W. H. & K. Bell. 1964. Predaceous insects, spiders and mites of Arkansas cotton fields. Ark. Agric. Exp. Sta. Bull. 690.

Discrete, A Computer Program for Fitting Discrete Frequency Distributions

Charles E. Gates[1]

ABSTRACT The computer program DISCRETE, which fits eight common discrete distributions, has been generalized to fit, in principle, any discrete distribution. The **quid pro quo** is that the user has to provide initial estimates of the parameters and expand two pre—existing subprograms in FORTRAN. These subprograms specify the likelihood function and calculate the expected probabilities for all cells using the final estimates. The analyses of five carefully selected entomological data sets show that the researcher has to be careful in defining the sampling unit to be neither too large nor too small and that sample size must be sufficiently large to permit discrimination between various distributions.

The fitting of discrete distributions plays an important role in entomological research. In addition to its importance in knowing the underlying discrete distribution for purposes of making non—linear transformations for a linear, additive model analysis of the data, the end user also needs information for purposes of determining appropriate sampling plans.

The statistical and entomological literature is rife with various and diverse discrete distributions and there is at least one book concerned solely with discrete distributions (Johnson and Kotz 1969). While some of these distributions are readily fitted, even by handheld calculators, others require much more complicated algorithms.

Materials and Methods

A FORTRAN computer program, DISCRETE, written to fit various discrete distributions has been developed (Gates & Ethridge 1972 and Gates et al. 1987). The impetus for developing this program arose from the desire to fit the negative binomial distribution (Fisher and Bliss 1953). Initially, there was some question as to which distributions to include in the computer program, DISCRETE. However, we heeded the advice of Katti (1966) who suggested that the five two—parameter distributions: negative binomial (Fisher and Bliss 1953), Poisson with zeros (Cohen 1960), Neyman type A (Douglas 1955), logarithmic with zeros (Chakravarti et al. 1967), and Poisson—bionomial (McGuire et al. 1957), should fit a set of discrete data about as well as any distribution. These five distributions cover the extremes with respect to kurtosis and skewness. We also added the following distributions: Poisson (Steel & Torrie 1980), binomial (Steel & Torrie 1980) and Thomas double—Poisson (Thomas 1949).

---

[1]Department of Statistics, Texas A&M University, College Station, TX 77843.

Table 1. Definition of the eight discrete distributions used by DISCRETE.

1.	Poisson	$f(x) = \dfrac{e^{-\mu}\mu^x}{x!}$	for $x = 0,1,2,...$
2.	Negative binomial (Bliss and Fisher 1953)	$f(x) = \dfrac{(k+x-1)!}{x!\,(k-1)!}\dfrac{p^x}{q^{k+x}}$	for $x = 0,1,2,...$
3.	Binomial	$f(x) = \dfrac{n!}{(n-x)!x!}p^x q^{n-x}$	for $x = 0,1,...,n$
4.	Thomas Double	$f(x) = e^{-m}$	for $x = 0$
		$f(x) = \sum\limits_{r=1}^{x} \dfrac{m^r e^{-m}}{r!}\dfrac{(r\lambda)^{x-r}e^{-r\lambda}}{(x-r)!}$	for $x = 1,2,3,...$
5.	Neyman Type A	$f(x) = \dfrac{m_2^x}{x!}e^{-m_1}\sum\limits_{j=0}^{\infty}\dfrac{j^x}{j!}(m_1 e^{-m_2})^j$	for $x \geq 0$ and $m_1, m_2 > 0$
6.	Poisson–binomial (McGuire et al. 1957)	$F_x = \alpha pn \sum\limits_{i=0}^{x-1}\dfrac{(x-1)!}{i!(x-1-i)!}(n-1)^{[x-1-i]}p^{x-i-1}q^{n-x+1}F_i$	

$n = 2, 3, 4, ...; x = 0, 1, ...$ NO
[j] indicates factorial moments, i.e.
$(n-1)^{[0]} = 1$
$(n-1)^{[1]} = n-1$
$(n-1)^{[2]} = (n-1)(n-2)$
$(n-1)^{[3]} = (n-1)(n-2)(n-3)$, etc.
α, p are the parameters.

	Now we define	$f(x) = F_x/x!$	for $x > 0$
	where	$f(0) = F_0 = e^{-\alpha}(1-q^n)$	for $x = 0$.
7.	Poisson with zeros (Cohen 1960)	$f(x) = 1 - \theta$	for $x = 0$
		$f(x) = \dfrac{\theta e^{-\lambda}\lambda^x}{(1-e^{-\lambda})x!}$	for $x = 1,2,3,...$
8.	Logarithms with zeros (Chakravarti 1967)	$f(x) = 1 - \lambda$	for $x = 0$
		$f(x) = \dfrac{\lambda\theta^x}{-x\ln(1-\theta)}$	for $x = 1,2,...$

Table 2. Theoretical means and variances of eight distributions of DISCRETE

Distribution	μ	σ^2
Poisson	μ	μ
Negative binomial	kp	μq
Positive binomial	np	μq
Thomas double–Poisson	$m(1 + \lambda)$	$m(1 + 3\lambda + \lambda^2)$
Neyman type A	$m_1 m_2$	$\mu(m_2 + 1)$
Poisson–binomial	αpn	$\mu[\, 1 + (n{-}1)p \,]$
Poisson with zeros	$\theta\lambda/(1 - e^{-\lambda})$	$\mu(1 + \lambda - \mu)$
Logarithmic with zeros	$\dfrac{-\lambda}{\ln(1 - \theta)} \cdot \dfrac{\theta}{1 - \theta}$	$\mu[\, \mu + \dfrac{1}{1 - \theta} \,]$

The eight discrete distributions fitted by DISCRETE are shown explicitly in Table 1, along with an explanation of the symbols used. Table 2 gives a convenient summary of the theoretical means and variances for these distributions. It is rather obvious on examining these theoretical means and variances why the variance/mean ratio exceeds unity, except for the Poisson and the binomial distributions.

The details of the computational methods utilized by DISCRETE are much too involved to be enumerated here. These computational details are available in the User's Guide for DISCRETE. Copies of this guide may be obtained from the author. DISCRETE itself is available in three versions: the portable mainframe version and two PC–compatible versions, the Microsoft and the PROFORT versions. The latter version requires a math co–processor chip. The mainframe version is free if you are on the BITNET. If you send me a formatted diskette and return postage, then the PC–compatible versions are also available *gratis*. I will support all versions, as long as the user has made no substantive programming changes. The three versions give identical results as far as is known.

DISCRETE is designed to respond with minimum user input. Less effort is required on the part of the user to have all distributions fitted rather than a subset (i.e., the default is set such that the distribution will be fitted rather than omitted). Goodness–of–fit is indicated by two criteria, the usual chi square goodness–of–fit test and a visual plot. The visual goodness–of–fit is illustrated by plotting the cumulative frequency distributions of the observed and the expected values for each of the eight distributions. DISCRETE is programmed to accept data in frequency data form (the standard input) or as an option, it will accept raw data (actual counts) and will construct the necessary frequency table.

Output includes the problem title, observed and expected frequencies for each distribution fitted, as well as chi–square, degrees of freedom, which cells are combined, (the default is to pool if cell expectation < 1), chi square probability, estimates of the parameters, etc. For the distributions fitted by iterative methods, some intermediate values are shown optionally. After all problems are fitted, a three–page summary table is given, including ' * ', ' ** ', or 'NS', depending on whether the calculated chi square values are significant at the 0.05, 0.01 levels or non–significant, respectively, final estimates of the parameters, etc.

A major, recent modification to DISCRETE permits the user, in principle, to fit any discrete distribution. The quid pro quo is that in order to do this, the user must provide the following, using FORTRAN: (a) initial starting values for all parameters, and (b) two subprograms for which the basic shells are provided. In providing initial starting values, the user will have available the mean and variance as well as the third and fourth moments, if moment estimators are desired. The two subprograms that have to be completed are: (a) a function giving the likelihood, or log likelihood, of the target distribution and (b) a subroutine evaluating the expected theoretical probabilities for all cells, based on final estimates of the parameters. Detailed instructions are provided for the user, with an illustrative example.

With the aid of DISCRETE (Gates et al. 1987), I analyzed several hundred entomological data sets. From these analyses I chose five examples, not because they are "typical", but because they are "atypical". These data sets are shown in Table 3 in detail, except for an abbreviated version of Data Set 1.

Data Set 1 represents a collection of pear psylla nymphs, *Psylla pyricola* (L.), (Homoptera: Psyllidae), (Harris 1972). The data were taken near Geneva, New York, on 30 June, 1970. This orchard was comprised of backcrosses of a *Pyrus ussuriensis* and *P. communis* hybrid to *P. communis*. Thus, the orchard trees were genetically heterogeneous. The sampling procedure consisted of taking 10 randomly selected spurs (each spur consisting of six to eight leaves) from each tree in the orchard. The psyllae were separated from the foliage in the laboratory by means of a Berlese–Tullgren funnel. The data shown here have been combined across leaves and clusters by individual trees.

The second set of data is from a study of the pecan weevil, *Curculio caryae* (Horn), (Coleoptera: Curculionidae), (Ring 1975). Random samples of pecan nut clusters composed of approximately 100 nuts were taken twice a week from Aug. 5 to Aug. 29 and once a week from Sept. 1 to Nov. 1 in 1975. With rare exceptions the number of weevils per nut appear to have resulted from a single bout of oviposition by one adult female. The pecans were from native, rather than cultivated, trees located near Hamilton, Texas.

The third set of data is from a study of the sorghum midge, *Contarinia sorghicola* (Coquillett), (Diptera: Cecidomyiidae), (M. E. Merchant personal communication). This insect is considered a key pest of sorghum and is the most damaging of all sorghum pests in the United States. The sorghum midge is a small fly that lays eggs in the flowers of sorghum plants. Larvae cause yield loss by feeding on the developing seed. The sampling units used here are the counts of sorghum midges on a panicle (seed head). Systematic sampling was used, wherein every 50 paces a single plant was sampled. The data were taken from a field near Corpus Christi, Texas, during the summer of 1987.

Table 3. Five interesting data sets (blanks represent 0 frequencies).

Count	Distribution[a]				
	1	2	3	4	5
0	12	9313	46	1549	1169
1	11	240	25	1	144
2	12	409	4		92
3	15	606	0		54
4	12	496	1		29
5	8	153			18
6	8	28			10
7	2	14			12
8	0	1			6
9	1	1			9
10	2	1			3
11	2				2
12	4				0
13	2				0
14	2				1
15	1				0
16	0				0
17	0				0
18	0				0
19	1				1
20	2				
21	1				
22	1				
23	1				
24	0				
25	3				
26	0				
27	0				
28	0				
29	0				
30	0				
31	0				
32	0				
33	0				
34	3				
35	2				
36	0				
37	1				
42 – 495	54				
Totals	163	11,262	76	1550	1550

[a]Harris 1972, Ring 1978, Merchant 1988, Lin 1985, and Lin 1985, respectively. See text for full details.

The 4th and 5th data sets were taken from a large study on the Southern pine beetle, *Dendroctonus frontalis* Zimmerman, (Coleoptera: Scolytidae), in Southeast Texas (Lin 1985). The Southern pine beetle infests all species of southern yellow pine. The adults mine under the bark of the pine trees and ultimately girdle the trees. The pines must be killed for the brood the develop. Once a tree is colonized, then depending on site, stand, and weather conditions, many adjacent pines also may be colonized. The damage is very high; in 1976 the economic loss was $14 million in stumpage alone in state and private forests, (Lin 1985). The data I report here consists of the number of "spots" of infestations in 5' by 5' geographic "rectangles", which are roughly 8 km. square. Initial counts are from aerial photographs, but the counts were verified by ground crews before a new spot was considered infested. Data were collected monthly for the 5 years: 1975–1977, 1982–1983. Data set 4 is for the month of January, 1982, while data set 5 is for September, 1976.

Results and Discussion

No known discrete distribution can be fitted to Data set 1, Table 3. Most discrete distributions are defined in terms of combinatorials, if not factorials, and numbers of the order of magnitude of 495! are extremely difficult to handle. The basic difficulty here is obviously that far too much aggregation of the data has taken place; the sampling unit is much too large. The values reported in Table 3 consist of six to eight leaves from ten separate clusters composited. Undoubtedly, the data would be analyzable statistically if the original data for individual leaves, if not clusters, were available.

The second set of data in Table 3 is readily analyzable by DISCRETE. Unfortunately, of the eight discrete distributions available in DISCRETE, none gave a good fit. The reasons for this are two–fold: Tests of goodness–of–fit are very powerful for this data set, with over 11,000 observations and 11 filled cells. The basic difficulty in fitting this data set is that there is a great excess of zeros (9,313), compounded with an excess (or deficiency) of values at three pecan weevils per nut. Two of the eight distributions are specifically designed for an excess of zeros, the Poisson and the logarithmic distributions with zeros. Apparently, the additional peak prohibits fitting these distributions. The expected values for all fitted distributions, except the binomial, are shown in Table 4. The chi square goodness–of–fit values are also given in Table 4; these range from 168 to 36,025. The Poisson–binomial achieved the minimum chi square value of the fitted distributions; the second best fit was the Poisson with zeros distribution (chi square = 246). In none of the fitted distributions would I consider the fit adequate for either statistical or biological purposes. In actuality, the observed data may represent a mixture of two or more distributions.

Data set 3, is in a real sense, the opposite of Data set 2 in that all eight of the discrete distributions from DISCRETE fit the data. This occurrence is fairly common with only four cells, but this particular data set is one of very few cases in which five cells have permitted all eight distributions to fit. There is also an obvious explanation here. There are only 76 observations scattered over five cells (four filled cells). The power of the chi–square goodness–of–fit tests will

Table 4. Expected values for data set 2, pecan weevils per nut, from Table 3. Expected values rounding less than 0.1 are not shown.

Class	Observed Frequency	Expected Frequencies[a]						
		P	PZ	NB	TDP	NTA	PB	LZ
0	9,313	6658.9	9313.0	9239.1	9078.7	9371.0	9209.3	9313.0
1	240	3499.1	337.9	875.2	464.3	345.0	285.7	865.6
2	409	919.4	483.7	400.2	679.7	464.9	592.2	369.5
3	606	161.0	461.6	231.3	514.7	423.3	622.9	210.3
4	496	21.2	330.4	147.7	280.3	296.6	348.8	134.7
5	153	2.2	189.2	99.7	131.8	174.1	113.1	92.0
6	28	0.2	90.3	69.6	60.5	92.0	43.4	65.4
7	14	0.1	36.9	49.9	28.2	46.7	27.0	47.9
8	1		13.2	36.4	13.1	23.7	12.2	35.8
9	1		4.2	26.9	6.0	12.2	4.4	27.1
10	1		1.6	86.1	4.7	12.6	2.9	100.8
Chi square		36,025	246	2,113	451	365	168	2,414

a

P = Poisson	PZ = Poisson with zeros
NB = Negative binomial	TDP = Thomas double poisson
NTA = Neyman type A	PB = Poisson–binomial
LZ = Logarithmic with zeros	

be very low for data sets such as these. The solution here is also exactly opposite to the solution Data set 2. Here evidence is that the sampling unit for sorghum midges was not large enough under 1985 growing conditions. Perhaps pairs of sorghum plants or all sorghum plants within a linear 0.5 m of row, for example, should have been combined to give the basic sampling unit.

Data set 4 is another non–analyzable data set, but for an entirely different reason than Data set 1. Here the counts of Southern pine beetle infested spots are 1,549 zeros and one 1. Thus, it is not possible to fit any discrete distribution. Technically speaking, one can obtain a perfect fit to any single parameter distribution, such as the Poisson, but the fit would not be meaningful.

Data set 5 was selected as it is one of the few distributions that was fitted by the negative binomial, but none of the other seven distributions in DISCRETE; however, this should be qualified. In my experience, and in particular, in analyzing these entomological data sets, the negative binomial fit the data far more often than any other distribution, perhaps more than all others combined. Because of the technical requirements of many of the discrete distributions, it is not often possible to obtain a fit at all. An iterative method may fail to converge, for example, because of the nature of the data. This even happened in about 10% of the current cases for the negative binomial. The expected values and other summary statistics are shown in Table 5 for all distributions in DISCRETE except the binomial.

In summary, the quantitatively–minded entomologist needs to be cognizant of the following: The sampling units have to be "just the right size". If sampling units are too large, the data will not be analyzable for technical reasons (495! is not manageable). If the sampling units are too small, then there may be too few filled classes to satisfactorily analyze the data (if at all). From a practical perspective in which insect populations may vary enormously from season to season; what is the researcher to do? My recommendation is that the sampling unit be designed to be on the "low" side, or in other words, just about right for a high population year. Samples can be combined but never can be correctly partitioned if they are initially too large. The other important point brought out by the data sets illustrated above is that the researcher must take enough samples (combined samples, in the final analysis) such that the goodness–of–fit tests have some statistical power to distinguish between various common discrete distributions. Undoubtedly, the power is a function, not only of total sample size, but also the number and the way cells are filled. For example, from the above analysis on Data set 3, substantially more than 76 observations would be needed to permit the user to eliminate most of the eight distributions from contention; perhaps as many as 150 observations would be required.

Table 5. Expected values for data set 5, Table 3. Expected values rounding less than 0.1 are not shown.

		Expected Frequencies[a]						
Class	Observed Frequency	P	PZ	NB	TDP	NTA	PB	LZ
0	1169	785.3	1169.0	1165.3	1271.0	1151.7	1303.2	1169.0
1	144	534.0	82.3	169.4	22.3	109.2	1.9	178.1
2	92	181.6	104.9	78.9	54.3	112.7	14.1	74.0
3	54	41.2	89.2	45.2	66.5	81.0	50.9	41.0
4	29	7.0	56.9	28.3	55.4	47.0	92.4	25.6
5	18	1.0	29.0	18.6	35.9	24.4	67.5	17.0
6	10	0.1	12.3	12.6	20.2	12.2	2.1	11.8
7	12		4.5	8.8	10.8	6.0	4.3	8.4
8	6		1.4	6.2	5.9	3.0	5.9	6.1
9	9		0.4	4.4	3.4	1.4	4.8	4.5
10	3		0.1	3.2	1.9	0.7	1.9	3.4
11	2			2.3	1.1	0.3	0.2	2.5
12	0			1.7	0.6	0.1	0.3	1.9
13	0			1.3	0.3	0.1	0.2	1.5
14	1			0.9	0.2		0.1	1.1
15	0			0.7	0.1			0.9
16	0			0.5				0.5
17	0			0.4				0.4
18	0			0.3				0.3
19	1			0.2				0.3

a

P = Poisson	PZ = Poisson with zeros
NB = Negative binomial	TDP = Thomas double Poisson
NTA = Neyman type A	PB = Poisson–binomial
LZ = logarithms with zeros	

References Cited

Bliss, C. T. & R. A. Fisher. 1953. Fitting the negative binomial distribution to biological data. *Biometrics* 9: 176–200.

Chakravarti, I. M., R. G. Laha, & J. Roy. 1967. Handbook of methods of applied statistics, Vol. 1. John Wiley, Ne York.

Cohen, A. C., Jr. 1960. An extension of a truncated Poisson distribution. *Biometrics* 16: 446–450.

Douglas, J. B. 1955. Fitting the Neyman type A (two parameter) contagious distribution. *Biometrics* 11: 149–158.

Gates, C. E. & F. G. Ethridge. 1972. A generalized set of discrete frequency distributions with FORTRAN program. *Intl. Assoc. for Math. Geo.* 4: 1–24.

Gates, C. E., F. G. Ethridge. & J. D. Geaghan. 1987. Fitting discrete distributions. User's documentation for the FORTRAN computer program DISCRETE. Texas A&M University, College Station.

Harris, M. K. 1972. Host resistance to the pear psylla in New York. Ph.D dissertation, Cornell University, Ithaca.

Johnson, N. L. & S. Kotz. 1969. Discrete distributions. Houghton Mifflin, Boston.

Katti, S. K. 1966. Interrelations among generalized distributions and their components. *Biometrics* 22: 44–52.

Lin, Shih–Kang. 1985. Characterization of lightning as a disturbance to the forest ecosystem in East Texas. M.S. thesis, Texas A&M University, College Station.

McGuire, J. U., T. A. Brindley & T. A. Bancroft. 1957. The distribution of corn borer larvae *Pyrausta nubilalis* (HBN.), in field corn. *Biometrics* 13: 65–78.

Ring, D. R. 1978. Biology of the pecan weevil, emphasizing the period from oviposition to larval emergence. M.S. thesis, Texas A&M University, College Station.

Steel, R. G. D. & J. H. Torrie. 1980. Principles and procedures of statistics. 2nd Ed. McGraw, New York.

Thomas, M. 1949. A generalization of Poisson's binomial limit for use in ecology. *Biometrika* 36: 18–25.

Calibration of Biased Sampling Procedures

Lyman L. McDonald[1] and Bryan F. J. Manly[2]

ABSTRACT   Sampling without a well defined frame potentially leads to biased inferences. We consider procedures developed for estimation of selection functions in the study of natural selection and procedures for correction of size—biased samples to give a general theory for correction of biased samples. Applications are made to the correction of selection bias in samples of insects collected by trapping.

Consider a population of experimental units to which an experimenter wishes to make statistical inferences. On each unit, a vector of r random variables, $\underline{X} = \{X_1, X_2, ..., X_r\}$, is defined by the experimenter and inferences are to be directed to parameters of the natural probability density function, $f(\underline{x})$, of the random variables. The population size, N, will often be finite, but in those cases we will assume that a probability density function, $f(\underline{x})$, gives an adequate representation of the distribution of $\underline{X}$.

A subset of the population is obtained by means of a sampling mechanism which is under the control of the experimenter. For example, "light traps" might be set for one night in an orchard in an attempt to obtain individuals of a certain species of insect living in the orchard at that point in time. Each element of the random vector $\underline{X}$ is recorded for each individual in the subset. The subset is defined to be an *unbiased sample* if the members of the subset have the same distribution, $f(\underline{x})$, as the original population. Otherwise, the sample is *biased*, and the *biased probability density function* of the sample is the function, $g(\underline{x})$.

We will assume that under the sampling mechanism the probability of selection of a member of the population with $\underline{X} = \underline{x}$ depends only on $\underline{x}$ and is proportional to the *selection function (weighting function)*, $w(\underline{x})$. The actual probability of selection is denoted by $p = k \cdot w(\underline{x})$. It is then a well known result that the probability density function of $\underline{X}$ for the subset of individuals selected is proportional to $w(\underline{x})f(\underline{x})$. When this second proportionality constant is needed in formulas, it will be denoted by

$$E_f(w(\underline{x})) ,$$

the expected value of the selection function with respect to the natural probability density function, $f(\underline{x})$. The biased distribution is

---

[1]Departments of Zoology and Statistics, Box 3332, University of Wyoming, Laramie, Wyoming, 82071, U.S.A.

[2]Department of Mathematics and Statistics, University of Otago, Dunedin, New Zealand

$$g(\underline{x}) = w(\underline{x})f(\underline{x})/E_f(w(\underline{x})). \tag{1}$$

The relationship between the proportionality constants is

$$P = k \cdot E_f(w(\underline{x})) \tag{2}$$

where P can be interpreted as the average probability of selection of individuals for the biased sample, or the proportion of the population selected in the biased sample. In general, P cannot be estimated unless the sizes of the populations are known.

We consider the case when there is no well defined sampling frame. The number of units in the population is unknown, the size of the sample is a random variable depending on the outcome of the sampling mechanism, an individual in the population with $\underline{X} = \underline{x}$ is selected with probability proportional to $w(\underline{x})$ and the observed data are from the biased distribution, $g(\underline{x}) = w(\underline{x})f(\underline{x})/E_f(w(\underline{x}))$. The objective of the analysis is to estimate the selection function, $w(\underline{x})$, and then use the selection function to obtain unbiased estimates of parameters in the natural distribution, $f(\underline{x})$. The general theory for unequal probability sampling in classical finite sampling theory (e.g., Cochran 1977) is not applicable to the class of problems under consideration because the selection probabilities are estimated and there is no sampling frame.

Biased distributions have been studied extensively in the statistics literature under the name of weighted distributions and are widely applicable in scientific studies (see Manly 1985 for a summary of procedures for the estimation of selection functions and their use in the study of natural selection). All areas of science must deal with biased samples and numerous special procedures have been developed to "correct" the analysis. For example, Southwood (1978) reviews ecological methods which are particularly applicable to the study of insect populations, identifying potential biases, and when possible, gives techniques for correction of bias. A complete review of the literature is beyond the scope of this paper. One new mathematical theorem is utilized, but the major contribution of the paper is the combination of known mathematical results into a general theory for "correction" of selection bias in samples.

The next section contains a brief review of the relevant mathematical theory of biased distributions, and the theory for estimation of parameters in the natural distribution when the selection function is known (or estimated). An application is given to "size–biased" sampling with one random variable X, where $w(x) = x$. We present two experimental designs where it is possible to obtain sufficient data for estimation of the selection function. Required assumptions for use of the estimated selection function in the correction of biased data are discussed. The case is considered when individuals in the population can be assigned to one of k classes (i.e., X is a multinomial random variable with k classes). For example, the population might consist of a polymorphic collection of insects where the morphs are present in unknown proportions in the "natural" distribution. If the sampling mechanism selects individuals with unequal probability from the different morphs, then methods are outlined for estimation of the selection probabilities (e.g., Manly 1985) and correction of the observed "biased" distribution of morphs. The theory is illustrated with application to a sample of snails (*Cepaea nemoralis*) collected by Cain & Currey (1968).

The last section contains theory concerning the case when there are several quantitative variables X_1, X_2, ..., X_q which may influence the probability of selection (e.g., Manly 1985). In addition, there may be variables Y_{q+1} ..., Y_r which are of interest to the researcher and are correlated with those variables which directly influence the probabilty of selection. The theory is illustrated with application to field collection of tsetse flies (*Glossina swynnertoni*) by Glasgow (1961).

Mathematical Theory

The point estimators can be justified from sampling theory where there is a finite population of N elements (Hanruav, 1962). Let y_i denote the value of one of the random variables in the vector $\underline{X}$ for the ith member, and let $Y = \Sigma y_i$ denote the total of the values of the random variable over all members of the population.

Theorem 1. Assume that every member of the population of size N has positive probability of selection by the sampling mechanism and that n individuals are selected. Further assume that the selection function and the proportionality constant, k, are known so that given the ith member of the biased sample with $\underline{X}_i = \underline{x}_i$, the probabilities of selection

$$p_i = k \cdot w(\underline{x}_i), i = 1, 2, ..., n,$$

can be computed. Unbiased estimators of the population size, N, and the total, Y, are given by

$$\hat{N} = \sum_{i=1}^{n} (1/p_i), \text{ and} \tag{3}$$

$$\hat{Y} = \sum_{i=1}^{n} (y_i/p_i) . \tag{4}$$

Proof. The proof follows an argument used extensively in the literature. Note that $\hat{Y} = \sum_{i=1}^{N} (y_i/p_i)I_i$, where the sum is over the entire population and the indicator function, I_i, is 1.0 if the ith item is encountered and 0.0 otherwise. In repeated sampling the I_i's are the random variables, and the y_i's are a set of fixed numbers regardless of whether or not they are ever measured. Then, $E(I_i) = p_i$, and $E(\hat{Y}) = Y$. With $y_i = 1$, we see that $E(\hat{N}) = N$.

The variance of $\hat{Y}$ is given by the expression

$$\text{var}(\hat{Y}) = \sum_{i=1}^{N} (y_i)^2(1-p_i)/p_i + 2\sum_{i<j}^{N}\sum^{N}(y_iy_j)(p_{ij}-p_ip_j)/p_ip_j, \tag{5}$$

where p_{ij} is defined to be the joint probability that both the ith and jth individuals are selected by the sampling mechanism for the biased sample. If $y_i = 1$ for i = 1, 2, ..., N, then $\hat{Y} = \hat{N}$ and $\text{Var}(\hat{N})$ can be obtained from equation (5) by substitution of 1 for y_i. Define an indicator function I_{ij} to be 1 if both the ith and jth individuals are in the biased sample and to be 0 otherwise. Define $p_{ii} = p_i$, for i = 1, 2, ..., N. Note that the expected value of the sum $\left(\sum_{i=1}^{n}\sum_{j=1}^{n}(1/p_{ij})\right)$ is equal to N^2.

Theorem 2. An unbiased estimator of the variance of $\hat{N}$ is given by

$$\text{var}(\hat{N}) = (\hat{N})^2 - \sum_{i=1}^{n}\sum_{i=j}^{n}(1/p_{ij}). \tag{6}$$

In the case that selection of the ith individual is independent of the selection of the jth individual, then $p_{ij} = p_ip_j$, and the unbiased estimator of $\text{Var}(\hat{N})$ in Theorem 2 reduces to

$$\text{var}(\hat{N}) = \sum_{i=1}^{n}(1/p_i^2) - \sum_{i=1}^{n}(1/p_i). \tag{7}$$

Applying this theory to the total Y, the unbiased estimator of $\text{Var}(\hat{Y})$ is

$$\text{var}(\hat{Y}) = (\hat{Y})^2 - \sum_{i=1}^{n}\sum_{j=1}^{n}(y_iy_j/p_{ij}) \tag{8}$$

where again p_{ii} is defined to be equal to p_i. If the selections of the ith and jth individuals are independent then equation (8) reduces to

$$\text{var}(\hat{Y}) = \sum_{i=1}^{n}(y_i/p_i)^2 - \sum_{i=1}^{n}(y_i^2/p_i). \tag{9}$$

The "ratio" estimator (the weighted mean of the observed values of the random variable Y_i)

$$\hat{\mu}_y = \hat{Y}/\hat{N} = \sum_{i=1}^{n}(y_i/w(\underline{x}_i))/\sum_{i=1}^{n}(1/w(\underline{x}_i)) \tag{10}$$

is proposed for the mean of the random variable Y in the natural distribuiton, where the weights are the reciprocals of the probabilities of inclusion in the biased sample. Note that the proportionality constant cancels from this weighted mean. Thus, even if the proportionality constant is not known or estimated, the weighted estimator $\hat{\mu}_y$ can still be computed.

Estimation of variances is difficult. The simultaneous probabilities p_{ij} may be small and the estimators of p_{ij} are subject to high sampling variance. Variances of the estimators in this theory are usually estimated by using repeated independent replications of the basic sampling mechanism (e.g., Kaiser 1983). An ad hoc alternative which should give approximate estimates of variances and standard errors is illustrated in the examples.

It is interesting to note that the above estimator of the total $Y = \sum_{i=1}^{n} y_i$, its variance and unbiased estimator of the variance have exactly the same formulas as derived in Horvitz & Thompson (1952).

Size–Biased Field Sampling

Size–biased sampling (Cox 1969) is an important case of the general theory because it is often possible to design studies so that the probability of inclusion of elements in the biased sample is proportional to some measure of "size", e.g., sampling of fibers in yarn (Cox (1969)), and line intercept sampling of vegetation (McDonald (1980), Kaiser (1983)). In forestry, the probability of a tree being "sampled" by the Bitterlich plotless technique for estimation of basal area coverage is proportional to $x =$ basal area of the tree (described in Grosenbaugh 1952). This example illustrates another interesting application of the general theory. The probability of selection is $p = k \cdot x$, where the proportionality constant k is evaluated for the particular angle and units used. Interest is in estimation of the total basal area $X = \sum_{i=1}^{n} x_i$ and the estimator is

$$\hat{X} = \sum_{i=1}^{n} (x_i/kx_i) = n/k.$$

Thus, basal area is never actually measured on the trees in the biased sample. Only the total number of trees in the sample is recorded and adjusted for the proportionality constant. If other random variables were also measured on a tree counted in the biased sample then those variables are selected with probability proportional to basal area of the tree. Basal area would have to be measured to adjust estimators as in equation (10) above.

The fact that items are selected with probability proportional to x, the variable of interest, means that it is not necessary to measure x on any item in order to estimate the total of that random variable over the population. This interesting application has been discovered and utilized in many areas of science (e.g., stereology (Weible 1980), and geology (Davy & Miles 1977)).

If X is a positive univariate random variable, $w(x) = kx$, and the indicated moments exist, then

$$E_g(x^r) = E_f(x^{r+1})/E_f(x), \text{ and}$$

$$E_g(x^r y^s) = E_f(x^{r+1} y^s)/E_f(x) ,$$

where Y denotes a correlated random variable also measured on the members of the biased sample (Cox, 1969). In particular, let μ_x and μ_y denote the means, σ^2_x and σ^2_y denote the variances and σ_{xy} denote the covariance of X and Y, respectively, in the natural distribution. The means of the observed values in the biased sample (in terms of parameters of the natural distribution) are:

$$E_g(x) = \mu_x + \sigma^2_x/\mu_x, \text{ and}$$

$$E_g(y) = \mu_y + \sigma_{xy}/\mu_x .$$

These formulas clearly show the magnitude of the bias when individuals are selected by a size–biased procedure. However,

$$E_g(1/x) = 1/\mu_x, \text{ and}$$

$$E_g(y/z) = \mu_y/\mu_x .$$

Using the last two results, Cox (1969) proposed the harmonic mean of values of x in the biased sample as the estimator of the mean of the natural distribution μ_x and the weighted mean of observed values of y for the natural mean μ_y of correlated random variables, i.e.,

$$\hat{\mu}_x = n/\sum_{i=1}^{n} (1/x_i), \text{ and} \tag{11}$$

$$\hat{\mu}_y = \sum_{i=1}^{n} (y_i/x_i)/\sum_{i=1}^{n} (1/x_i) . \tag{12}$$

These formulas can be obtained from equation (10) for the case $w(x) = x$.

For another example of size–biased sampling, consider encounter sampling of mobile populations McDonald et al. (1987)). Let x_i be the length of time spent in a certain activity ('life', life stage, nesting, egg laying, care of young, fishing, camping, shopping, 'behaviors', etc.) by a finite population of N individuals. Assume that the probability of encounter of the ith individual is proportional to the time x_i spent in the study area, i = 1,2,...,N. If the value of x_i can be measured for each animal encountered then estimation formulae for the totals (X,Y) and means (μ_x,μ_y) are the same as in the above example of size–biased sampling.

Two Designs for Calibration of Biased Sampling Procedures

Design One.

This design is in common use, but a formal treatment does not appear to be available. Assume that an expensive, or time consuming, sampling procedure exists which can be used to yield an unbiased sample of size n_0 from the natural population with distribution, $f(\underline{x})$. Denote the unbiased sample by $\{\underline{x}_{01}, \underline{x}_{02}, ..., \underline{x}_{0n_0}\}$. The objective is to calibrate a procedure which is inexpensive or quick, but potentially biased. Assume the probability of selection by the inexpensive method is proportional to the selection function, $w(x)$, and that a biased sample of size n_1 is available with distribution $g(\underline{x})$ proportional to $w(\underline{x}) \cdot f(\underline{x})$. Denote the potentially biased sample by $\{\underline{x}_{11}, \underline{x}_{12}, ..., \underline{x}_{1n_1}\}$. Estimation of $w(x)$ is mathematically equivalent to estimation of the selection function in the study of "natural selection" (Manly, 1985).

Under the assumption that the selection function $w(x)$ is the same for a future application of the biased sampling procedure, the estimate $\hat{w}(\underline{x})$ can be evaluated for each of the individuals in the biased sample to obtain the selection values $\{\hat{w}(\underline{x}_1), ..., \hat{w}(\underline{x}_n)\}$. Reciprocals of the selection values are used in equation (10) to provide estimators of the means in the natural distribution, $f(\underline{x})$. Also, the selection values can be used to "adjust" the biased sample. If the vector $\underline{x}$ occurs with frequency $f_{\underline{x}}$ in a future biased sample of size n then, the "corrected" frequency of occurrence for the natural probability density function is $(f_{\underline{x}})/\hat{w}(\underline{x})$. In the case that every vector is unique, the frequencies are equal to 1 and the corrected frequencies are $1/\hat{w}(\underline{x})$. It will be easier to interpret the corrected frequencies, say $f_{\underline{x}}^{*}$, if they are computed by the equation

$$f_{\underline{x}}^{*} = (f_{\underline{x}}/\hat{w}(\underline{x}))\{n/\underset{\underline{x}}{\Sigma}(f_{\underline{x}}/\hat{w}(\underline{x}))\}, \tag{13}$$

so that they sum to the original size of the biased sample. The mean of the corrected sample is the weighted mean of equation (10) and initial simulations indicate that one can obtain an approximate analysis if standard errors are computed on the corrected sample as if it were an unbiased sample from the population.

If P, the proportion of the population selected in the biased sample, is known then the proportionality constant k can be estimated from the equation

$$P = \hat{k} \cdot [(\underset{\underline{x}}{\Sigma} \hat{w}(\underline{x}))/n_1] . \tag{14}$$

In this case the quantity $\hat{p}_x = \hat{k} \cdot \hat{w}(\underline{x})$ can be interpreted as the absolute probability that an individual with $\underline{x}$ will be in the biased sample.

Design two.

The second design is for a situation where no procedure exists for obtaining an unbiased sample from the population as is required in Design One. Additional requirements must be satisfied in order for this design to be applied:

1. The sampling mechanism can be applied repeatedly (at least twice) such that given an individual with $\underline{X} = \underline{x}$, the selection function, $w(\underline{x})$, is exactly the same on all occasions.
2. Individual experimental units (e.g., animals, leaves, etc.) can be marked and returned to (left in) the population such that the biased distribution, $g(\underline{x})$ of the marked units remains unchanged until the next sample is collected.
3. The probability of survival until the next sampling occasion is the same for all marked individuals.

For example, in an attempt to obtain an unbiased sample from an animal population, the assumptions require that:

a. It is possible to capture animals and mark them such that the occasion of the last recapture is known,
b. the probability of an animal being returned alive to the population is the same for all marked animals,
c. the probability of survival until the next sampling occasion is the same for all marked animals,
d. the probability of capture on any occasion is proportional to the selection function, $w(\underline{x})$.

The design depends on "sampling the biased sample" with the probability of selection proportional to the same selection function, $w(\underline{x})$. That is, we are assuming that the "bias" is the same on both occasions and wish to estimate the selection function so that the first sample (or future samples) can be corrected.

The assumptions imply that the distribution $g_1(\underline{x})$ of the "first" biased sample, $\{\underline{x}_{11}, \underline{x}_{12}, ..., \underline{x}_{1n_1}\}$, is proportional to $w(\underline{x}) \cdot f(\underline{x})$.

If the same selection function is applied a second time to this distribution, then the biased sample "from the biased sample", $\{\underline{x}_{21}, \underline{x}_{22}, ..., \underline{x}_{2n_2}\}$ is distributed as $g_2(\underline{x})$ proportional to $[w(\underline{x})]^2 f(\underline{x})$.

The distributions $g_1(\underline{x})$ and $g_2(x)$ differ only by the selection function, $w(\underline{x})$. This is mathematically equivalent to the situation in Design One. Again, the procedures reviewed by Manly (1985) apply for estimation of the selection function based on data from $g_1(\underline{x})$ and from $g_2(\underline{x})$. Adjustment of the biased samples follows the same procedure outlined in Design One, admittedly with more assumptions.

In some cases, it may be possible to calibrate the biased procedure by sampling from a population where the distribution $f(\underline{x})$ is known. For example, large numbers of insects might be reared in the laboratory, released into the natural habitat and sampled by the biased procedure. Or, a complete enumeration might be made of plants in a small part of the population. Care must be taken to insure that the distributions of the "natural" and the "known" populations are identical and that the probability of selection is the same for both groups.

Design Two is in fact sampling from a known population, i.e., the marked members of the first biased sample. It has been used to evaluate sampling bias dependent on categorical variables (e.g., Cain & Currey 1968). For categorical variables, the procedures for correction of the biased sample have in effect been used by Cain & Currey and probably others. However, the analysis for making unbiased statistical inferences about parameters of the population is not clearly specified even in this simplest case. We analyze a small subset of Cain & Currey's data as an example below.

Categorical Variables

Consider the case where the individual in the population can be assigned to one of m classes. Assume that the "first" element of the vector $\underline{X}$, is x a categorical random variable which designates the class to which the individual belongs, $x = 1, 2, ..., m$. For example, the population might consist of a polymorphic collection of insects where the morphs are present in unknown proportons, f_x, $\{x = 1, 2, ..., m\}$. Assume that the sampling mechanism of an "inexpensive, or quick and easy" but potentially biased procedure is such that the selection function is proportional to $w(x)$. The probability of selection in the biased sample depends only on x, the class to which the individual belongs and the proportionality constant, k. In addition to the class, suppose that random variables $Y_2, ..., Y_p$, are recorded for individuals selected.

In Design One, we assume that a sampling mechanism is available which will yield unbiased estimates of the (marginal) frequencies, f_x, for the natural probability density function of X. Data might be arranged in the following table (the observed frequencies a_x denote the number of individuals assigned to class x in the sample of size n_0 from the unbiased procedure, and d_x denote the number of individuals assigned to class x in the sample of size n_1 from the biased sampling procedure):

class (x)	unbiased	biased
1	a_1	d_1
2	a_2	d_2
3	a_3	d_3
.	.	.
.	.	.
.	.	.
m	a_m	d_m

$$\text{total} = n_0 \qquad\qquad \text{total} = n_1 \quad .$$

Following theory developed in the second section, the distribution for X in the biased sample is g_x proportional to

$$w(x) \cdot f_x \ . \tag{15}$$

Manly (1985, p. 97) considers the problem of estimation of the selection coefficients $w(x)$, $x = 1, 2, ..., m$, in the context of natural selection experiments. The selection coefficients are estimated by the ratios

$$\hat{w}(x) = (d_x/a_x)/\sum_{j=1}^{m} (d_j/a_j) \ . \tag{16}$$

The selection coefficients are standardized to sum to 1.0 and can be interpreted as the relative probabilities that the "next" individual will be selected from class x by the biased sampling procedure.

Manly (1985) derives the following equations on the assumption that the d_x and the a_x values are Poisson distributed, using the Taylor series method. Let

$$\chi_x = [w(x)]^2 [(1/E(d_x)) + (1/E(a_x))] \ , \tag{17}$$

then

$$\text{bias}(\hat{w}(x)) = w(x) \cdot (\sum_j \chi_j) - \chi_x \ , \tag{18}$$

$$\text{var}(\hat{w}(x)) = (w(x))^2 (\sum_j \chi_j) + (1-2w(x))\chi_x, \text{ and} \tag{19}$$

$$\text{cov}(\hat{w}(x), \hat{w}(x')) = w(x) \cdot w(x') \ (\sum_j \chi_j) - w(x)\chi_x - w(x')\chi_{x'} \tag{20}$$

for $x \neq x'$.

If the frequencies f_x are known for each of the classes, maximum likelihood estimates with estimated biases and variances are given by (Manly, 1985, pp. 94–95). In addition, if the biased procedure is used to resample the biased samples with large sample sizes then the methods developed by Manly for fitting log–linear models may be applicable.

Example 1. Collection of snails on the Marlborough Downs.

In an extensive study of the snail *Cepaea nemoralis* on the Marlborough Downs, Cain & Currey (1968) evaluated the randomness of their collection procedures by releasing marked individuals and comparing the morph frequencies in their subsequent collections by human observers with the known frequencies released. We will consider this as a Design Two study. It is not clear from their paper, but we assume that the snails were first collected by the (potentially biased) collection procedure, marked, and released into the same habitat to be resampled by the same procedure.

Snails collected were of three colors (yellow, pink, and brown) in two habitat types (downland grass and nettlepatches). Summary data from Cain & Currey's table 2 are:

downland grass	yellow	pink	brown
released	720	525	845
recaptured	196	148	163 ,

and

nettlepatches	yellow	pink	brown
released	742	498	877
recaptured	283	192	322 .

The selection coefficients and their standard errors computed by equations (16) and (19) are:

	yellow	pink	brown
downland grass	0.364 (0.024)	0.378 (0.025)	0.258 (0.020)
nettlepatches	0.336 (0.020)	0.340 (0.022)	0.324 (0.019) .

The selection coefficients for the morphs are not significantly different from each other in the nettlepatch habitat when compared by the χ^2 test of homogeneity (Cain & Currey, 1968). However, in the downland grass habitat, the χ^2 test of homogeneity indicates that they are significantly different at the 0.05 level. The same conclusion would be reached based on the above estimates and their standard errors.

In section 3.1 of their paper, Cain & Currey report the following totals for snails collected in 1962:

	yellow	pink	brown	sample size
downland grass	671	457	871	1999
nettlepatches	556	363	678	1597 .

When "corrected" for the biased sampling by use of equation (13) the estimated morph frequencies in the population are:

	yellow	pink	brown	sample size
downland grass	573.2 (28.7%)	376.0 (18.8%)	1049.8 (52.5%)	1999
nettlepatches	548.8 (34.4%)	354.1 (22.2%)	694.1 (43.5%)	1597 .

The corrected percentages agree with those given by Cain & Currey for the downland grass habitat. However, neither the corrected nor the uncorrected percentages agree for the nettlepatches. It is not important for this illustration, but it appears that a typographical error occured in one of their tables. We would agree with their apparent decision to correct the frequencies in both habitats even though the estimated selection coefficients are not significantly different in the nettlepatch habitat. The discrepancies actually observed are in the same direction for both habitats with brown snails being more difficult to collect while yellow and pink snails have approximately the same selection values.

Cain & Currey continued to analyze their capture–recapture study for evidence of natural selection by predators. No data are given on other variables such as weight, or length of snails in the 1962 sample. However, purely for illustration, consider the following hypothetical data on a character for the downland grass snails:

morph	mean	biased frequencies	corrected frequencies
yellow	3.7	671	573.2
pink	3.8	457	376.0
brown	4.3	871	1049.8
estimated pop. mean	3.98		4.03 .

Equation (10) allows the estimate 4.03 to be computed in the usual way from the corrected frequencies. The standard error of the corrected mean can be approximated by use of standard formulas on the corrected sample frequencies. In this hypothetical case, the standard error of the corrected mean is approximately 0.333.

Quantitative Variables

Manly (1985, pp. 55–72) has recently summarized results for estimation of selection functions for quantitative variables. See his text for references and citations to the original research in the literature. He reviews cases where a random variable(s) follows the univariate normal, multivariate normal, gamma or beta distributions and has considered robust models for selection functions. Unfortunately, the general model for non–normal multivariate distributions, does not appear to apply to the present problem.

In this paper, only a brief outline of the use of the existing models will be given by illustrating the case when both the unbiased and biased samples are from normal distributions with a single random variable. Assume that the variable X has a normal distribution with mean μ_0 and variance V_0 in the population and that the probability of selection of an individual with $X = x$ is proportional to the selection function

$$w(x) = \exp\{\ell x + mx^2\} , \tag{21}$$

where ℓ and m are constants. Under these assumptions, the "biased" sample will also be normally distributed with mean μ_1, and variance, V_1, where

$$\mu_1 = (\mu_0 + \ell V_0)/(1 - 2mV0) , \text{ and} \tag{22}$$

$$V_1 = V_0/(1 - 2mV_0) . \tag{23}$$

These equations can be solved for the constants in the selection function to yield

$$\ell = (\mu_1/V_1) - (\mu_0/V_0) , \text{ and} \tag{24}$$

$$m = [(1/V_0) - (1/V_1)]/2 . \tag{25}$$

It is also known that if the distribution of X is normal in both samples then the selection function must be of the form $w(x) = \exp\{\ell + mx^2\}$.

Consider two samples which differ by the selection function $w(x)$. In Design One, an unbiased sample is also available from the population. In Design Two, a second biased sample is available from the first biased sample. Denote the usual sample means and variances by $(\hat{\mu}_0, \hat{V}_0)$ and $(\hat{\mu}_1, \hat{V}_1)$ for the two samples respectively. The reciprocal of the sample variance should be adjusted slightly when used to estimate the reciprocal of the corresponding parameter. For the two samples, $j = 0, 1$, let

$$\hat{B}_j = (n_j - 3)/n_j \hat{V}_j , \tag{26}$$

with estimated variance

$$\mathrm{var}(\hat{B}_j) = 2(\hat{B}_j)^2/(n_j{-}5) \,, \tag{27}$$

and let

$$\hat{a}_j = \hat{\mu}_j\hat{B}_j \,, \tag{28}$$

with estimated variance

$$\mathrm{var}(\hat{a}_j) = (\hat{B}_j/n_j) + (\hat{\mu}_j)^2(\mathrm{var}(\hat{B}_j)) \,. \tag{29}$$

The estimators of ℓ and m are -

$$\hat{\ell} = (\hat{a}_1{-}\hat{a}_0) \,, \tag{30}$$

with estimated variance

$$\mathrm{var}(\hat{\ell}) = \mathrm{var}(\hat{a}_1) + \mathrm{var}(\hat{a}_0) \,, \tag{31}$$

and

$$\hat{m} = (\hat{B}_0{-}\hat{B}_1)/2 \,, \tag{32}$$

with estimated variance

$$\mathrm{var}(\hat{m}) = [\mathrm{var}(\hat{B}_0) + \mathrm{var}(\hat{B}_1)]/4 \,. \tag{33}$$

Once the selection function is estimated, it will be of interest to plot its graph and to test hypotheses concerning the significance of the constants. Adjustment of the biased samples will follow the methods outlined above and illustrated in example 2.

Example 2. Field sampling of tsetse flies.

Manly (1977, 1985) illustrated the methods for estimation of selection functions in the study of natural selection with analysis of data collected and analyzed by Glasgow (1961) on female tsetse fly *Glossina swynnertoni*. Two samples of flies were collected. The first, class 0, were collected as puparia in the field then held in a laboratory until the flies emerged. The length X of the middle part of the fourth longitudinal wing vein was recorded (millimeters/0.0349) as a measure of size. In addition, adult flies were collected in the bush and the same measure of size recorded. Apparently, the collection of adult female flies in the field is difficult. The amount of wing fray was used as an approximate measure of age for individuals collected in the field. Flies with "perfect" wings were assigned to class 1. As the amount of fray increased, flies were assigned to classes 2–5 with class 5 showing extreme fray of the wings. Data calculated from Glasgow's (1961) figure 4 were analyzed by Manly (1977, 1985). The first two classes are:

size (X)	class 0	class 1
41.25 – 42.25	0	1
– 43.25	31	15
– 44.25	156	47
– 45.25	359	137
– 46.25	411	179
– 47.25	418	164
– 48.25	243	62
– 49.25	84	16
– 50.25	9	2

We assume that sample 0 was an unbiased sample of the population of puparia (an assumption implicit in Glasgow's analysis) and hence the size measurement on class 0 is an unbiased sample of the natural distribution of X at age zero. Natural mortality and the field collection procedure for adult female flies combine to generate an unknown selection function, $w(x)$, which yielded the "biased" sample of individuals in class 1. The objective is to estimate $w(x)$ so that information collected in future biased samples of class 1 adult flies can be corrected to age zero.

Using the above frequency distributions and the mid–points {41.75, 42.75, ..., 49.75} of the intervals on X, the following basic statistics are computed:

$$\hat{\mu}_0 = 45.97, \ \hat{V}_0 = 1.971, \ \hat{\mu}_1 = 45.85, \text{ and } \hat{V}_1 = 1.641 \ .$$

The intermediate statistics are: $\hat{B}_0 = 0.5064, \ \hat{a}_0 = 23.280, \ \hat{B}_1 = 0.6065, \ \hat{a}_1 = 27.806.$ Finally, the estimated constants in the selection function with standard errors are:

$$\hat{\ell} = 4.526 \ (\text{s.e.} = 1.779) \ , \text{ and}$$

$$\hat{m} = -0.05 \ (\text{s.e.} = 0.0194) \ .$$

The estimates are large compared to their standard errors and we conclude that there is a significant bias in the field sample of class 1 adult females when compared to the size of individuals emerged from the sample of puparia. Considering equations (24) and (25), the bias depends both on the mean and the variance of the sizes of emerging flies. The estimated selection function (proportional to the probability of selection) is

$$\hat{w}(x) = \exp\{(4.526)x - (0.05)x^2\} \ .$$

In the following table, the value

$$\hat{w}(x)/\hat{w}(45.75)$$

has been computed so that the relative probability of selection of a female fly with size equal to the mid–point of the middle interval is 1.0:

mid–point of interval	relative probability of selection
41.75	0.547
42.75	0.739
43.75	0.903
44.75	0.999
45.75	1.000
46.75	0.906
47.75	0.742
48.75	0.550
49.75	0.369 .

From this table and under the assumptions made, it is seen that individuals in the smaller and larger size categories have roughly one–half the chance of survival and capture in the biased field sample of adults when compared to the central values. If the selection function is assumed to apply to a future field sample than data could be corrected by the use of equation (13). Inferences would then apply to the adult females emerging from puparia. Standard errors computed by the standard formula do not take all sources of variation into account, namely, the "variance due to estimation of $w(x)$". It does take sampling variation into account and we conjecture that estimates will be approximately correct.

In this illustration, the observed frequencies for the class 1 females collected in the field may be corrected by equation (13), where x is the mid–point of the corresponding interval. The estimated mean length of the middle part of the fourth longitudinal wing vein at age zero based on the adjusted field sample is 45.964. The approximate standard deviation for the population of sizes computed from these corrected data is 1.418. Comparable statistics computed directly from class 0 females emerging from puparia are 45.97 and 1.404.

References

Cain, A. J. & J. D. Currey. 1968. Studies on *Cepaea* III. Ecogenetics of a population of *Cepaea nemoralis* (L.) subject to strong area effects. *Phil. Trans. Roy. Soc. Lond.* B, 253: 447–482

Cochran, W. G. 1977. Sampling Techniques. Wiley, New York.

Cox, D. R. 1969. Some sampling problems in technology. pp. 506–527. In New Developments in Survey Sampling. [eds.] N. L. Johnson and H. Smith. Wiley–Interscience, New York.

Davy, P. J. & R. E. Miles. 1977. Sampling theory for opaque spatial specimens. *J. Roy. Statist. Soc.* B, 39: 56–65.

Glasgow, J. P. 1961. Selection for size in tsetse flies. *J. Anim. Ecol.* 30: 87–94.

Grosenbaugh, L. R. 1952. Plotless timber estimates – new, fast, easy. *J. Forestry* 50: 32–37.

Hanurav, T. V. 1962. Some sampling schemes in probability sampling. *Sankhyā A.* 24: 421–428.

Horvitz, D. G. & D. J. Thompson. 1952. A generalization of sampling without replacement from a finite universe. *J. Amer. Statist. Assoc.* 47: 663–685.

Kaiser, L. 1983. Unbiased estimation in line–intercept sampling. *Biometrics* 39: 965–976.

Manly, B. F. J. 1977. The estimation of the fitness function from several samples taken from a population. *Biom. J.* 19: 391–401.

Manly, B. F. J. 1985. The Statistics of Natural Selection. Chapman and Hall, London.

McDonald, L. L. 1980. Line–intercept sampling for attributes other than coverage and density. *J. Wildl. Mgt.* 44: 530–533.

McDonald, L. L., M. A Evans, D. L. Otis & E. Becker. 1987. Encounter sampling of fisherman, campers and other wild animals. Technical Report, Department of Statistics, University of Wyoming, Laramie, Wyoming.

Southwood, T. R. E. 1978. Ecological Methods. Chapman and Hall, London.

Weible, E. R. 1980. Stereological Methods: Vol. 2 Theoretical Foundations. Academic Press, London.

Arthropod Sampling Methods in Ornithology: Goals and Pitfalls

Michael L. Morrison[1], Leonard A. Brennan[1], and William M. Block[1]

ABSTRACT Methods used to sample arthropods in ornithological studies were reviewed; most studies likely obtained only a partial and potentially biased view of the arthropod community. Results of several studies, including our own, showed that data obtained on both absolute and relative abundance of arthropods varies widely among different sampling techniques. Such differences indicate that several different methods, rather than the single method (e.g., sticky traps) usually used in ornithological studies, must be used if more than a cursory and potentially biased examination of an arthropod community is necessary.

The objective of this paper is to critically review and evaluate some methods used by ornithologists to assess arthropod communities. Such an evaluation includes quantification of the arthropods available to birds, as well as those actually consumed by them. We will focus on the assessment of arthopod availability, this being one of the least understood and controversial areas of sampling methods in studies of animal ecology. Accurate quantification of resource availability is essential to detailed analyses of energy requirements of animals, and has direct implications for the evaluation of reproductive success, survival, competitive interactions, as well as more applied areas such as the role of birds in the regulation of arthropod populations.

That birds eat arthropods is common knowledge (e.g., see Morse 1971). The ultimate goal of many avian ecologists who study bird–arthropod relationships, however, is to determine what impact birds have on their prey and vice versa (e.g., Morris et al. 1958, Campbell et al. 1983). But not every study can possibly answer this ultimate question; few even try. Rather, ornithologists attempt to partition their work among numerous smaller studies whose objectives are to provide at least some information that may clarify the bird–arthropod question. One point here is critical in placing the quality of "ornithological insect data" in proper perspective: one must keep the objectives of the researcher in mind when evaluating the arthropod data obtained. We will discuss the general goals of ornithological studies that included arthropod sampling, including the types of data that are necessary to meet these goals. Further, we will compare the data obtained from an intensive, community–level survey of arthropods with the types of data obtained in ornithological studies. Finally, we will present data we collected during a comparison of three arthropod–sampling methods.

---

[1]Department of Forestry and Resource Management, University of California, Berkeley, CA 94720.

Availability

First, we should define what we mean by availability before discussing how arthropods have been quantified. Not all arthropods in the foraging area of a bird are available for consumption by the bird for several reasons. The size, life stage, toxic properties, coloration, spatial and temporal activity patterns, and other characteristics of arthropods all influence whether or not they will be eaten by birds. For example, arthropods that are too small or large, unpalatable, and/or camouflaged will probably not be eaten.

Thus, we should recognize that arthropod abundance indicated simply as biomass may not reflect the actual energy available to birds (see review by Martin 1986). Although total abundance or biomass may give a sufficient description of the arthropod community for some purposes (discussed beyond), such measures likely lack the information that is necessary for a detailed examination of many biological relationships (e.g., predator–prey interactions, competition). We will refer to this point throughout our paper.

Goals

When evaluating the usefulness of any data set, we should consider what the goals of the researcher were in conducting the study. For example, reproductive success of certain birds has been related to the biomass of arthropod prey (see Martin 1986, 1987). Here, the critical variable was apparently biomass, and the methods used to obtain the data, while simple, were apparently adequate to achieve the study objective. Given that birds synchronize their breeding period to coincide with peak or near–peak levels of prey abundance (e.g., Morse 1971), it is not surprising that these studies were able to show a relationship between prey abundance and reproductive success. Using identical sampling techniques among years may give an adequate description of the relative abundance of at least major taxa over time. Determination of actual availability may be unnecessary if the abundance of many alternate prey simply swamp the more subtle relationships that likely exist. Such relationships may not hold over the long–term, however, because arthropod populations will likely go through different, species–specific, population fluctuations.

Avian ecologists have conducted numerous studies on the possible role of intra– and interspecific competition in determining the species composition and abundance (i.e., "community structure") of birds in an area (e.g., see Martin 1986). Most of these studies compared behavior––primarily habitat use and foraging– –in assessing the existence and nature of competition. One way to assess the role of competition is to relate a decrease in the fitness of individuals of one species to the consumption of food by individuals of one or more other species (see Martin 1986). Although habitat can be readily measured (albeit with proper sample sizes and procedures; see Block et al. 1987), assessment of the prey resources is difficult due to the ephemeral nature of the arthropod populations. We believe that quantification of the arthropod community must go beyond simple measures of overall biomass to achieve an accurate assessment of competitive relationships. Here, the use of arthropods relative to their availability is critical to assess. The same would be true for studies examining the use of prey by even single species of birds. Thus, the methods used must be able to sample the range of different taxa comprising the arthropod community.

Sampling Methods

Very few studies are available that have conducted intensive sampling of an arthropod community in temperate North America while simultaneously analyzing the effectiveness of several sampling techniques. Such studies provide valuable information to ornithologists designing sampling techniques, and further, provide an indication of the effectiveness of single—method studies in sampling the arthropods available to birds. Below we describe one of the few intensive comparisons of sampling techniques, and relate the results to those generally obtained by ornithologists interested in the arthropod community.

Voegtlin (1982) used eight sampling techniques to describe the arthropods occupying the canopy of mature conifer in the Cascade Mountains of Oregon. Our summarization of his data raises several major points of direct concern to ornithological studies incorporating arthropod sampling. First, no single method used by Voegtlin accounted for more than about 40% of the total taxa captured (Table 1).

Table 1. Frequency of capture of arthropods by abundance categories, H. J. Andrews Experimental Forest, Oregon (summarized from Voegtlin 1982).

Method	Abundance (No. of individuals)				
	1	2–10	11–100	>100	Total
	Percent captured by one method only (n = 335)				
Sticky trap	6.9	18.0	12.4	4.7	42.0
Black light	15.3	7.1	2.9	0.4	25.8
Pitfall trap	0.7	1.6	0.9	0.2	3.3
Filtration	0.0	0.7	0.4	0.7	1.8
Other methods[a]	0.7	0.2	0.7	0.0	1.6
Subtotal[b]	24.1	28.2	17.7	6.1	74.4

[a]Methods accounting for < 1% of total captures not displayed individually.
[b]% of total captures (n = 450).

Sticky traps were by far the most effective technique in number of taxa sampled (although they missed about one—half of the taxa captured overall). Blacklights were about half as effective at sampling the community as sticky traps, but were far more successful than the remaining methods used. Blacklights, of course, sample primarily adults of nocturnal taxa.

Voegtlin's data show that sticky traps are the most effective technique at sampling above the family level (Table 2). Sticky traps, however, captured only 64% of the total number of families sampled. Blacklights were particularly effective at capturing lepidopterans (100%) and trichopterans (89%) relative to sticky traps (8% and 44%, respectively). In contrast, sticky traps captured 90% of the coleopterans and 81% of the diperans, while blacklights captured only 23% and 22% of these families, respectively.

Table 2. Frequency of capture of arthropod families by different trapping methods, H. J. Andrews Experimental Forest, Oregon (summarized from Voegtlin 1982).

	Sticky trap	Trunk sticky trap	Pitfall trap	Tullgren	Filtration	Vacuum	Blacklight trap	Cookie
Collembola 4 families	25	25	100	75	75	25	0	75
Ephemeroptera 2 families	100	0	0	0	0	0	0	0
Orthoptera 1 family	0	100	100	0	0	0	0	0
Isoptera 1 family	100	0	0	0	0	0	100	0
Plecoptera 3 families	100	0	0	0	0	0	0	0
Psocoptera 8 families	63	13	38	13	25	13	13	13
Thysanoptera 3 families	100	0	0	0	67	67	0	0
Hemiptera 5 families	80	0	0	0	0	20	60	0
Homoptera 9 families	89	33	0	0	44	44	11	0
Neuroptera 4 families	100	50	25	0	0	0	75	0
Coleoptera 30 families	90	17	23	13	3	10	23	3
Trichoptera 9 families	44	0	11	0	0	0	89	0
Lepidoptera 13 families	8	0	0	0	8	8	100	0
Diptera 32 families	81	25	22	0	6	3	22	0
Hymenoptera 15 families	100	47	20	13	7	33	20	7
Acari 31 families	0	3	23	68	42	16	0	68
Araneae 14 families	100	0	0	0	0	0	0	0
Overall Orders(n=17)	88	53	53	29	53	59	59	29
Families(n=184)	64	16	18	17	16	13	26	15

These results have several clear implications for arthropod sampling in ornithological studies. First, no single technique adequately samples the arthropod community. Not surprisingly, various methods designed to capture specific types of arthropods (e.g., filtration) are necessary to develop a thorough assessment of the community. Although when taken alone these specialized methods do not account for a large proportion of the total community, taken together they account for a substantial segment of the taxa present. It is important to note, however, that most of these methods do not give accurate assessments of actual arthropod densities.

The results of the blacklight sampling may be especially important to ornithologists. Although blacklights are used at night, they capture animals that are consumed by birds during the day: birds frequently feed on both the larvae and adults of lepidopterans (e.g., Campbell et al. 1983). Using only sticky traps— —as is often the case in ornithological studies— —would greatly underestimate the abundance and types of lepidopterans possibly available to birds. These data indicate that more research into the use of blacklights in ornithological studies is warranted (e.g., types of arthropods captured by blacklights that are eaten by birds).

As part of our on—going research on bird foraging behavior, we compared three methods of sampling bark—inhabiting arthropods: trunk—attached pan traps and up— and down—opening bark traps described by Moeed & Meads (1983). Except for Collembola, results for the up— and down—opening traps were similar (Table 3). The single exception, however, could result in markedly different conclusions regarding the potential availability of arthropods to birds. Data for pan traps differed substantially from those for the other traps, as seen for Araneae, Collembola, Diptera, and Hymenoptera. Pan types are, of course, much less selective than the other bark traps. For example, pan traps captured about 30% more families and 60% more individuals than the bark traps in our samples. Here again, we see that the method(s) used can greatly influence conclusions concerning the absolute and relative abundances of arthropods. Our comparison does not indicate, however, how closely the data approximated the true density of arthropods in the area.

Our review of published ornithological studies that included arthropod sampling showed that virtually all studies used only one method for sampling arthropods; examples of some of these studies are given in Table 4. Sweep netting and sticky traps were the most frequently—used methods. Several of these studies were attempting to describe the use of specific taxa by birds (e.g., lepidopteran larvae), and were apparently successful in doing so. However, most of the studies were attempting to describe the overall arthropod community. Given the results of Voegtlin (1982), our data presented herein, and what is known generally about arthropod sampling techniques (e.g., Morris 1960), it is unlikely that most ornithological studies obtained more than a cursory and biased glimpse of the arthropods available to birds. Most sampling methods are, by the nature of their design, biased towards particular types of arthropods.

Table 3. Percent captures by trap type for arthropod sampling at Blodgett Forest, El Dorado Co., California during fall–winter 1985–86.

Taxa	Trap Type (%)[a]		
	Down–opening	Up–opening	Pan
Araneae	30.5	49.0	11.1
Thysanura	0.4	0.2	0.8
Collembola	22.0	1.6	0.5
Orthoptera	7.5	2.4	2.8
Isoptera	0.2	0.0	0.0
Psocoptera	1.0	1.8	0.8
Hemiptera	1.0	2.4	1.3
Homoptera	0.8	1.6	0.7
Neuroptera	0.2	0.4	0.3
Raphidioptera	0.4	0.2	0.8
Coleoptera	2.5	2.6	2.2
Trichoptera	0.2	0.2	0.2
Lepidoptera	7.3	15.7	15.8
Diptera	7.5	6.8	32.7
Hymenoptera	17.8	14.9	29.9

[a] No. of individuals captured: down–opening = 478, up–opening = 502, pan = 602.

In conclusion, we feel ornithologists must clearly identify their goals in sampling arthropods, and then adequately justify the methods used to achieve these goals. Ornithological studies that sample arthropods are usually trying to relate some measure of abundance to some aspect(s) of bird biology (e.g., abundance, reproduction, foraging behavior). Consequently, ornithologists probably do not need to know absolute arthropod densities; some reliable (i.e., unbiased) index of relative abundance will probably suffice. Our review and data indicate, however, that the use of only one or two sampling techniques may not adequately characterize an arthropod community.

Table 4. Examples from published studies which illustrate how ornithologists sample arthropods for ecological investigations.

Bird foraging group	Sampling method	Source
ground–foraging and foliage–gleaning breeding insects	sweep–nets and emergent traps for aquatic–	Orians & Horn (1969)
cone–probing of cones for larvae	systematic dissection	Gibb (1958)
foliage and twig gleaning	clipped foliage samples	Gibb (1960)
foliage and twig gleaning	collected frass from larvae	Kluyver (1961)
foliage and twig gleaning	sweep–net samples	Root (1967)
foliage and twig gleaning	clipped and fumigated twigs and foliage; frass collection from larvae	Tinbergen (1960)
ground–foraging	sweep–net samples (dead leaf litter)	Fischer (1981)
dead–leaf gleaning (forest understory)	examination of dead leaves	Gradwohl & Greenberg (1982)
dead–wood excavating	cube–shaped samples of bark and wood from standing dead trees	Cambridge & Knight (1972)
bark–foraging wood–excavating	bark cores	Otvos (1965)
sediment probing	sediment core samples rinsed through screen	Buchanan et al. (1985)
foliage gleaning (forest understory)	sticky traps	Blake & Hoppes (1968)
foliage gleaning	visual counts of leaves	Graber & Graber (1983)
foliage gleaning	sticky traps	Hutto (1981)

Acknowledgements

We thank D. L. Dahlsten and K. A. With for reviewing an earlier draft, J. A. DeBenedictis for arthropod identification, and L. Merkle for manuscript preparation. C. J. Ralph assisted with development of the bark traps.

References Cited

Blake, J. G. & W. G. Hoppes. 1986. Influence of resource abundance on use of tree—fall gaps by birds in an isolated woodlot. *Auk* 103: 328–340.

Block, W. M., K. A. With & M. L. Morrison. 1987. On measuring bird habitat: influence of observer variability and sample size. *Condor* 89: 241–251.

Buchanan, J. B., L. A. Brennan, C. T. Schick, M. A. Finger, T. M. Johnson & S. G. Herman. 1985. Dunlin weight changes in relation to food habits and available prey. *J. Field Ornithol.* 56: 265–272.

Cambridge, W. F. M. & F. B. Knight. 1972. Factors affecting spruce beetles during a small outbreak. *Ecology* 53: 830–839.

Campbell, R. W., T. R. Torgersen & N. Srivastava. 1983. A suggested role for predaceous birds and ants in the population dynamics of western spruce budworm. *For. Sci.* 29: 779–790.

Fischer, D. H. 1981. Wintering ecology of thrashers in southern Texas. *Condor* 83: 340–346.

Gibb, J. A. 1958. Predation by tits and squirrels on the Eucosmid *Ernarmonia conicoland* (Heyl.) *J. Anim. Ecol.* 27: 375–396.

Gibb, J. A. 1960. Populations of tits and goldcrests and their food supply in pine plantations. *Ibis* 102: 163–208.

Graber, J. W. & R. R. Graber. 1983. Feeding rates of warblers in spring. *Condor* 85: 139–150.

Gradwohl, J. & R. Greenberg. 1982. The effect of a single species of avian predator on arthropods of aerial leaf litter. *Ecology* 63: 581–583.

Hutto, R. L. 1981. Temporal patterns of foraging activity in some wood warblers in relation to the availability of insect prey. *Behav. Ecol. Sociobiol.* 9: 195–198.

Kluyver, H. N. 1961. Food consumption in relation to habitat in breeding chickadees. *Auk* 78: 532–550.

Martin, T. E. 1986. Competition in breeding birds: on the importance of considering processes at the level of the individual. *Current Ornithol.* 4: 181–210.

Martin, T. E. 1987. Food as a limit on breeding birds: A life—history perspective. *Ann. Rev. Ecol. Syst.* 18: 453–487.

Moeed, A. & M. J. Meads. 1983. Invertebrate fauna of four tree species in Orongorongo Valley, New Zealand, as revealed by trunk traps. *New Zealand J. Ecol.* 6: 39–53.

Morris, R. F., W. F. Cheshire, C. A. Miller & D. G. Mott. 1958. The numerical response of avian and mammalian predators during a gradation of spruce budworm. *Ecology* 39: 346–355.

Morris, R. F. 1960. Sampling insect populations. *Ann. Rev. Entomol.* 5: 243–264.

Morse, D. H. 1971. The insectivorous bird as an adaptive strategy. *Ann. Rev. Ecol. Syst.* 2: 177–200.

Orians, G. H. & H. S. Horn. 1969. Overlap in foods and foraging of four species of blackbirds in the potholes of central Washington. *Ecology* 50: 930–938.

Otvos, I. S. 1965. Studies on avian predators of *Dendroctinous brevicomis* Le Conte (Coleoptera: Scolytidae). *Can. Entomol.* 97: 1184–1199.

Root, R. B. 1967. The niche exploitation pattern of the blue–gray gnatcatcher. *Ecol. Monogr.* 37: 317–350.

Tinbergen, L. 1960. The natural control of insects in pinewoods. I. Factors influencing the intensity of predation by songbirds. *Arch. Neerl. Zool.* 13: 265–343.

Voegtlin, D. J. 1982. Invertebrates of the H. J. Andrews Experimental Forest, western Cascade Mountains, Oregon: a survey of arthropods associated with the canopy of old–growth *Pseudotsuga menziesii*. Oregon State Univ. For. Res. Lab. Spec. Publ. 4.